Agriculture, Environment and Society

Agriculture, Environment and Society

Contemporary Issues for Australia

Edited by

Geoffrey Lawrence, Charles Sturt University

Frank Vanclay, Charles Sturt University

Brian Furze, University of New England

First published 1992 by
THE MACMILLAN COMPANY OF AUSTRALIA PTY LTD
107 Moray Street, South Melbourne 3205
6 Clarke Street, Crows Nest 2065

Associated companies and representatives
throughout the world

National Library of Australia
cataloguing in publication data

Agriculture, environment and society: contemporary issues for Australia

Includes index.
ISBN 0 7329 1258 X.
ISBN 0 7329 1257 1 (pbk.).

1. Agriculture — Social aspects — Australia. 2. Agriculture — Environmental aspects — Australia. 3. Environmental impact analysis — Australia. 4. Australia — Rural conditions. I. Lawrence, Geoffrey, 1950– . II. Vanclay, F.M. (Francis M.). III. Furze, Brian, 1957– .

306.3490994

Set in Plantin and Helvetica by DOCUPRO, Australia
Printed in Hong Kong

CONTENTS

NOTES ON CONTRIBUTORS

Ross Annels is currently a postgraduate student at the Australian National University. He was previously a Research Assistant in the Science Policy Research Centre, Griffith University, Brisbane.

Neil Barr is a Policy Research Officer in the Sustainable Agriculture Unit, Victorian Department of Food and Agriculture, Bendigo.

Lia Bryant is currently a PhD student at Flinders University, Adelaide. She was previously Rural Research Officer with the South Australian Department of Agriculture.

David Burch is a Senior Lecturer in science policy in the Division of Science and Technology, Griffith University, Brisbane.

Andrew Campbell is the National Landcare Facilitator and is based in the School of Agriculture and Forestry, University of Melbourne.

John Cary is a Senior Lecturer in agricultural economics and extension, School of Agriculture and Forestry, University of Melbourne.

Shankariah Chamala is a Senior Lecturer in extension, Department of Agriculture, University of Queensland, Brisbane.

Peter Cock is a Senior Lecturer in sociology in the Graduate School of Environmental Science, Monash University, Melbourne.

Stephen Dovers is a Research Officer in the Centre for

Resource and Environmental Studies, Australian National University, Canberra.

Bruce Frank is a Senior Lecturer in extension, Department of Agriculture, University of Queensland, Brisbane.

Brian Furze is a Lecturer in sociology, Department of Social Science, University of New England, Armidale.

Ian Gray is a Lecturer in sociology, School of Humanities and Social Sciences, and a Key Researcher, in the Centre for Rural Social Research, Charles Sturt University, Wagga Wagga.

Richard Hindmarsh is a PhD student, in the Division of Science and Technology, Griffith University, Brisbane.

Geoffrey Lawrence is Associate Professor of sociology, in the School of Humanities and Social Sciences, and Director of the Centre for Rural Social Research, Charles Sturt University, Wagga Wagga.

Stewart Lockie is a Masters (Honours) candidate in the School of Humanities and Social Sciences, Charles Sturt University, Wagga Wagga.

Peter Martin is a Lecturer in rural environmental management, in the Faculty of Agriculture and Rural Development, and is affiliated with the Centre for Extension and Rural Development, University of Western Sydney — Hawkesbury, Richmond.

Ian Reeve is a Project Director with The Rural Development Centre, University of New England, Armidale.

Roy Rickson is Associate Professor of sociology and Deputy Dean (Research), Division of Environmental Studies, Griffith University, Brisbane.

Sharman Stone is a consultant rural sociologist, based in Melbourne. She previously worked for the Rural Waters Commission, Victoria.

Shane Tarr is a Lecturer in politics at the University of Adelaide.

Frank Vanclay is a Lecturer in sociology in the School of Humanities and Social Sciences, and a Key Researcher, in the Centre for Rural Social Research, Charles Sturt University, Wagga Wagga.

Chris Watson is a Consultant Environmental Scientist, based in Canberra. He was formerly an Experimental Scientist with the CSIRO Division of Soils, Black Mountain Laboratories, Canberra.

PREFACE

The deterioration of the environment as a direct consequence of agricultural practices is arguably Australia's most pressing ecological problem. Fortunately, it is a problem beginning to be addressed by governments, scientists, conservationists and farmer organisations.

More often than not, however, the debate about the causes of degradation has been one framed by narrow technocratic concerns. Many analysts continue to believe that poor landuse strategies (overstocking, overploughing, overirrigating and inappropriate use of agrichemicals) in combination with production-limiting natural features of the agricultural landscape (shallow soils, low rainfall, frequent droughts) best explains environmental degradation.

Similarly, discussion of potential solutions has been constrained by an uncritical acceptance of current features of the Australian economy and of the social and political life of its people. Questions which are asked in the form 'how can farmers increase output while adopting more environmentally friendly techniques?', 'how can we best educate farmers to introduce soil conservation measures?' or 'what incentives can the government provide to farmers to reduce individual pollution levels?' appear, at first glance, to be technologically and politically 'neutral'. They are not. The assumptions underlying the construction of such questions are that agriculture can, and should, pursue its present trajectory; that the provision of more information will lead to changes in farmer behaviour; and that the state's role should not extend beyond that of providing a 'guiding hand' in assisting farmers in adopting more environmentally sound production methods.

Each of these assumptions arises from a particular view of society — one which assumes such features as the normality of a balanced economy, the continuation of an agricultural system based on increased production, rational action on the part of farmers, and a state free from class bias.

Social scientists have begun to recognise — and hold up for public scrutiny — the ideological underpinings of much of the debate about the causes of, and potential solutions to, environmental degradation in Australia. The first step has been to move beyond the orthodox, but largely unhelpful, approaches of those who either have a stake in the continuation of the present system of agricultural production, or possess a constricted biophysical approach to issues concerning agriculture and the environment. The next step requires acknowledgement that any explanation of the problem of environmental degradation must be one located within an analysis of social structure. A final step is to attempt to link the agricultural and the environmental to the social in a manner which provides an holistic account of what might otherwise be considered as 'technical' problems, and in a way which will allow potential solutions to be developed on an integrated basis.

If these steps are not followed, partial solutions (at best) will be forthcoming. If Australia is to begin its transition to an economically secure future — one characterised by rural social development, agricultural sustainability and ecological vitality — then many of the previously held views about what constitutes 'growth', 'development', 'efficiency' and 'productivity' will have to be carefully scrutinised.

If the environmental problems of rural Australia are profoundly *social* in origin, then it behoves those who are responsible for rural policy, education, environmental management — or who work the land for a living — to examine the findings of social scientists for insights into the evolution of better systems of production and social organisation.

An enormous amount of research remains to be undertaken before a clear picture emerges of the social origins of environmental degradation in Australia's farming systems. This book attempts to bring together the best information currently available: chapters have been commissioned from Australia's most highly regarded social scientists involved in the exploration of issues concerning agriculture and the environment.

Expertise has been drawn from a wide variety of academic disciplines, including social history, environmental management, sociology, social ecology, agricultural extension and political science, while other contributors have had a wide experience in rural regions as counsellors, researchers and government officials. The breadth of interest of contributors is revealed most clearly within individual chapters of the book.

For thematic continuity, the book is divided into three parts. Part 1 sets the scene for later discussion of sustainable options by providing an assessment of the sociopolitical development of capitalist agriculture in Australia. It is thought to be less than constructive to attribute blame to previous generations of farmers and politicians for today's environmental problems. What is required is an appreciation of how past decisions about land use were moulded and constrained by Australia's particular role in the global order. Analysing Australia's past and present role is a key to understanding the continuation of environmental problems.

Contributors in Part II focus upon the social bases of rural life and discuss the forms of rural social organisation (and bureaucratic structure) which have evolved to deal with issues of degradation. Various 'models' such as Total Catchment Management, Landcare and Whole Farm Planning are evaluated for their potential to move Australian agriculture toward sustainability.

In Part III, the questions raised are: 'will the development of a sustainable agriculture be possible given the constraints and demands of the economic system?' and 'will a more sustainable agriculture be one which demands a change in rural social organisation?'. Needless to say, there are no clear answers to these questions. Some contributors are convinced that farmers — for reasons of economic survival — will be forced to abandon many non-sustainable practices; others believe sustainability will be impossible without a fundamental realignment of the society–nature relationship and reject outright the present agricultural trajectory linked to corporate (and especially transnational) capital.

The book does not attempt to address every problem which presently affects the agricultural environment and the people dependent upon farming as an occupation. Rather, it looks at selected issues critically and from the perspective of the social sciences. Contributors weave a line between the political, the economic, the social and the environmental so as to highlight the past, the present and the emerging relationship between agriculture, environment and society. This is the first book in Australia to provide an analytical assessment of that relationship. The book does not attempt to 'force' an overall conclusion. It will have served its purpose if it instils a critical focus in the minds of readers, engenders concern for the future of Australian agriculture, stimulates discussion about future options and provides a basis — even if on a small scale — for public action. We hope it will act as a catalyst in generating and shaping research questions which will be of importance to Australian agriculture and rural society in the 1990s.

The book developed as a logical outcome of discussions between members of Charles Sturt University's Centre for Rural Social Research. One of the centre's main areas of interest is the social aspects of agriculture, the environment and sustainability. Individually, and jointly with colleagues of the Centre and from other universities, members have begun to address many of the important issues discussed in this book. The editors acknowledge the support received from Centre members and, in particular, thank Helen Berry for time spent wordprocessing final (and not so final) drafts. We also thank Peter Debus of Macmillan and Jenni Coombs for final assistance in preparing the manuscript.

This book is appearing at a time when the country's most productive agricultural region (and the region in which we live), the Murray–Darling Basin, is being labelled as 'an environmental disaster area' which boasts a river system described as 'the world's longest sewer'. There is growing acceptance among state officials that the existing problems of degradation in the Basin — and throughout Australia — will be virtually impossible to reverse. In such circumstances it is easy to be disenchanted with governments, just as it is easy to be outraged by stories of continued and purposeful environmental destruction. What is needed, however, is a *constructive* basis for discussions about the future of the agricultural environment in Australia. We believe this collection may go some small way towards achieving development of a rational basis for debate. We dedicate this book to our children in the hope that they — and their children — might live in a future Australia where environmental and social justice considerations are elevated above the increasingly narrow economic (profit-making) justifications for the present forms of use of Australia's limited natural resources.

Geoffrey Lawrence
Frank Vanclay
Brian Furze

Acknowledgements

The authors and publishers are grateful to the following for permission to reproduce the following copyright material:

Figure 4.1, State Library of Victoria;
Figure 4.2, City of Hamilton Art Gallery;
Figure 4.3, Private Collector;
Figure 6.4, Australian Conservation Foundation;
Figure 6.5, Department of Conservation and Environment, Victoria.

1 THE HISTORY OF NATURAL RESOURCE USE IN RURAL AUSTRALIA: PRACTICALITIES AND IDEOLOGIES

STEPHEN DOVERS

The European occupation of Australia in the late eighteenth century heralded a period of revolutionary change in the physical, ecological and human character of the continent, a revolution that is far from over. From the earliest encounters through to the voyage of James Cook, the reports of explorers had given a firmer shape to the rumour and myth of *terra australis incognita* — the unknown south land — and the British Empire knew just enough to perceive a use, albeit the unglamorous one of penal colony.

The British colonists claimed possession of what they defined as *terra nullius* — literally, empty land — looked upon as simply to be occupied, with no one to conquer or to consider. Thus they effectively denied the existence of a human culture 50 000 or more years old. For the later use of the land and resources, which was a trial and a battle, this was perhaps significant. What the settlers could have learnt from the Aboriginal people had they so attempted is hard to say. The Aborigines knew the landscape and survived in it; some of their knowledge certainly would have been useful to settlers. It has taken us two centuries to realise that the original occupants of the continent did not merely wander the landscape and take from it what they could, but that there existed both a mythic and technical culture which was tied reciprocally to the land, and that they did *manage* the land. Aboriginal society had its impacts on the ecosystems, particularly through the use of fire, although these pale in comparison with those of the later European occupation. The new settlers did not occupy an empty wilderness but the human landscape of Aboriginal Australia (see Chapter 4).

The denial of the previous culture, combined with ignorance and perhaps the hubris of an expanding imperial society, led the colonists to view the country also as *terra sine scientia* — land with no knowledge. Despite being confronted by what was 'more a new planet than a new continent' (Rolls 1985) it was taken that there was nothing to learn from the people or the landscape. The cultural baggage that was imported with the colonists — ideological and technical, political and practical — merely had to be applied. The story of rural resource use in Australia is the story of realising the impotence of much of this baggage and the discarding, invention and reshaping of the contents of the tool kit of land management. Again and again realities — soils, drought, economic crises, technological limitations and chance — were to disturb and overcome the way people thought things were or believed they should be; the prevailing ideology overcome by practicality.

The 'new planet' certainly was strange to the newcomers. So different from the soft landscapes of home, its age, climate and evolution had left their mark. There were unimaginable plants and animals, shallow and deficient soils, and above all a pattern of rainfall and runoff which defied understanding. This resource endowment, Biophysical Australia, is a dominating backdrop against which European Australia has evolved, and continues to evolve.

CONFUSED BEGINNINGS AND EMERGING VISIONS

The British settlement in Australia must be counted as one of the most unusual beginnings in the history of the European expansion across the globe — a motley collection of military personnel, civilians and convicts, strangers in a strange land of which they knew virtually nothing of any utility, and who were very much thrown upon their own resources. The initial task was one of survival, with longer-term hopes and plans postponed. In fits and starts, against drought, floods and pests, and contained within the immediate surrounds of Sydney Cove, a rudimentary Australian agriculture was established, easing hunger but not sufficient for trade. For these early years, the small patches of available alluvial soil were the limits of production. Tradable goods came primarily from sealing and whaling, later supplemented by cutting red cedar (*Toona australis*) which was a spur to further settlement in those areas where it occurred. So the great alchemy of European Australia, turning natural resources into capital quickly via the route of export, began not on land but in the sea. The difficulties of establishing farming in the new environment are perhaps now hard to imagine but are illustrated by the fact that it took 28 years of settlement to discover a practical method of removing the troublesome eucalypt stumps from cleared land so as to allow ploughing; or that food shortages at Sydney Cove were only eased in 1810 by imports from Britain and the new settlement in Tasmania. (For a description of the history of farming in Australia, see Wadham 1967; Davidson 1981.)

The realities of Biophysical Australia were already becoming painfully apparent and set the stage for a battle between these and the imported perceptions, beliefs and ideologies of Britain. Western landscape sensibilities were in the throes of change, with the wild, romantic and picturesque gaining ground against the traditional preference for settled rural scenes. Utility was an overriding concern. The perceptions and visions of the Australian environment varied enormously in the first few decades of occupation, from the profoundly optimistic to the devastatingly pessimistic. Perhaps a critical role here was played by the variability of this environment itself, especially rainfall, with a green and promising tract of land within a short space of time becoming, under the influence of season and drought, far less alluring. Even as the unknown environment became better known, the swings in opinion remained. One outcome of the strangeness of Biophysical Australia, and one which would continue for many years, was a constant desire to improve upon the array of plants and animals which nature had provided. Plants and animals from elsewhere were to be introduced — by accident, for farming, for sport, for visual appeal — to the continent one after another. Some were to have little impact, some failed to establish, some formed the basis of modern farming, and still others caused ecological and economic nightmares.

The constraints of topography were first effectively broken in 1813; movement followed across the Blue Mountains in 1815, and prospects of expansion came with subsequent exploration. The great period of exploration in Australia in the first half of the nineteenth century was a mixture of the heroic, the foolish and the pragmatic. Epic journeys driven by high hopes and curiosity often preceded shorter journeys, filling in the maps and assessing the country. When Captain Mark Currie in 1823 recorded the first impressions of Monaro in southern New South Wales, he gave a flavour of the more pragmatic reasons for exploration when describing the country he saw and commenting on its 'interesting nature connected to sheep grazing' (Currie 1825). The land was being discovered and much of it was soon put to use more quickly in fact than many in power thought seemly.

Lands were granted to settlers, the military and to ex-convicts in a controlled manner. The successes were only modest. The allure of new lands was strong and a pattern of expansion soon became set. The term 'squatter' was originally used for unsavoury characters who led doubtful existences on the periphery of settlements, but the title soon transferred to those who took a mob of cattle, or less often sheep in the early days, into the new lands and took possession of the most suitable land they could. Squatting, the rural land use to dominate Australia for many years, had begun, and was proceeding apace, at times even preceding the official explorers. This was a mixed blessing for officialdom: on the one hand welcome expansion was occurring, but on the other this expansion was becoming uncontrollable. In 1824, a process of surveying 19 counties was begun, an area where pastoral use was to be permitted and controlled. Spanning that part of current New South Wales from Batemans Bay in the south to the Manning River in the north and west through Wellington and Yass, the counties were defined in 1831. The squatters had by then already gone beyond the boundary. Elsewhere, settlements had begun, in Tasmania in 1803, and later in Queensland, South Australia, Western Australia and Victoria. This phase of rapid, opportunistic expansion began the pioneering phase of rural resource use, the underlying ideology of which was resource exploitation with scant concern for resource or environmental management.

Despite the rapid expansion of pastoralism, the problem that soon arose was that while the colony absorbed the new produce for a time, there was very much a limit on this capacity. The expanding pastoral industry needed a product and a market. The development of the product, wool, was to be critical to the history of rural Australia.

CAPITAL, EMPIRE AND SQUATTERS

Wool was, in retrospect of course, just what the colony needed: a saleable commodity which could be produced with reasonably low labour levels, did not require a great deal of land management, and could be transported and stored. It was a commodity tailor-made for the squatters. The story of wool has been often told: it was an interplay of economic imperative, outstanding characters, and diverse inventiveness and application. The chief ingredients were new blood lines via Spanish and Saxony stock, a few characters (notably the Macarthurs at first), and the squatters and their land. It was a hard task, and often slow, but woolgrowing succeeded and the product could be sold. It was the impetus for the birth of an export-oriented capitalist agriculture in Australia. In the words of the historian W.K. Hancock (1930: 12): 'Here was Australia's opportunity, the grand occasion offered to the most wretched of colonies to pay its ransom and win its way to freedom and self-respect. Wool made Australia a solvent nation, and, in the end, a free one.'

The colonial economy was boosted by the many things which could be achieved following the solvency gained on the sheep's back. The freedom was not simple, however. The fact that Australia had shown an ability to pay its way attracted what admittedly was much needed investment, but, as Hancock notes, once wool was seen as successful, English capital took control over Australia's destiny. Immigrants with capital set up as squatters; so did some large companies created for the purpose, such as the Australian Agricultural Company which began in 1824, with half a million acres on the Liverpool Plains and in New England. The actual level of foreign control can be questioned, especially from the late 1800s onwards when family ownership of land and Australian-based capital increased (Tsokhas 1990). Tsokhas has questioned the presumption that Australia's wool industry remained an adjunct controlled by Imperial policy and capital.

The squatters and their sheep, as a land use, possessed attributes which were well suited to time and place. The first was the nature of the land — relatively open country of savanna woodland that could be grazed without the effort of clearing, and mild winters which obviated the need for stock shelter. Second, apart from shepherding, the labour input required was low, which was a great advantage. Third, the product was of relatively high value, and could be transported to the distant end markets (initially in England, but soon elsewhere as well). The 'squattocracy' which emerged in this era was monopolistic, politically powerful, often exploitative and did not fully utilise the land resource. However, as Williams (1975) notes, the 'squattocracy' was also perhaps ideally suited to the prevailing environmental, economic and social conditions. The effect of this land use on the resources of Biophysical Australia is difficult to gauge. Limited clearing occurred, and doubtless the fire regime of the Aborigines changed, but the effect of these is unclear. The impact of grazing on the native vegetation and thus also on the soils and streams was certainly significant — Rolls (1985) estimates that major changes occurred in the species composition and distribution of the grass and shrub layers within as little as six years. The impact on humans was mixed. The rapid pastoral expansion was disastrous for Aboriginal Australia. For the most part, the squatters themselves prospered, but living conditions for their employees or servants were less favourable — perhaps not so bad compared to the alternative of industrialising, urban Britain, but according to some observers (for example, Lhotsky 1835), quite brutish none the less.

Whatever its faults, squatting opened up the country and provided the beginnings of a background of understanding of Biophysical Australia. Great breakthroughs were rare, but the maps were filled in, stock were improved, and the tool kit of land management started to acquire some particularly local additions. The period from about 1830 to 1860 was the golden age of squatting, when grazing expanded rapidly. The squatters' hold on the land was to last for many years to come, but it was, in the latter part of this period, increasingly under attack.

RUSHES AND SELECTORS

Attacks on the squattocracy were often planned, with the aim of settling more people on the land. Before these attempts gathered any force, the scene was disturbed by, as it were, chance, with the discovery of gold and the ensuing gold rushes from 1851 onwards. The scale of this disturbance and the resulting changes are simply illustrated by the massive relative increases in population. Australia's non-Aboriginal population grew from just over 400 000 to over 1 000 000 during the 1850s, and in Victoria, where the gold rushes were so concentrated, the increase was from about 80 000 to over half a million people. This Victorian growth was to see the concentration of capital in Melbourne, something which had a strong influence for many years to come.

Gold produced social and economic changes which may never have otherwise happened, and doubtless brought on many others prematurely. The environmental impact in the mined areas was often devastating, particularly in terms of tree clearance and alterations to streams. This impact was such that criticisms and debates began emerging in the public arena (see Powell 1976). The mining population created a demand for food and fibre (and numerous other services, both legal and otherwise). This was a great impetus for agriculture and for development generally. However, the most lasting impact came when the rushes peaked, and increasing numbers of people looked elsewhere for occupation and income (and for a political voice resulting in the emergence of colonial legislatures during the 1850s). The selectors became the critical mass for a challenge to the squatters which had been brewing for some time. The obvious outlet for the pressure of people was on the land, changing the broadacre and shifting pastoralism of the squatters to a denser, more populated farming system. This impact of gold was to mirror an interaction between mining and agriculture that was to occur often. Unlike today's methodical searching for minerals, early discoveries were more often happenstance, commonly following early pastoral use (see Heathcote 1975). With larger developments (for example, Broken Hill, Charters Towers, Mount Isa), the mining activity had significant impacts on the environment and on the farming systems nearby.

The reasons for wanting more people who would use the land varied. For those without occupation and without the hoped-for riches from gold the desires were immediate and obvious. For governments, the reasons were not so obvious, but were very powerful: unease over the potential and actual political power of the squatters; a desire to increase and diversify agricultural production; the need to accommodate the surplus population, all played a part. Overriding these 'practical' influences, and in many ways arising from them, there existed a theoretical and conceptual belief in the inherent rightness of a more closely settled, agrarian population — the image of a society of humble but adequate farms peopled by a hard-working and independent (but nevertheless by definition obedient) 'sturdy yeomanry'. A theoretical basis of this ideal was the influential 'systematic colonisation' of E.G. Wakefield in his *A Letter from Sydney* in 1829, and its essence is captured by Williams (1975: 63): 'The land needed to be used and civilised, and that, of course, would best result from its settlement by a stable and sedentary society of farmers'. So despite the fact that the size of the urban population was overtaking the rural by the 1850s, it was to the country that hope was directed, as the angry mob in front of the parliament in Melbourne in 1860 showed in their slogan 'Every man a vote, a rifle, and a farm'.

Moves for encouraging smaller-scale and more diverse agricultural holdings which had been attempted right from the start almost consistently failed. (For coverage of the various pushes for more intensive land use, see King 1957; Williams 1975.) The most notable was in South Australia, where a Closer Settlement pattern had been instilled from the start. The more organised means to this desired end was to be 'free selection'. Against the stern opposition of the squatters, Robertson's Crown Lands Alienation and Crown Lands Occupation Acts of 1861 passed into law in New South Wales, and there was an attempt at land reform of massive proportions. These Acts allowed the 'selection' of portions of land, held under a squatters' lease or not, from 40–320 acres (16–130 hectares). Similar schemes were introduced in the other colonies, with similar intent. Prices were not great and payment was typically over a period of time, and many took advantage of the opportunity to purchase. The impact was to redraw the tenure map in record time; there was less impact on the production pattern. In New South Wales between 1861 and 1891, the amount of alienated land rose from around 3 million hectares to over 20 million, with an attendant 21 000 new farms and 199 000 extra people on the land. This, however, was not as great a transformation as had been expected. Deceit and false claims abounded. The squatters retained control of a great many areas through selection by 'dummies' and by selecting around the key watering points ('peacocking'). The selectors, for the most part, had no hope of economic survival on the small blocks allowed, even when later these were enlarged. The ones who survived were usually lucky, especially astute, or managed to select or purchase additional land. Many simply ran sheep or cattle in the same manner as had the squatter but on a smaller scale and less profitably.

In line with the intensification of land use, the impact of the free selection era on the environment greatly exceeded that of the previous years. People struggling to make farming pay on unviable holdings placed extra pressure on the vegetation and soil. Increasing attempts at cultivation, spurred on by the need for produce and new technologies such as the Ridley wheat stripper (1840s) and the celebrated stump-jump plough (1870s), required clearing, and so 'ringbarking' was born. Increasing competition for land and more intensive use saw fencing become a greater necessity, with fences, before the advent of wire (from the 1870s), requiring large amounts of timber. Part of the various settlement schemes was a requirement for residency. More significantly, 'improvements' had to be undertaken. One standard improvement was tree clearance (it was only in the 1980s that a last vestige of official encouragement for clearance in the federal tax system was removed). By the end of the nineteenth century, it was generally believed that tree removal and ringbarking for agriculture had a greater impact on forests than commercial forestry (Bolton 1981). It was later in the period of free selection that soil erosion was first noticed on a large scale. The impact of agriculture on the native vegetation of Australia has been profound. The most well-known changes have been in the forests, and these have been large. However, changes of often greater magnitudes occurred in the woodlands, shrublands and grasslands of the drier areas (see Australian Surveying and Land Information Group 1990).

Despite increasing land parcel sizes and continual amendment of the land laws, by the mid-1880s it was becoming obvious that the dream of a closely settled and frugally prosperous farming society was not being achieved. Surprisingly, many survived, often with outside work. How they did so before technology and transport made smaller-scale farming profitable has been described as 'one of the mysteries of Australian agriculture' (Davidson 1981). The legacies of free selection were longlasting: the redrawn tenure survives still in many places, while the struggling selector became a part of the national folk myth. In many areas, the families which survived now provide the leading farmers in their districts. Also, importantly, the failure of many to succeed was an impetus to research and to the development of technology and management skills which were to be basic to later farmers. Of these, critical ones were to include fertilisers, plant breeding and water supply. In the 1890s, but not fully implemented in various parts of the country until much later, 'Closer Settlement' was initiated by government. This involved, at its most basic level, the purchase or resumption of land and its subdivision for sale. Other, more specific efforts at encouraging small-scale farming were also tried, including the colonial government's design of the fledgling Queensland sugar industry from the 1860s onwards. The impetus for intensification of land use would continue and, in fact, speed up.

BOOM, BUST AND DROUGHT

There was great optimism in the 1880s despite the failures described above. Rural production was increasing, while economic growth in the cities, especially Melbourne, was a source of confidence (as well as capital). This was not to last. Financial crashes, recession and above all drought were on the horizon.

Agriculture was being boosted by a number of factors. Management regimes continued to become more sophisticated, helped by the creation of research and training institutions such as Roseworthy Agricultural College (in 1881), and by the increasingly active agricultural departments of state bureaucracies. A belief in the rewards of science was emerging, as was a perception that the state could usefully play a leading role. Improved varieties of plants, the benefits of fertiliser use, and the wider availability of wire fencing, including the new barbed wire, were significant. Farrer's wheat breeding experiments, culminating in the release of the Federation strain in 1901, had a great influence, as did the development of subterranean clover in the first decade of the twentieth century. From the

1870s onwards, the advent of refrigeration was of immense significance to Australia, benefiting the beef and dairy industries in particular with increased trade possible in perishable goods. In the case of dairying, this was boosted by new separator and butterfat testing technologies. A result of these innovations was renewed clearing and development of the more forested, higher rainfall areas. The development of irrigation increased following the Deakin Royal Commission on Water Supply in the mid-1880s, and was to play an ever-increasing role in rural Australia. Bore-drilling for groundwater began in New South Wales in 1880 and increased rapidly allowing the stocking with sheep of drier areas hitherto little used.

The financial boom of the 1880s in the cities (especially Melbourne) created a pool of capital, some of which supported rural industries. There were, however, dry years over this period, resulting in a series of droughts between 1891 and 1901. These droughts, combined with the depression of the 1890s and the rising damage to the resource base by rabbits, caused farming, and much else, to suffer. The droughts were not the first, of course, but they had a major overall economic impact for the first time, providing some important lessons.

Rural development and production rose quickly from 1902 to the beginning of World War I, aided by improved techniques, and particularly by the influence of governments, both through settlement schemes and through public expenditure on infrastructure and services in rural areas. Closer Settlement proceeded apace, and more than 10 000 farms each over one million hectares in size had been allotted (Williams 1975). This period also saw the emergence of the 'home maintenance area' concept, which formed the basis of the area of land allotted for farm settlement. Intended to enable a reasonable living for the farm family, the areas were all too often inadequate, something quite quickly recognised. But the optimism of the period, justified by good prices and by improvements in productivity, swept aside many such doubts.

FURTHER DELAYS, SLOW ADVANCES

World War I proved both a boon and an obstacle to agriculture. Interruptions in trade created problems, but prices in general rose with the war demands between 1914 and 1919. Governments began playing a stronger role, with a wool marketing scheme and the Commonwealth Wheat Board being established. Farming flourished (Wadham 1967). The inflated prices of this period created much optimism, and this, in combination with the continuing ideology of intensification and the problems of what to do with the returned servicemen, saw the soldier settlement schemes begin. Heavily subsidised allocations of land on attractive terms occurred around the country, from the semi-arid areas to small irrigation blocks. As Williams (1975: 94) explains, 'the idea of giving returned servicemen a block of land and financial assistance to start farming was apparently the highest and most desirable reward which society could think of.' Certainly this was so, but other economic and political reasons counted as well. This phase of new settlement, on top of the Closer Settlement, was to place a great deal of new pressure on the land resource. It was in the 1920s that the perception of soil erosion began to spread, and warnings by experts were issued, including a prophetic report on dryland salinity in south-west Western Australia. In general, the experts found that they were welcome bearers of news on how to lift yields, but unwelcome pessimists when they highlighted effects of poor management practices (Bolton 1981). The age of the scientific expert and adviser had arrived, buoyed by demonstrated successes like sub-clover, superphosphate, the control of prickly pear and of livestock diseases. Agriculture was supported by government as it was perceived to be contributing significantly to the 'national interest'. Following the establishment of the Council for Scientific and Industrial Research (the precursor to the CSIRO) in 1926 by the Prime Minister, Stanley Bruce, much attention was given to the farm sector. If Australia rode on the sheep's back, then the sheep and its pastures have been similarly supported by the CSIR and CSIRO.

Changes in underlying beliefs and ideologies are difficult to date, but around this time in Australia new ones were emerging. The pioneering, exploitative and unplanned phase of rural resource use was being usurped by one which was based on government leadership and the calculated use of resources (based on scientific assessment and the perceived interests of the nation and the economy). The roots of change went back several decades and related to new attitudes toward management, research, government policy and legislation.

The depression of the early 1930s brought this post-war era of prosperity to an end. Lack of capital and collapsing markets prompted some attempts at amelioration, including ill-fated ones to expand production of wheat, as well as more effective ones involving price-setting measures and direct subsidies. Prices rose in the late 1930s, and some recovery was made. One impact of the depression was to focus attention more firmly on improving productivity, including the research and management necessary for this (Wadham 1967). Cropping practices evolved rapidly to sounder procedures, while the catchcry of 'pasture improvement' was heard more often. Fodder conservation was begun in earnest, and, most significantly of all, machinery fed by fossil fuel was replacing human and animal energy. Innovations at this time, such as the bulk handling of wheat, which replaced the more labour-intensive bagging, were indications of the new scale and style of agriculture to come. However, this momentum was not allowed to reach its potential. War held at bay much of the potential for change and expansion in rural Australia. These forces were to be unleashed fully in the period following World War II.

MARKETS, PRICES, PEOPLE AND TECHNOLOGY

World War II had a more profound impact on rural Australia than did the Great War. Strange and sometimes conflicting demands, the result of the urgent needs of war, were placed on a farm sector robbed of much of its labour force. Shortages of fuel and the slow rundown of plant and infrastructure took their toll, a situation not helped by the impact of a drought in 1944–45. Throughout the war, Australian agriculture generally managed to maintain production, but by the end was in a semi-exhausted state.

The immediate post-war years saw farmers competing and losing against the other demands of post-war reconstruction, with much-needed inputs being denied to farmers. When the wire, fuel, tractors and so on reached rural Australia farmers were well poised to take advantage of new improvements in technology and management. In the expanding post-war global economy, Australia as a food and fibre producing nation had its role defined. With domestic population increasing, both through natural increase and mass immigration, the domestic market grew rapidly. As after World War I, soldier settlement schemes were implemented, with returned servicemen once again being rewarded with land. They became the shock troops in the latest manoeuvre aimed at intensification of rural land use.

Rural production was thus poised for expansion. The most telling factor was the realisation of many techniques and factors which already existed, but had been delayed. Pasture improvement, weapons against diseases in livestock, fertilisers, improved crop varieties — all these counted. So did the removal of the immense pressure of the rabbit in the 1950s with control being achieved by the *Myxoma* virus, one of the most famous services of science to Australian agriculture. Also, management was very much a focus, both physically and economically. Agricultural economics was emerging as influential, and the 'science' of agriculture was coming into its own (helped on by widely read volumes such as Wadham and Wood 1939).

By far the most critical factor was the inrush of energy from fossil fuels, which, although beginning decades before, was only now realised fully. What could be done in weeks by human and horse could be done in a day on a tractor or bulldozer. The scale of possible cultivation and harvesting was magnified greatly, transport and storage of produce was revolutionised, and there was a renewed period of large-scale clearing utilising heavy machinery and drag-chains. This occurred, significantly, in marginal areas and in previously unused areas, such as the brigalow lands of southern Queensland and the black soils plains of northern New South Wales. The full cultivation of these regions awaited the heavy machinery which became available in the 1950s and 1960s. The longer-settled regions also underwent change. The need to cultivate crops (typically oats) to provide feed for draught animals had been obviated. With greater cultivation ability, the more universal use of leguminous plants in rotation schemes and greater use of artificial fertilisers, landscapes were altered. Changes were especially important on lands which had been based on a typical rotation including fallow, wheat crop and oats. In northern Australia, the introduction of *Bos indicus* cattle and the later development of new mixed breeds increased beef production rapidly, placing new pressure on northern ecosystems (see Chapter 7). These examples indicate that no simple chronology of land use development can be applied in Australia, as many fits, starts and changes occurred in different places with the introduction of new methods and technologies.

Away from the farms themselves, the country towns were rapidly changing. Many factors were involved, but the increasing use of the private car was the most significant. Increased personal mobility and the greater mobility and availability of goods and services saw the demise of many smaller towns and service centres and the greater uniformity of settlements across rural Australia (see Davidson 1976).

Taken together, these factors equalled a much greater rate of physical influence on Biophysical Australia, an inescapable outcome of increases both in the scale of farming effort and in the productivity of this effort. However, management and amelioration of the impacts had advanced somewhat since the last phase of intensive development earlier in the century. Attention had been drawn to degradation of the resource base (see, for example, Ratcliffe 1947). Soil conservation services which emerged before the war were slowly instilling the principles of a more appropriate land management regime. Nevertheless, the expansion of production continued, with increased irrigation development (notably the Snowy Mountains Scheme with its overtones of national purpose). Resource management amounted mostly to damage control rather than preventative measures as problems emerged in the wake of new activity. That great alchemy of making capital from resources was beginning to be exposed as an equation between production and the conservation of the resource. The challenge of reaching this balance continues today.

Table 1.1 Rural Production, 1950–51 and 1991–92

	1950–51	1991–92
Wheat (million ha)	4.7	8.1
Other crops (million ha)	3.4	13.7
Sown pasture and grasses (million ha)	7.2	30.0
Beef cattle (million)	10.4	21.6
Sheep (million)	109	152

Source: Australian Bureau of Agricultural and Resource Economics (1989 and various publications).

The 1950s and 1960s saw the assimilation of Australian agriculture and primary production far more tightly into an international system of trade. Less and less tied to Britain, this period saw America and Japan become more significant as destinations for wool, wheat, meat and minerals. The returns were greater, yet the potential for disturbance to the national economy from swings in the international marketplace increased. Although prices of major commodities rose and fell, Australian farmers generally managed to shift production in a manner which allowed continued output increases. Similarly, droughts came and went, disturbing but not stopping the advance of production. The scale of post-war increase is worth noting (see Table 1.1), particularly keeping in mind the fact that over this period farm employment has fallen from 474 000 to 407 000.

CONCLUSION

The post-war boom (or series of booms and busts, but with the booms dominating for three or four decades) has faded. The story of the evolution of rural Australia continues. Many people's vision of 'how things should be' is based on the characteristics of the best years of this more recent phase. While the characters and specific arguments of this story have certainly changed, the plots and the principal themes are often surprisingly familiar. The climate and the soils, the people, the markets and the technology continue to mix together to form the stream of the story, with new twists to old plots at every turn, often not without paradox. The 150 year-long push for more intensive settlement has been replaced by the slogan of 'get big or get out'. Diversification into new, intensively farmed products (stimulated by favourable tax treatment for city and overseas investors) and the continuing 'alternative settlement' movement run counter to this and so complicate the story. Capitalist export agriculture which began with English capital a century and a half ago is now being driven by East Asian capital — something perceived as undesirable by many people.

The result of this story — of ideologies overcome by practicalities, and interruption by climatic, political or economic spasms — is today's pattern of resource use of rural Australia. Modern capitalist agriculture in Australia is characterised by production-intensive, owner-operated enterprises, concentrating on a comparatively few products for export markets (Lawrence 1987). It rests upon a resource base which is relatively unimproved and has a high propensity for degradation (Heathcote 1987; Scott 1987; Laut 1988). In the space of two centuries, farming has grown to cover 63 percent of the Australian landmass, the bulk of which is represented by unimproved grazing enterprises (see Boyden *et al.* 1990). In this form, agriculture makes a vital contribution to the Australian economy, although its relative contribution has fallen in recent decades.

What cannot be doubted is that today's agricultural sector is beset by special problems: often lower than average incomes; restricted employment opportunities and falling relative employment; isolation and related social problems; and a testing dependence on a world market not known for its stability and predicability. The rural landscape is in many places highly degraded, with over half the landmass requiring treatment for land degradation (Woods 1984). The practicalities of Biophysical Australia make degradation a crucial issue for land managers. The confusions and emerging visions with which the story began have not disappeared. The resource endowment, the innovations, as well as the players in their political and economic settings, continue to interact.

In the two decades before and after 1900, the pioneering ideology was slowly replaced by one which stressed the rational assessment and use of resources. The rural environment became a resource that was useful to European Australia and was to be well managed to provide ongoing economic benefits. While this set of beliefs is still dominant, another ideology is now challenging this and is a cause of much conflict for rural industries. Concern for resources has been replaced to an extent by concern for a broader entity — the environment. This adds a further complication to the problem confronting rural Australia. These problems are, as they have always been, economic, social, political and environmental. We are now in a phase of introspection, doubt and internationalisation in a changing world of unsure markets and stressed environments. While rural Australia is being forced to redefine its role, most of the outcomes of the debates about appropriate resource use are simply not able to be predicted.

This chapter has been about hindsight. Hopefully the lessons of hindsight, so briefly sketched here, can illuminate why the present ecological, social and economic landscapes of rural Australia have the features they do. Yet one great lesson should be kept in mind — while we can expose causes, we should shy away from appropriating blame to those in the past. This is easily and often done, but it is both unfair and unfruitful. Geographer Joseph Powell sums this up well in discussing the age of pastoral expansion in Australia:

> . . . its inhabitants were required to participate in a new era of Western progress long before there had been time to achieve more than a rudimentary grasp of the continent's most crucial ecological characteristics.
>
> They were then deeply engrossed in a variety of political and economic issues, and obviously, neither they nor any of their fellow colonists can be blamed for activities which had the full sanction of their times (Powell 1976: 3, 31).

The history of rural Australia has seen many cases of degradation both of the resource base and the native biota. It also has extended at times to resident human populations (of all races). Despite some early warning signs, both the state of knowledge and the nature of production (which existed within a framework of nation-building capitalism) prevented the introduction of measures to address environmental destruction. Degradation continues, but the plea of ignorance is no longer admissible and becomes less so each year. Uncomfortably, then, this forgiveness of blame cannot be extended to current generations, engrossed as we may be in our own economic 'crises'. We now possess the knowledge — if not the wherewithal — to achieve a more sustainable future.

REFERENCES

Australian Bureau of Agricultural and Resource Economics (1989) *Commodity Statistical Bulletin, December 1989,* Australian Government Publishing Service, Canberra.

Australian Surveying and Land Information Group (1990) *Natural Vegetation: Australia's Vegetation in the 1780s* and *Present Vegetation: Australia's Vegetation in the 1980s,* 1:5 million map series, AUSLIG, Canberra.

Bolton, G. (1981) *Spoils and Spoilers: Australians make their Environment 1788–1980,* Allen and Unwin, Sydney.

Boyden, S., Dovers, S. and Shirlow, M. (1990) *Our Biosphere Under Threat: Ecological Realities and Australia's Opportunities,* Oxford University Press, Melbourne.

Currie, M. (1825) 'Journal of an Excursion South of Lake George in New South Wales in 1823', in B. Field, *Geographical Memoirs of New South Wales,* John Murray, London.

Davidson, B. (1976) 'History of the Australian Rural Landscape', in G. Seddon and M. Davis (eds), *Man and Landscape in Australia: Towards an Ecological Vision,* Australian Government Publishing Service, Canberra.

Davidson, B. (1981) *European Farming in Australia: An Economic History of Australian Farming,* Elsevier, Amsterdam.

Hancock, W. (1930) *Australia,* Ernest Benn, London.

Heathcote, R. (1975) *Australia,* Longman, London.

Heathcote, R. (1987) 'Pastoral Australia', in D. Jeans (ed.), *Australia — A Geography,* Volume 2, *Space and Society,* Sydney University Press, Sydney.

King, C. (1957) *An Outline of Closer Settlement in New South Wales,* Government Printer, Sydney.

Laut, P. (1988) 'Changing Patterns of Land Use in Australia', in Australian Bureau of Statistics, *Year Book Australia 1988,* Australian Government Publishing Service, Canberra.

Lawrence, G. (1987) *Capitalism and the Countryside,* Pluto Press, Sydney.

Lhotsky, J. (1835) *A Journey from Sydney to the Australian Alps Undertaken in the Months of January, February and March, 1834.* Reprinted as A. Andrews (ed.), (1979), Blubber Head Press, Hobart.

Powell, J. (1976) *Environmental Management in Australia 1788–1914. Guardians, Improvers and Profit: An Introductory Survey,* Oxford University Press, Melbourne.

Ratcliffe, F. (1947) *Flying Fox and Drifting Sand,* Halstead Press, Sydney.

Rolls, E. (1985) *More a New Planet Than a New Continent,* CRES Paper 3, Centre for Resource and Environmental Studies, Australian National University, Canberra.

Scott, P. (1987) 'Rural Land Use', in D. Jeans (ed.), *Australia — A Geography,* Volume 2, *Space and Society,* Sydney University Press, Sydney.

Tsokhas, K. (1990) *Markets, Money and Empire: The Political Economy of the Australian Wool Industry*, Melbourne University Press, Melbourne.

Wadham, S. (1967) *Australian Farming 1788–1965*, Cheshire, Melbourne.

Wadham, S. and Wood, G. (1939) *Land Utilization in Australia*, Melbourne University Press, Melbourne.

Williams, M. (1975) 'More and Smaller is Better: Australian Rural Settlement 1788–1914', in J. Powell and M. Williams (eds), *Australian Space Australian Time: Geographical Perspectives*, Oxford University Press, Melbourne.

Woods, L. (1984) *Land Degradation in Australia*, Australian Government Publishing Service, Canberra.

2 AN ECOLOGICALLY UNSUSTAINABLE AGRICULTURE

CHRIS WATSON

All is not well on the farm! Australians generally are well aware of the financial hardships now being suffered by many farmers, particularly since the dramatic drop in both wool and wheat prices in 1990. There is also some awareness of land degradation problems such as soil erosion and salinity, yet very few people see these as the symptoms of an underlying problem. That problem is that the whole farming system, as currently practised, is fatally flawed and ecologically unsustainable. It is a system supported by the use of non-renewable oil and fertiliser. One result is an insidious decline in the health of the landscape itself — its soil, its flora and fauna and, finally, the well-being of its human custodians.

Since the arrival of the first European settlers in 1788 stock-raising and cropping have increasingly grown to become a national enterprise geared to supply food and fibre to distant cities at home and abroad. An export–import economy is now taken for granted. Farmers have now become isolated even within their own nation. Today, our predominantly urban population — and a rapidly increasing one at that — has little inkling that the nation's heartlands are being farmed unsustainably. The change in atmospheric gas concentrations, heralding a likely global warming, has alerted scientists to the fact that something is amiss, planet-wise. Unfortunately it seems that very few urban Australians are able to relate this notion of 'biotic crisis' (Price 1986: 91) to the Australian landscape. Australia is a land where ecosystems evolving over millennia have been decimated in two short centuries. As much as ever it continues to be relentlessly exploited for its short-term productive capacity. No wonder the quality of life for farmers and urban dwellers alike is declining.

DEGRADATION OF LAND AND STREAM

> Australia's soil resources are being mined without recognition of the fact that they are non-renewable. (CSIRO 1990:12)

All over the world as societies of hunter–gatherers were replaced by agricultural settlements a staggering price has been paid in soil loss. This happened some 10 000 years ago in many parts of the world, but only two centuries ago in Australia. Often this goes unreported because, once soil has gone, people forget that it used to be there. Yet 2000 years ago in Greece, Plato was writing of serious soil degradation as a result of farming practice: 'the rich soft soil has all run away leaving the land nothing but skin and bone' (*Timaeus and Critias*, translated by Lee 1971: 131).

Australian soils have similarly suffered wherever European settlers denuded the land of its original vegetation consisting largely of perennial trees, shrubs and grasses. For example, once the current cereal–sheep belt of south-eastern Australia was stripped last century of its native vegetation for cropping regimes, soil deterioration was rapid (Walker 1986). The degradation continues unceasingly in various guises.

We need to look beyond the obvious gully scars and billowing soil clouds generated by the wind. Usually we find that these readily visible phenomena are the culmination of insidious forms of soil decline that have been occurring over decades. Wind and water erosion are important but soil salinity, acidity and nutrient decline are considered to be of major concern. Each of these three facets of soil degradation signals unsustainable and improper land use — and it is these which should be considered in detail.

SOIL SALINITY

The clearing of hills and slopes of their perennial tree, shrub and grass cover (often a legacy of the ringbarking enthusiasm of last century) changes an area's hydrology. Less evapotranspiration occurs and consequently more rainfall enters the watertable. Watertables rise, bringing dissolved salt present in the soil to the surface and making the land unsuitable for agricultural use. Already 800 000 hectares are affected and the area is increasing; in turn, streams have become some 4 to 20 times more salty than pre-existing levels (Williamson 1990). Downstream users — such as Adelaide residents at the mouth of the Murray–Darling Basin — bear the brunt of deteriorating water quality. One obvious remedy is to regenerate the salt patches and replant trees on higher land. As farmers are often reluctant or financially unable to fence off this land from grazing (to protect the seedlings) and are likely to continue cropping, they need incentives to change.

Irrigation, too, in the semi-arid Australian clime almost always creates saline soils and streams. It may take decades but sooner or later salty waters rise close to the surface, suffocating the plant roots. The finale is a bare salt-encrusted soil, plus perhaps a few gaunt trees. Only about 6 percent of Australia's irrigated land has a subsurface drainage system to prevent accessions to the watertable (Watson 1986), hence a dire future awaits most of the 1.5 million hectares of Australia's irrigated land. Drainage schemes to drain rising watertables often proposed as simple solutions themselves are flawed; the disposal of the saline effluent is usually a little like a game of 'musical chairs'. This poses an intractable problem in our main irrigation areas. For instance, the extensive Murray and Murrumbidgee schemes are many hundreds of kilometres from the sea so that seepage into downstream agricultural lands is virtually inevitable. Also, disposal into the sea itself may severely affect marine life. Widespread installation of effective drainage networks is unlikely because of the sheer scale of disposal problems and cost involved. Our irrigation areas have become a liability. The lack of drainage is one issue — but irrigation inevitably also demands profligate use of energy, fertilisers and biocides. If larger and larger areas of land are not to become salinised most of our irrigation schemes must be drastically modified or phased out, beginning with the areas of shallow saline watertables. Flooding of fields must be replaced with methods that use water more sparingly and changes have to reflect the severe environmental side-effects of applying water. Clearly we need a new kind of political and social thinking in addressing this problem.

SOIL ACIDITY

This problem has come to the fore in recent years and is now estimated to affect some 8 million hectares, mainly in the south-eastern corner of Australia (CSIRO 1990). Its presence bespeaks a gross misuse of land. Its extent is also an indictment of a recent generation of agricultural scientists who have not foreseen the result of decades of use of 'sub and super' (subterranean clover or other legumes fertilised by superphosphate) on our cropping and grazing lands. These scientists have focused too narrowly on yield–response curves and this has prevented them from properly evaluating the chemical imbalances brought about by increased nitrogen production via the clover root nodule. Even now, costly lime amelioration is being recommended. One is inclined to view this as a bandaid measure. Research attention should rather be given to land use systems using deep-rooted grass or tree species that will utilise excess nitrogen. The concept of 'improved' pastures is now something of a misnomer. Chemical fertilisation has brought about soil acidification over a time scale of decades. Our task in reversing this trend is formidable.

SOIL NUTRIENT DECLINE

The loss of essential plant nutrients both in eroded soil and in rural produce itself is a threat to the very sustainability of our systems of land use. It is difficult to estimate the quantity of nutrients from the soil of an average paddock eroded by water and wind. However, the amounts are often likely to be large especially in cultivated and overgrazed country. By contrast, it is easy to calculate the amounts taken off the farm each year in the produce (Table 2.1). Cumulatively, they are significant. These nutrients are virtually all exported to distant markets; some 95 percent of wool ends up overseas and so does over 80 percent of our wheat. There is thus little possibility of recycling any waste. The sewage of Cairo or Tokyo is a long way from the Australian wheat–sheep belt and, in any case, most of Australia's human wastes are pumped into the sea rather than being recycled. Nitrogen, a very large component of the food and fibre exported, can be restored by the use of legumes but may result in leaching and soil acidification.

Table 2.1 Major nutrients removed annually in wheat and wool, Australia 1985–86 to 1989–90

		NUTRIENT							
	Production (million tonnes)	Nitrogen %	Nitrogen Amount ('000 t)	Phosphorus %	Phosphorus Amount ('000 t)	Potassium %	Potassium Amount ('000 t)	Sulphur %	Sulphur Amount ('000 t)
Wheat	14.7	2.0	294	0.3	44	0.4	59	0.2	29
Greasy wool	0.9	10.0	90	0.02	0.2	1.5	14	2.0	18

Source: Calculated from Australian Bureau of Statistics data.

Replacement of other major nutrients such as phosphorus and potassium is not so easy. Most Australian soils used for cropping have needed repeated dressings of artificial fertiliser containing phosphorus. Sooner or later all soils that are cropped or grazed will require dressings of other major and minor nutrients. The continued addition of artificial fertiliser is, of course, neither sustainable nor affordable in the long term. Furthermore, manufacturing processes and transportation over long distances entail great fossil energy cost. Even at the present prices for oil this is a major burden on farmers. Moreover, soluble nutrients in fertilisers are often leached into groundwaters or washed into streams, creating hazards to water supplies. The emerging incidence of algal blooms in waterways is almost certainly the result of increasing phosphate levels arising largely from agricultural sources.

The very concept of an export-oriented rural production system must therefore be questioned. Grains, in particular, rapidly diminish the soil's reserve of nutrient. The existing monocultural emphasis on wheat (with wool) will have to be replaced by more diverse rotations. Less cropping must be demanded and the land fertilised primarily by wastes and excrement from regionally based populations.

A MUCH DIMINISHED BIOLOGICAL DIVERSITY

> The history of European settlement is a monumental ecological disaster. (Margules 1989: 2)

The ecological disaster to which Margules refers is the systematic replacement of Australia's biologically diverse flora and fauna with monocultural export-oriented systems of agriculture and plantation forestry. These simplified systems of land use are intrinsically unstable on ecological grounds. They are almost entirely based on introduced plants and animals, whose diverse and 'wild' ancestors have all but disappeared from their overseas locations. It is inevitable that monocultures become vulnerable to disease — especially as large fields are sown with the one variety of crop, or grazed by the same breed of animal. The use of pesticides then becomes the norm. Both plant and animal breeders are committed to producing a never-ending stream of (initially) disease-resistant plants or animals; but we have on hand only a limited supply of germplasm in artificial gene banks. One alternative to this is the use of polycultures and rotations.

In the wheat–sheep belt of southern Australia, for example, the original woodland was more or less eliminated so that cultivation could take place and crops or pastures sown. Virtually overnight, complex ecosystems (with their array of soil insects and microorganisms) disappeared except for remnants, usually on road verges. The land has indeed become a production line for wheat and wool. Naydler (1989: 30) calls it 'an industrial landscape, a huge outdoor factory'!

We must learn to regard the remnants of native vegetation as sacred groves — seedbanks for regeneration of the surrounding farmed land. Expanded areas of native habitat are essential for a myriad of functions such as stabilising the land, preventing salinisation, providing predators for pests on adjacent farmed areas, a source of genetic material for breeding purposes (allowing for the continued evolution of flora and fauna over the aeons of time). We are not talking about 'conservation' of existing wildlife areas but of the imperative to expand native habitat as a land use policy. The prospect of rapid climatic change makes the task of regeneration doubly urgent (Busby 1988). At the same time we must foster the conservation of linear corridors (Hobbs *et al.* 1990) to allow movement of wildlife and see them as the precursors and lifeline of a basically diversified landscape.

Although there has been little actual clearing of vegetation in Australia's large arid zone, or across northern Australia, grazing of cattle and sheep and the growth in numbers of feral animals has brought about serious land degradation. Gardener *et al.* (1990) document the deleterious changes under grazing of the belt of tropical woodland stretching across northern Australia. Recently the rate of degradation has begun to accelerate from the grazing impact of crossbreed cattle, introduced as being able to withstand better the climatic vicissitudes. It is not commonly appreciated that all of Australia's arid and semi-arid country (some 70 percent of the continent) is seen to be in a process of degradation or 'desertification' (Boyden *et al.* 1990). Salient indicators are the loss of the very shallow (and far more fertile) surface layer of soil together with the lack of regeneration of perennial vegetation. We therefore need both to control the spread, and population levels of, the introduced domestic animals and to control the ubiquitous feral animals as well. The latter task is formidable; the scale of the problem is exemplified by the havoc caused by the rabbit in the arid lands of the south.

Exotic plants and feral animals are seriously competing with the remaining biota over most of the continent. McIntyre (1990) believes that the effects of invasive plants are particularly severe in the east, largely because of the more intense human impact and exploitation.

EXHAUSTIBLE ENERGY SUPPLIES

> We cannot base our agriculture on fossil fuels and expect that agriculture to outlast the supply of this resource.
>
> (Weiskel 1989:103)

Australian per capita energy use is amongst the highest in the world (Boyden *et al.* 1990) and its use in the agricultural sector is no exception: mechanised farming is dependent on oil. It is not only used for powering machinery but for the long haul transportation of fertilisers, pesticides and produce.

Fertilisers and pesticides themselves embody considerable

fossil energy in their manufacture, while their application often hides a declining soil fertility. Increases in production and even maintenance of existing levels of production often simply reflect the high rates of fertiliser and biocide used. These inputs are partly replacing the non-seed functions of plants such as vigorous root systems and the generation of compounds to repel pests. Almost 60×10^{15} joules (60 PJ) of energy are now used annually on Australian farms (Australian Bureau of Agricultural and Resource Economics (ABARE) 1991a). This quantity of energy is around 50 percent above that used 15 years ago. Yet this is only the beginning of the energy input–output chain. For example, some three times more energy is required before rural produce gets 'to the table' (Watt 1984).

The undue dependence of our agricultural and forestry systems on fossil energy, particularly oil, makes them extremely vulnerable to shortages and price rises; for example, many farmers were severly affected by the large increases in oil prices during the Gulf Crisis of 1990, which came as it did at a time of rapidly falling wool and wheat prices. Increasingly, oil prices will be dictated by those few countries which hold the major reserves (Hall 1990). The whole economies of oil-dependent countries like Australia can therefore be readily destabilised. For this reason, if for no other, Australian agriculture must change to practices that minimise dependence on oil. At present, although minimum tillage has reduced the energy used for cultivation in much of the cereal belt, this has been offset by an increasing use of herbicides. Present programs of breeding and introducing herbicide-tolerant crop species offer only very short term solutions since they simultaneously create continuing dependence on fossil fuel energy and ecological destruction. The use of nitrogenous fertilisers — very energy demanding in their manufacture (Stout 1990) — is to be deplored. Instead, we need to use perennial legume-based rotations to rejuvenate the soil after cropping. These rotations can be grazed and represent a far more energy-conserving system than annual crops.

In our quest to reduce fuel use there has to be a radical change in the distribution of our population — away from cities into the productive lands themselves. Only then has the huge amount of fuel necessary for transporting farm inputs and outputs a chance of being minimised. For this and other fundamental reasons such as the feasibility of recycling sewage, human communities need to live close to their source of food (without, of course, occupying vast tracts of the most productive farmland).

AUSTRALIA AND THE PROSPECT OF RAPID CLIMATIC CHANGES

> The population of the earth is such that the next big climatic event will produce problems of a magnitude never seen before.
>
> (Bryson 1988: 14)

Most scientists agree that major climatic changes will occur globally as the result of human-induced changes, mainly from the use of fossil energy and the razing of forests and other vegetation. We should hardly be surprised that climatic effects have begun to appear, given the surge in population, world-wide: 'profound changes in the biosphere and the geosphere have been the hallmark of [humankind's] recent and rapid emergence as the dominant species of the earth and of the accelerating growth of our global technological civilization' (Price 1986: 91).

Australia, with its high per capita energy consumption, is adding disproportionately to the burden of 'greenhouse gas' world-wide and hence to global warming and other climatic effects. In addition to carbon dioxide emissions from the burning of oil, agricultural operations also result in the venting of gases such as methane and nitrous oxide (Ehrlich 1990). Ruminants such as cattle and sheep, with which Australia is abundantly stocked (some 152 million sheep and 24 million beef and dairy cattle, ABARE 1991b), are a major source of methane, while the growing use of artificial nitrogenous fertiliser is a source of nitrous oxide.

Climatic change in Australia is likely to induce a significant shift in rainfall distribution with heavier downpours and an increase in the likelihood of severe floods and droughts (CSIRO 1990). For example, the cereal belt of southern Australia, especially in Western Australia, could become drier, adding yet another impact to an already ecologically impoverished region.

On the east coast of Australia higher rainfall intensities are likely to increase the already serious soil erosion problem. We also need to be concerned about the effect of climatic change on the 'tiny islands' of native habitat that remain in a 'sea of wheat'. Land restoration is essentially dependent on revegetation with native species, many of which may not be able to adjust to the projected rate of temperature rise. In other words, the ever-continuing life-giving evolutionary process is put in jeopardy.

Unless global change occurs in a manner which will cut the emission of greenhouse gases, Australia can expect rapid climatic change in rural areas. There have been calls world-wide for a 20 percent reduction in carbon dioxide emission over the next 15 years but very few countries are taking it seriously. Even though the Australian government has given its endorsement in principle, energy use continues to increase (ABARE 1991a). Moreover, no government has yet called for a reduction in domestic animal numbers and in fertiliser use for the express purpose of cutting emissions of methane and nitrous oxide. Few governments have addressed the most fundamental culprit of all — the rapidly increasing human population.

DOMINATION OF WORLD MARKETS AND TECHNOLOGY

> The farmer today has become a hired-hand of industrialised society.
> (Fukuoka 1987: 17)

Australian rural production by and large is not geared to local needs but does the bidding of overseas buyers. As mentioned earlier we produce and export virtually all our wool and four-fifths of our wheat, while in recent decades we have sent away large amounts of timber from native forests as woodchips. The wholesale shipping abroad of produce is seen by some as a form of exploitation by the present major economies and transnational corporations. Lawrence and Vanclay (1992) have referred to the process as 'environmental imperialism' while Coombs (1990) suggests we are in a period of 'neo-colonialism' as a nation progressively subordinated to the world economy. Wheelwright (1990) argues that we have the 'export or die' mentality and in the process since much of our export capacity comes from the land our regional ecosystems become impoverished. The benefits flow to big cities and foreign markets, which either readily ignore or are ignorant of environmental deterioration such as soil nutrient diminution and loss of biological diversity. The impact on local human communities, let alone on the generations to come, is rarely considered by those consumers in distant marketplaces. Customers worldwide have become used to the great range of goods coming from afar. The bonanza may not continue (see Crawford and Marsh 1989). Since the onset of industrialisation in the 1700s and the concomitant rise in the world's population from one to five billion people, the decimation of the earth's biota by farming has been massive (Richards 1986). The situation worsens, yet mainstream economists still see the world as a 'single' global trading system and as one with abundant resources or where the possibility always remains of satisfactory substitution. Such presuppositions are now increasingly being shown to have no ecological justification.

Rather than supplying needs of food and fibre in ecologically benign ways, current practices may be viewed as ephemeral —

propped up on fossil inputs of one form or another and relying on a depauperate gene pool not specifically adapted to the soils and climate.

The science and technology serving our industrialised and market-oriented society have proved no panacea for the land — quite the reverse. The training of scientists has become designed to produce specialists who see 'a caricature of the world that is designed to exclude everything' outside the immediate issue under investigation 'in order to produce a "clean" result' (Ornstein and Ehrlich 1989: 198). Goldsmith (1989) regards this pursuit of science and technology, to which modern society is so committed, as leading inevitably to the destruction of all living things. Australian science is no exception — our agricultural and forestry science is largely geared to supporting the existing rural production systems and has so far shown little perception of the complexities of ecological sustainability.

POPULATION GROWTH

> We are creating a giant, crowded 'monoculture' of human beings.
> (Ehrlich and Ehrlich 1990: 143)

The very concept of ecologically sustainable systems of land use assumes that human numbers and demands must be kept in check. Most of the earth's biological systems are reeling under the impact of burgeoning numbers of people and their domestic crops and stock, plus profligate lifestyles; the worldwide human increase is over 90 million per year and in Australia it is around 250 000 per year. Australia's population is rising rapidly, growing now at a rate of 1.5 percent per year according to the Australian Bureau of Statistics.

It is salutary to reflect on the level of the Aboriginal population at the time of the European invasion — estimated to be around one million (Thorne and Raymond 1989). This number is an indication of a sustainable human population level for the continent with ecosystems assumed to be in good health without the impact of domestic stock or crops and no input of fossil resources. On this basis, today, the long-term 'carrying capacity', with so many ecosystems severely degraded (and some completely obliterated), would only be in the hundreds of thousands. This is not to suggest, of course, that the Australian population should fall to this level. But it is to suggest that there is a crucial need to provide a sustainable future for the present level.

Population control in Aboriginal culture was a basic principle in the maintenance of the ecosystems themselves (Gebbie

1981). Moreover, Gebbie's investigations show that the people were healthy. He argues, in particular, that mothers were well prepared both physically and mentally for child-rearing. In contrast, the health of the modern Australian leaves a lot to be desired; not only is there the stress of city life, where epidemics can readily spread, but we also suffer from degenerative diseases due to diets that bear no relation to the food on which we evolved for almost all of our evolutionary history (Crawford and Marsh 1989; Cohen 1989). Meats from domesticated animals with their high proportion of saturated fats may be less desirable nutritionally than the flesh of game, fish and other sources which provide unsaturated fats essential for the development of healthy nervous and vascular systems (Hill and Hurtado 1989). Australians are in a double bind: not only are we grazing, unsustainably, many millions of domestic stock for the food (and clothing) of rising populations both here and overseas, but the meat itself contributes to the onset of degenerative diseases.

CONCLUSION

> In Australia there is not the continuity and congruency of land, population, history, tradition and language that knit together a people's soul.
> (Jose 1985: 314)

The evidence of 'terminal disease' (in ecological terms) in the biological health and productivity of the Australian landmass is overwhelming. In just two centuries, European settlers have imposed a land use system of domesticated crop, pasture and stock — a system seriously flawed because of its reliance on exhaustible inputs and genetic simplicity. This replaced a relatively stable and biologically diverse environment on which probably less than one million humans depended. Now, not only is the land asked to supply food and fibre for over 17 million local inhabitants (with numbers growing rapidly), but it is farmed year by year to produce enough food for, say, 50 million or more and clothes for perhaps another 300 million people overseas. Such an operation can only last while: (1) oil remains available and affordable; (2) soil fertility and supply of nutrients last; (3) there remains suitable germplasm for breeding purposes; and (4) the predicted climatic perturbations do not take a heavy toll.

Australian society has hardly yet glimpsed the gravity of the situation. We are still in the grip of an economic system which is consumer oriented, based on the export of minerals and rural products in return for the import of manufactured goods (and such is our level of consumer demand that we can't even balance our 'economic books' — overseas debt is presently well over the $100 billion mark). Our citizens are mainly city dwellers (many of recent overseas origin) who have no roots in the land or attachment to it. There is but a small commitment 'to place'. It will be difficult to promote the message of the ecological reformers. Goldsmith (1988: 177) says that modern society will have to 'methodically and systematically deindustrialise and decentralise'. Ornstein and Ehrlich (1989: 198) call for a 'conscious cultural evolution' in which human thinking embraces the long time-frames involved. Certainly, as a species, we do have the capacity to understand our past and the future implications of our action or inaction. Knowledge is not hard to gain for Australians. What has so far eluded us is putting it to work intelligently to serve this country with its own unique ecology.

If Australia is to move toward ecological sustainability in which humans have a place of pride and dignity, there will have to be a marked change in our sense of values, and a major reorganisation of our living patterns. The first step must be to question existing levels of population increase and to reduce dramatically the populations of domestic farm animals. The society will, perforce, put self-reliance first and live in geographical 'bioregions' where international trade in rural produce will be greatly reduced.

REFERENCES

Australian Bureau of Agricultural and Resource Economics (1991a) *Projections of Energy Demand and Supply, Australia 1990/91 to 2004/5*, Australian Government Publishing Service, Canberra.

Australian Bureau of Agricultural and Resource Economics (1991b) *Agriculture and Resources Quarterly* 3 (1).

Boyden, S., Dovers, S. and Shirlow, M. (1990) *Our Biosphere Under Threat: Ecological Realities and Australia's Opportunities*, Oxford University Press, Melbourne.

Bryson, R. (1988) 'Civilisation and Rapid Climatic Change', *Environment Conservation* 15: 7–15.

Busby, J. (1988) 'Potential Impacts of Climate Change', in G. Pearman (ed.), *Greenhouse: Planning for Climatic Change,* CSIRO, Melbourne.

Cohen, M. (1989) *Health and the Rise of Civilization*, Yale University Press, New Haven.

Coombs, H. (1990) *The Return of Scarcity: Strategies for an Economic Future*, Heinemann, London.

Crawford, M. and Marsh, D. (1989) *The Driving Force: Food, Evolution and the Future*, Heinemann, London.

CSIRO (1990) *Australia's Environment and its Natural Resources: An Outlook*, CSIRO, Canberra.

Ehrlich, A. (1990) 'Agricultural Contributions to Global Warming', in J. Leggett (ed.), *Global Warming: The Greenpeace Report*, Oxford University Press, Oxford.

Ehrlich, P. and Ehrlich, A. (1990) *The Population Explosion*, Simon Schuster, New York.

Fukuoka, M. (1987) *The Natural Way of Farming: The Theory and Practice of Green Philosophy*, Japan Publications, Tokyo.

Gardener, C., McIvor, J. and Williams, J. (1990) 'Dry Tropical Rangelands: Solving One Problem and Creating Another', *Proceedings of the Ecological Society of Australia* 16: 279–286.

Gebbie, D. (1981) *Reproductive Anthropology — Descent Through Woman*, John Wiley and Son, New York.

Goldsmith, E. (1988) *The Great U-turn: De-Industrialising Society*, Green Books, Devon.

Goldsmith, E. (1989) 'Towards a Biospheric Ethic', *The Ecologist* 19 (2): 68–75.

Hall, C. (1990) 'Sanctioning Resource Depletion: Economic Development and Neo-Classical Economics', *The Ecologist* 20: 99–104.

Hill, K. and Hurtado, A. (1989) 'Hunter–gatherers of the New World', *American Scientist* 77: 437–443.

Hobbs, R., Saunders, D. and Hussey, B. (1990) 'Native Conservation: The Role of Corridors', *Ambio* 19: 94–95.

Jose, N. (1985) 'Cultural Identity', in S. Graubard (ed.), *Australia: The Daedalus Symposium*, Angus and Robertson, Sydney.

Lawrence, G. and Vanclay, F. (1992) 'Agricultural Change and Environmental Degradation in the Semi-Periphery', in P. McMichael (ed.), *Food Systems and Agrarian Change in the Late Twentieth Century*, Cornell University Press, Ithaca.

Lee, H. (1971) Translation of Plato's *Timaeus and Critias*, Penguin, Baltimore.

McIntyre, S. (1990) 'Invasion of a Nation: Our Role in the Management of Exotic Plants in Australia', *Australian Biologist* 3 (2): 65–74.

Margules, C. (1989) 'Introduction to Some Australian Developments in Conservation Evaluation', *Biological Conservation* 50: 1–11.

Naydler, J. (1989) 'The Work of a Gardener', *Resurgence* 132: 29–31.

Ornstein, R. and Ehrlich, P. (1989) *New World, New Mind: Moving Toward Conscious Evolution*, Doubleday, New York.

Price, R. (1986) 'Global Change — Geological Processes Past and Present', *Episodes* 9 (2): 91–94.

Richards, J. (1986) 'World Environmental History and Economic Development', in W. Clark and R. Munn (eds), *Sustainable Development of the Biosphere*, Cambridge University Press, Cambridge.

Stout, B. (1990) *Handbook of Energy for World Agriculture*, Elsevier, London.

Thorne, A. and Raymond, R. (1989) *Man on the Rim: The Peopling of the Pacific*, Angus and Robertson, Sydney.

Walker, P. (1986) 'The Temperate Southeast', in J. Russell and R.

Isbell (eds), *Australian Soils: The Human Impact*, University of Queensland Press, St Lucia.

Watson, C. (1986) 'Irrigation', in J. Russell and R. Isbell (eds), *Australian Soils: The Human Impact*, University of Queensland Press, St Lucia.

Watt, M. (1984) 'An Energy Analysis of the Australian Farm System', *Energy in Agriculture* 3: 279–288.

Weiskel, T. (1989) 'The Ecological Lessons of the Past: An Anthropology of Environmental Decline', *The Ecologist* 19: 98–103.

Wheelwright, T. (1990) 'Should Australia be Striving to Become More Competitive in the International Marketplace?', in A. Gollan (ed.), *Questions for the Nineties*, Left Book Club, Sydney.

Williamson, D. (1990) 'Salinity — An Old Environmental Problem', in Australian Bureau of Statistics *Yearbook Australia, 1990*, Australian Government Publishing Service, Canberra.

3 AGRICULTURAL PRODUCTION AND ENVIRONMENTAL DEGRADATION IN THE MURRAY–DARLING BASIN

GEOFFREY LAWRENCE and FRANK VANCLAY

The Murray–Darling Basin (MDB) is Australia's most productive agricultural region; it is also experiencing severe environmental degradation. Although by world standards farms are large and efficient, employing advanced agro-industrial inputs to increase the volume of output available for export, existing production regimes are presently viewed as being non-sustainable. Yet Basin producers are structurally bound to the present system of farming. They have quite significant investments in plant and equipment and, in the context of general and widespread debt, are seeking to trade out of their difficulties by employing more of the same technologies and methods.

While it is certainly true that subsidies by governments in competing countries and world agricultural surpluses — all in the context of global recession — have resulted in falling real prices for Australia's primary products, there is little likelihood that trade reform will provide any immediate relief. Basin producers face a situation of falling terms of trade and economic uncertainty. Both militate against the adoption of more sustainable agriculture practices and result, instead, in even greater pressure on the environment.

At the macro level, the Australian economy is precariously placed in a world where income from the sale of raw materials is no longer sufficient to pay for the imports of manufactured goods. The export of even more rural products — based on the application of the latest biotechnological and other scientific inputs which increase efficiency and productivity levels — is considered to be part of the solution to falling terms of trade in agriculture. The other is to 'value add'.

This chapter considers both of these options and concludes that neither, in its present form, is likely to result in the development of a more environmentally sound agriculture in the MDB. The reason is that the structure of the Australian economy is such that the intensification of agricultural production is viewed by increasingly non-interventionist governments as an essential ingredient to enhanced export earnings. It is argued that with a compliant state unable (in relation to its own fiscal problems) and unwilling (in relation to transnational capital) to direct the process of capital accumulation in agriculture, there will be continuing environmental degradation within the Basin.

THE MURRAY–DARLING BASIN — AN OVERVIEW

The MDB covers an area of inland Australia comprising one-seventh of the nation (see Figure 3.1). With over 11 000 kilometres of waterways draining an area of one million square kilometres, the Basin is the fourth longest river system in the world. The area is noted for its flatness, low rainfall, extremely low runoff and great variability of water flow. European settlement of the inland was considered to be hampered by unpredictable climatic patterns and a relatively unproductive vegetative cover. The attempted solution was to clearfell most of the country and to build a series of dams, locks and irrigation channels — including the diversion of melting snow from the Great Dividing Range — to provide a constant water supply to inland farmers. Today, rural industries in the Basin — predominantly broadacre wheat cropping, extensive wool and beef production and intensive horticulture — account for one-third of the nation's total agricultural output. The Basin contains 12 percent of the nation's population, 40 percent of its farms (including one-quarter of its dairy farms), over half its sheep, lambs and cropland and three-quarters of its irrigated farmland. Annual production is currently about $10 billion (Crabb 1988; Cook 1989; MDB Ministerial Council 1990).

Soils in the Basin are thin and fragile, necessitating continuous and heavy applications of phosphatic and nitrogenous fertilisers to boost plant growth (Ockwell 1990). Recurring patterns of droughts and floods together with continuous cultivation have exposed soils to eroding winds and rains. There have been 9 major droughts and at least 30 severe droughts in the past century. The estimated cost of the 1982–83 drought was $2.5 billion which, with multiplier effects taken into account, cost the nation $5.0 billion in lost revenue (see Ockwell 1990; *Australian Farm Journal* November 1991). The 1991–92 drought is expected to reduce average wool cut by about 6 percent and wheat production by some 27 percent (*Australian Farm Journal* November 1991).

Farm families own and operate approximately 90 percent of the Basin's agricultural establishments. On average, there is

Figure 3.1 The Murray–Darling Basin

less than one hired worker per farm (Lewis 1990). Approximately one-third of farms have a member or members working off the farm as a means of supplementing income from agriculture (see Lawrence 1987).

In line with Australia-wide trends in broadacre production, the top 25 percent of farmers produce about half the gross value of output and do this with half the land and 40 percent of the industry's capital. This contrasts with the bottom 25 percent who produce only 10 percent of the industry's output using 15 percent of industry capital (Ockwell 1990). The government has viewed the former as those most likely to succeed and provides opportunities such as cheap credit to allow expansion; it views the latter as being in need of 'structural adjustment' to become larger and more productive or to leave the industry altogether (Musgrave 1990). In concert with normal market pressures, the Rural Adjustment Scheme has operated to decrease the number of rural holdings and to increase the size of the remaining properties (Bell and Pandey 1986). An expanding share of income is being obtained by larger farmers, with an increasing number of otherwise unviable producers (those with less than $10 000 per annum income from agriculture) continuing in farming through pluriactivity and/or poverty level incomes. It is acknowledged that:

> the tendency for farming to require greater entrepreneurial and technical skills will accelerate [and] the trend towards a bipolar distribution of farms will continue with increasing proportions of large and sometimes corporate farms at one end of the spectrum and small specialist and hobby farms at the other (Department of Primary Industries and Energy (DPIE) 1989: 21).

Consistent with findings in the US (see Buttel and La Ramee 1987), it is the 'middle' farmer with limited capital and isolated from opportunities for off-farm work in the larger regional centres who is under the most pressure to 'structurally adjust'. The emerging bipolar distribution (Bureau of Rural Resources 1991) is consistent with tendencies toward concentration and centralisation of capital in agriculture (Lawrence 1989).

Farmers who have been deemed the most successful have sought to improve their productivity as a means of offsetting declining terms of trade (Wonder and Fisher 1990). However, many who have borrowed to purchase new, more productive equipment or for farm expansion have been burdened with high interest payments. Farm debt has grown considerably — in line with the long-term tendency of farmers to substitute capital for labour. Debt repayment levels are now so high that they represent a barrier to further growth in farm capital stock, potentially threatening productivity improvements (Powell and Milhouse 1990).

Most land in the MDB is held under freehold tenure, although, in Australia generally, less than 13 percent of land is held privately — with the majority of leased lands being in the arid pastoral regions. With freehold tenure, landholders are able to buy and sell at will. The state retains rights to tax and to impose land use restrictions. Under the system of leasehold, lands are leased to pastoralists at below-market rates (but calculated on the carrying capacity of the land). For nominal rents, farmers are expected to maintain and improve the land. For many farmers, the land is most often 'improved' by clearing it! In fact, in past decades statutory requirements generally defined 'improvement' as clearing the land of trees to make it suitable for grazing activities. Despite penalties for intentionally degrading the land, there is little incentive for farmers to adopt more sustainable practices (see Campbell and Dumsday 1990).

In the irrigation areas — which past governments have developed and funded as a means of improving Australia's agricultural output — governments have imposed land tenure and production controls in accordance with closer settlement principles of the Home Maintenance Area (HMA) (one farm — one family). Land is held under both freehold and restricted title with the HMA set at 32.4 ha for fruit tree and vine production and 304 ha for field crop (particularly rice) and pasture production. An individual is prohibited from owning more than 140 percent of an HMA. Each HMA has an associated water entitlement designed to provide sufficient water for irrigation (see Committee to Examine Changes . . . 1988).

It is estimated that some 97 percent of the Basin's wool, 80 percent of its wheat, and 50 percent of its beef is exported (see Crabb 1988) and enters a volatile world market in a largely unprocessed form. In the mid-1980s, and again in the early 1990s, price variability and unpredictability has resulted in widespread economic difficulties among Basin farmers. With present depressed prices for wheat and wool, the average farmer in the Basin will receive a net farm income of $2100 in 1991–92. This is a 30 percent reduction on the previous year, with a further 21 percent reduction expected in 1992–93 (*The Australian* 26 June 1991; *The Bulletin* 16 July 1991). The terms of trade are expected to fall by 15 percent in 1991–92 to one of their lowest recorded levels. The economic conditions faced by farmers are considered to be as severe as they were in the depressions of the 1930s and 1890s. It is considered that one-third of producers will be removed from agriculture in the next decade (*The Bulletin* 16 July 1991). In line with the declining prospects for agriculture, land values have fallen, with prime grazing land near the Lachlan river selling for 30 percent of its pre-1990 level (*The Sydney Morning Herald* 6 June 1991). Some lands offered for sale are not being purchased (one reason why some banks have agreed to a moratorium on foreclosures). Farmers who might otherwise sell are therefore prevented from doing so. Instead, they remain in farming with little prospect of ever meeting high debt repayments. To help provide support, the number of rural counsellors has doubled in the past year (*The Weekly Times* May 1991) and some farmers are now able to apply for unemployment benefits.

Recent surveys have confirmed that rural producers are under quite severe financial pressure. In one Australia-wide survey, 32 percent of farmers considered their debt to be 'financially crippling' and indicated they were about to abandon farming, while some 27 percent complained of a deteriorating quality of life and indicated that family relations were suffering (ANOP 1991). In another two surveys of farmers within the Basin, over 70 percent believed their income was insufficient to maintain an acceptable lifestyle, less than half had taken more than a week's holiday in the past 12 months and farm operators reported working an average of 6.4 days per week and their spouses 5.8 days (on-farm versus off-farm was not disclosed). In comparison with city dwellers, family farm members were less likely to pursue social and cultural activities (Farmfacts 1991). Significantly, in terms of on-farm practices: the vast majority revealed that their machinery had been purchased in the late 1970s or early 1980s; only 6 percent had applied lime to their land as a means of redressing acidity (although agronomic knowledge indicates that 80 percent of the land would respond to liming); over 60 percent believed more labour was required to manage effectively their properties — but only 5 percent believed they could afford to hire labour; and approximately 70 percent reported static or declining crop yields and/or declining stocking rates. About one-quarter of respondents attributed productivity decline to their reduced ability to purchase the necessary inputs to farming (see Farmfacts 1991).

Despite evidence of the uncoupling of country towns from agriculture (Reeve and Stayner 1990) the economic fate of Basin communities remains quite heavily dependent on rural enterprise. For each dollar generated by agriculture, there is a further six dollars worth of consumption and employment generated in local towns (Beale and Fray 1990). The Australia-wide non-farm multiplier for agriculture is calculated at between 0.5 and 1.0 (Crofts, Harris and O'Mara 1988). The predicted fall of $4 billion in real gross farm output for 1991–92 will produce a fall of between $2 billion and $4 billion in non-farm output (Knopke and Harris 1991) with obvious consequences for regional areas.

One outcome of the deteriorating conditions of agriculture and the effects of the current recession on both regional and private economic activity has been the decline in service provision in smaller country towns. Towns of below 2000 in population are considered to be most likely to suffer reductions (Austin 1990a). Agribusiness rationalisation has reduced the number of farm machinery dealerships from 2400 in 1990 to 950 in 1991. Hundreds more are expected to close in the immediate future (Farmfacts 1991). Federal and state governments — whose agendas have moved away from supportive social welfare interventions to those of rationalisation and regionalisation of services (see Lawrence and Williams 1990) — have been responsible for the underfunding and under-resourcing of programs and services in inland Australia (Collingridge 1991). This is despite evidence that rural residents experience more severe health and welfare problems than people in cities (Lawrence 1987; Lawrence and Williams 1990).

ENVIRONMENTAL PROBLEMS IN THE MURRAY–DARLING BASIN

The MDB is one of Australia's most degraded environments. Aspects of degradation include salinisation, acidification, deteriorating soil structure, loss of topsoil through erosion, weed growth, water turbidity, destruction of wetlands, species decline and pollution from agrichemicals (O'Reilly 1988; MDB Ministerial Council 1990; Ecologically Sustainable Development Working Groups 1991).

There are rising salt levels in both irrigated and dryland farming systems. With the removal of tree cover, particularly in highland regions, and with the establishment of irrigation areas, groundwater levels have increased, bringing salts, from dissolving rock beds, to the soil surface (Vanclay and Cary 1989). Salts have also moved into the Basin's river systems (MDB Ministerial Council 1990), creating drinking water supply problems for population centres downstream. Salinisation caused by irrigation affects over 1200 square kilometres of the Basin and an even larger region is affected by dryland salinity (Crabb 1988). One-quarter of Victoria is salt affected as a direct result of tree removal. Productivity losses associated with salting are placed at $100 million per annum (Cook 1989). In the Kerang district of Victoria, one water course — Barr Creek — has recorded a salt reading of over 60 000 electroconductivity (EC) units. (The maximum level recommended in drinking water is 850 EC units and most plants cannot tolerate more than 1250 EC units; see *Time* 11 April 1988.) It is predicted that by the year 2040, some of the Basin's most productive irrigation districts will be unable to sustain fruit tree growth — the basis of their present economies (O'Reilly 1988).

Soil degradation is of particular concern. Soil formation rates throughout the Basin are so low that they are generally considered to be zero (Edwards 1988). The Basin's soil is, in other words, a non-renewable resource. For Basin conditions, the term 'tolerable soil loss' is not applicable. For each tonne of grain produced by Basin farmers about 13 tonnes of topsoil are lost (O'Reilly 1988). The generally accepted figure of soil erosion in the Basin by soil scientists is 5 mm per annum, or 50 tonnes per hectare per annum (Rickson *et al.* 1987). In fact, it has been estimated that 40 to 60 percent of Basin farmers are not adequately protecting their farms from soil erosion (Vanclay 1986). Heavy tillage has led to the loss of soil structure while each year, as a result of existing practices, some $144 million worth of nutrients are being removed from the soil and not being replaced (Fray 1991). The combined effects of salinity and soil degradation reduce agricultural production in the Basin by approximately $220 million per annum (MDB Ministerial Council 1990). The combined on-farm effects of current practices have been reduced yields, reduced crop varieties, death of soil organisms, less healthy soils and increasing costs for farmers as they seek ways to maintain viability (*Time* 11 April 1988). The cost of essential works to address environmental problems in the Basin has been put at $1.6 billion (estimated from figures in Crabb 1988).

Other concerns include the reduction in native plants and animals (10 plant species and 20 species of mammal are recently extinct; some 37 plant species and 22 species of birds and reptiles are on the endangered list) (MDB Ministerial Council 1989, 1990). Habitat destruction and wetlands degradation continue, which not only has implications for animal, plant and bird life, but will inevitably result in the loss of amenity value (see Lawrence 1990; MDB Ministerial Council 1990). According to the MDB Ministerial Commission's own calculations, native vegetation in the region is now only 50 percent of its pre-European level (and most of that is in the arid and semi-arid sections). Less than 9 percent of original vegetation remains in the sheep–wheat belt (MDB Ministerial Commission 1989).

Water quality in the city of Adelaide — where the Basin drains into the sea — has deteriorated rapidly in past decades (MDB Ministerial Council 1990). Over 1.3 million tonnes of salt are carried by the Murray River to Adelaide each year with waterborne salt levels in drier than normal seasons being in excess of World Health Organisation recommended levels (Beale and Fray 1990). Water quality is also affected by turbidity, industrial water and sewage, agrichemicals, weeds and algae (Crabb 1988).

MANAGEMENT OF THE MURRAY–DARLING BASIN

By the mid-1980s, there was recognition of the severity of the environmental problems of the Basin and of the failure of previous arrangements (such as the River Murray Waters Agreement) and organisations (such as the River Murray Commission) to address the problems (Crabb 1988; Junor 1989). A new initiative was the establishment of three bodies to manage the Basin: the MDB Ministerial Council, comprising political representatives of three state governments and the federal government; the MDB Commission, a bureaucratic arm; and the MDB Community Advisory Committee which provides opportunities for community involvement. The Ministerial Council sets goals for natural resource management and has identified the main goal as 'equitable, efficient and sustainable use of land, water and other environmental resources of the MDB' (MDB Ministerial Council 1990: 8). Two of the primary aims of the strategy are to restore degraded environments and to prevent further degradation in the Basin.

It is now considered that restoration of lands and river systems already degraded will be impossible (Junor 1989). Tree planting schemes have begun, with the federal government providing funding for 'the Decade of Landcare' and for tree planting (see Cameron and Elix 1991). However, while the Commonwealth wants to 'plant a billion trees' by the year 2000, over one billion trees have been removed from just two local government areas in Queensland since 1985, in keeping with that state government's land development policy (Beale and Fray 1990). It is estimated that 150 000 km^2 — approximately 15 percent of the Basin — requires replacement tree cover. This is equivalent to 12 billion trees costing at least $1.2 billion for the seedlings alone. This is considered a minimum to restore the Basin to a sustainable level (Beale and Fray

1990). However, despite efforts being made to address environmental problems, the cumulative effect of loss of tree cover in the past will ensure rising watertables and salinity problems for at least 50 years (Vanclay and Cary 1989; MDB Ministerial Council 1990).

Management issues in the Basin are complicated by administrative difficulties. The Basin comprises four state governments and a territory government, 33 government departments and authorities, 10 interdepartmental organisations and 256 local governments (*Time* 11 April 1988). Queensland, which covers 20 percent of the Basin and is the source of 25 percent of waters entering the Basin's river system, was not a signatory to the initial MDB agreement (Crabb 1988).

The MDB Ministerial Council's 'very limited powers' are seen to be a major barrier to success in addressing the problems of land and water use (Crabb 1988). The state governments are simply unwilling to concede powers to a trans-state authority and there are concerns about interaction between agencies. For example, local councils decry the 'paternalistic' attitudes of the federal government (see Osborne 1989) while the federal government notes the intractable nature of local government decision making (Cook 1989). Under the Australian Constitution, it is the state governments, rather than the federal government, which are responsible for land and water management. The former are noted for their disagreements, intransigence and parochial attitudes.

Priorities for research are also a concern. The MDB Commission is providing some $45 million for research into Basin problems. To date, sociological research has been accorded a very low priority, despite recommendations from the Community Advisory Committee (CAC) emploring bureaucrats to fund such research (MDBCAC 1991). At present, there is growing opinion that some Basin lands should be taken out of production and reafforestation implemented (CSIRO 1990; Cameron and Elix 1991). This is opposed by most landholders and by communities fearing declining revenues from land use changes.

One latest 'compromise' proposal to overcome salt problems is to construct a pipeline and to pump water containing 900 000 tonnes of salt per year from irrigation areas to the sea (Beale and Fray 1990). Critics consider this a 'bandaid' proposal at best — one which does nothing to confront the practices leading to environmental degradation. Moreover, such a 'solution' would appear to be deleterious to sea life (see Chapter 2).

CHANGES IN RESOURCE USE: ENVIRONMENTAL IMPLICATIONS

There is widespread recognition that existing agricultural practices must be modified (MDB Ministerial Council 1990) and in many cases abandoned (Beale and Fray 1990; Cameron and Elix 1991) before the problem of degradation is properly addressed. While farmers have been active in the Landcare movement and have argued that they are the best conservationists (see Gibson 1989) there is counter evidence to suggest that the combination of the generally small-scale nature of family farming, the cost/price pressures faced by the farming sector, and insensitive views about the environment in the past have combined to create most of the Basin's environmental problems (Junor 1989; Beale and Fray 1990). Land, water and environmental management practices are now considered to have been 'exploitative and not conducive to sustainable resource use' with 'current high levels of production having been achieved in many instances at the expense of resource(s)' (Cook 1989). The formation of the MDB Commission, the implementation of federal government policies such as the One Billion Trees Program and Landcare, and initiatives of some local councils are seen to be an acknowledgement that changes in attitudes and farming practices must take place.

The problem remains, however, that Australia is dependent upon the sale of raw materials for its export income. How then will it be possible to maintain the value of exports while improving the environment? Two initiatives form the basis of present government policy — the application of new agro-biotechnologies and 'value adding'.

BIOTECHNOLOGY

Biotechnology is being heralded as the most appropriate mechanism for both increasing agricultural productivity and overcoming many of the environmental problems associated with modern agriculture (such as the heavy use of pesticides and weedicides). Some consider biotechnologies will create the best opportunities for a sustainable future (DPIE 1989; Begg and Peacock 1990; Bureau of Rural Resources 1991).

Biotechnologies are expected to allow producers to reduce their levels of inputs (and hence costs) while achieving higher levels of output. Embryo technology, for example, may provide opportunities for transferring superior genes to existing cattle herds and sheep flocks at a lower per unit cost than normal breeding techniques. Vaccines created through biotechnology are considered to be superior to those obtained in conventional ways. Bovine somatotropin — a natural protein hormone produced through recombinant DNA technology — will allow more milk to be produced by dairy cattle from the same level of feed thereby increasing profits by lowering milk production costs (see Begg and Peacock 1990; Baumgardt and Martin 1991).

Experiments in Australian laboratories are designed to confer pest resistance on plants and so reduce or eliminate the need for chemical applications on Australian croplands. The creation of insect-resistant plant species may not only mean that fewer dangerous chemicals will be used in farming but also that the costs to farmers will be reduced. Biotechnologists are also working on ways to 'mop up' chemical pollution and to convert what are now waste materials from food manufacturing into new products.

Proponents estimate that biotechnologies may reduce the use of natural resources by between 40 and 60 percent, allowing farmers to move rapidly towards sustainable production (Begg and Peacock 1990). Threats to the further degradation of lands are expected to be averted through new genetic manipulations and applications which reduce input use and allow output increases without soil loss (Bureau of Rural Resources 1991).

Since biotechnologies are 'enabling technologies' they are likely to have different outcomes according to the purpose of their application. For Redclift (1990) biotechnology will fulfil its promise if it can encourage the development of a low-input, high-tech system of sustainable agriculture in which there are reduced applications of proprietary inputs. The hope, then, is that in line with growing public concerns for the environment, scientists will develop plants and animals with pest and disease resistance, salt tolerance and productivity-enhancing qualities which will overcome many of the problems associated with current agricultural practices (see Lowe *et al.* 1990; Baumgardt and Martin 1991). However, evidence from both Australia (see Chapter 16) and abroad (Lacy, Lacy and Busch 1988; Tait 1990; Busch *et al.* 1991) indicates that the biotechnological promise is, in the context of existing social arrangements, unlikely to be fully realised.

There are a number of concerns. First, environmentalists point out that if corporate capital is involved in the production and distribution of biotechnologies, the profit motive will distort both the basis of experimentation and the likelihood of benefits being distributed evenly amongst producers. Thus, the production of herbicide-tolerant plant species is not designed to free agriculture from chemicals but to have farmers purchase a proprietary package of herbicide and herbicide-tolerant seeds

(Kloppenburg 1988; Busch *et al.* 1991), something which will increase the dependence of farmers on the agrochemical industry and raise input costs for producers. Furthermore, with herbicide use continuing at high levels the possibility of chemical resistance among weeds is increased and there is a greater likelihood of groundwater pollution (Otero 1991).

Second, there is also no proof that genetically modified organisms will be environmentally benign. They may proliferate to occupy 'niches' in ecosystems thus displacing other organisms or produce substances toxic to other organisms. Here, the use of supposedly environmentally friendly genetically modified organisms may result in environmental decimation. Ironically, the new products may be even more dangerous than the dangerous chemicals they have been designed to replace (see Busch *et al.* 1991).

Third, if costs of the new products are reasonably high — which they are expected to be given that they will be corporate rather than state released products — the adoption of the new biotechnologies will be limited to the well-financed and usually larger farmers. That is, many of the possible environmental benefits (of reduced chemical applications) would not, in any case, be available to the often-struggling middle 'family' farmer. The very people who might have been most advantaged will inevitably fall behind, concentrating food production among those in the wealthier sector of farming. In the US, employment in farming is declining faster than virtually all other occupations. With existing trends heightened by biotechnology there will be fewer farmers (Busch *et al.* 1991). There is evidence that corporate-linked agriculture is no better, and is perhaps worse, than family farm agriculture in terms of environmental management (see Lawrence 1987; Strange 1988).

Byman (1990) considers it to be somewhat worrying that new technologies are being advanced as the answer to the problems of environmental pollution and oversupplied markets, when the applications of technologies have been the major cause of these problems. Redclift (1987), too, has argued that the future of the advanced societies such as those in the US, UK and Australia is premised upon the transformation of the environment — yet the transformation of the natural environment is occurring in a manner which reduces long-term productivity. The 'environmental contradiction' is viewed as the central contradiction of advanced capitalism (Redclift 1987; and see O'Connor 1990).

The global economy is dominated by transnational capital and it is the large, transnational agribusiness firms which are controlling biotechnological development in agriculture (Goodman, Sorj and Wilkinson 1987; Kloppenburg 1988; Otero 1991). Farming will exist, in its present form, only for as long as it can conform to the profit-making needs of firms supplying agricultural inputs and of firms involved in the food processing industry — those firms using either the direct products from farming or farming products converted for use for industrially produced 'biomass'.

VALUE ADDING IN AGRICULTURE

The second hope for the nation is to 'value add' to products before they leave Australia's shores. At present the $15 billion of agricultural goods Australia exports is currently converted into $80 billion abroad. It is argued that if these agricultural products were to be further processed in Australia, there would be higher levels of employment, higher levels of income, and enhanced foreign curency earnings (see Bureau of Rural Resources 1991).

According to the DPIE (1989: 7):

> Value adding is the essence of economic growth. Value adding is the means by which individuals and businesses meet their objectives to prosper and grow . . . if a country wants to trade for the purposes of economic growth without subsidies, it will only do so via industries, businesses and individuals who are able to compete successfully . . . Hence, value adding and the competitiveness of agribusiness are inexorably linked.

The opportunities are seen to be available for producers — such as those in the Basin currently producing largely undifferentiated food and fibre for world markets — to link with agribusiness in a manner beneficial to both parties. For agribusiness, the markets abroad are well known and already penetrated by branch firms, providing an easy entrée for those producers who seek agribusiness affiliation. For farmers, the sale of specific product lines which can be readily distinguished from those of competing producers will allow consumer brand identification; it is presumed that this will result in increased profits. With extra income, once-struggling farmers will be able to overcome debt problems and begin to undertake much needed environmental repair work. In this scenario, the further integration of family farm agriculture and international agribusiness will be a cornerstone to both improved environmental sustainability and the continuation of high export earnings — not from any increased volume of exports, but from the sale of higher value goods.

The positive environmental flow-on effects suggested above

are viewed as part of a healthy and prosperous agricultural sector. Would family farm agriculture be 'reinvigorated' by agribusiness? Agribusiness firms are renowned for their ability to organise their production and distribution activities in the input–supply and output–processing sectors without, as it were, getting their hands dirty on the farm (see Lawrence 1987; Mooney 1988; and Chapter 15). Market strength and management strategies enable agribusiness to leave the production risks with the farmer, while purchasing raw materials from the farmer as cheaply as possible. It is not on the farm where value is likely to be added but off the farm in food processing factories. The individual Basin farmer has little opportunity for value adding and product differentiation on the farm and is therefore unlikely to receive profits made by those involved in the processing industries. The question that remains is — can Australia benefit from the value adding activities of transnational agribusiness as it penetrates and helps mould family farming activities?

The answer would seem to be no. Foreign interests have determined that Australia is not the most appropriate location for value adding. For example, in 1988 five of Australia's top agricultural exporters were Japanese trading houses which sent abroad, in one year, approximately $7 billion of unprocessed food and fibre (*The Australian Financial Review* 15 March 1988).

Attempts by successive Australian governments to diversify the economy and to have foreign capital invest in food, fibre (and wider) manufacturing appear to have failed. In 1972 so-called 'elaborately transformed manufactures' (embodying high-tech processing and knowledge-intensive applications) comprised 13 percent of Australia's exports but this had fallen to 9 percent by 1986 (Fagan and Bryan 1991: 15). For the 1980s, Australia imported value added imports at a rate faster than both domestic growth in GDP and the export earnings of food and materials (Jones 1988). By the 1990s Australia had reverted to its 1930s economic base, selling 'simply transformed manufactures' (unprocessed or semi-processed raw materials) in exchange for manufactured goods.

This has placed Australia — and, of course, Basin producers — in a difficult economic position. Farming is, at best, a slow growing sector which is susceptible to world oversupply and deteriorating terms of trade. More importantly, Dunkley and Kulkarni (1990: 20) suggest:

> trade in [simply transformed manufactures] is unlikely to revive in the near future because of technological change raising global productivity, agricultural subsidies in major countries, a trend to

> self-sufficiency in developing countries, the emergence of new primary suppliers and possible reduction in demand for [some rural] products for ecological reasons.

Australia's manufacturing industry primarily constitutes branch plants of foreign transnational companies. It is being progressively locked out of Asia–Pacific markets because of cheaper production costs overseas, particularly in South-East Asia. There is no reason to believe that local or foreign agribusiness firms will discover advantages in food and fibre processing in Australia that they have been unable to obtain elsewhere. Labour in Asian countries is cheaper than in Australia so it is likely that raw materials will continue to be sent abroad in unprocessed form. This has been begrudgingly admitted by the federal government. According to the DPIE (1989: 15):

> in considering the question of adding value to Australian agricultural products, it is reasonable to argue that the value adding activity will often take place outside Australia (by companies that may or may not be Australian owned), and that this activity will be initiated by companies positioned near the retail end of the channel rather than near the raw material end.

Without tariff protection which has provided support for Australia's 'infant industries', there are few incentives for firms to move beyond simple semi-processing activities. Significantly, the processed foods area — that described as providing the best opportunities for value adding (Bureau of Rural Resources 1991) — now forms a declining proportion of total food exports (Wettenhall 1991). Bulk agricultural commodities constitute approximately 70 percent of Australia's exports (*Australian Farm Journal* June 1991) and are expected to continue to do so (DPIE 1989; Austin 1990b).

Neither biotechnology nor value adding appear capable of providing a basis, at least in the short term, for the development of a more sustainable form of agricultural production in the MDB.

ENVIRONMENTAL DEGRADATION IN THE BASIN: TOWARDS A SOCIOLOGICAL EXPLANATION

Increased mechanisation, product specialisation, farm amalgamation, the farming of marginal lands and monocultural production regimes in the MDB have led to major degradation of the physical environment. However, these characteristics and tendencies of Basin agriculture cannot be considered outside the context of agricultural policy, the socioeconomic structure of farm production and the place Australia occupies in the global order. While it is somewhat ambitious and certainly theoretically difficult to combine macro- and micro-sociological processes in a manner which explains environmental degradation, Buttel and Gertler (1982) have produced a model for the US which links a number of the above features and which has been adapted here to explain the causes of environmental deterioration in the MDB (see Figure 3.2).

Degradation of the MDB can be traced to the past role Australia has played, and the role it continues to play, in the international economic order. The nation is semi-peripheral in that it exhibits features of the economically powerful 'centre' nations (in terms of living standards, level of technology, social infrastructure and corporate presence) as well as features of the 'periphery' (reliance upon extractive industries for foreign capital inflow, balance of payments problems, little corporate research and development, dependent economic status) (Clegg 1980; Boreham *et al.* 1989).

Its settler colonial history was one marked by export of primary products from owner-operated freehold farms and from leased lands (see McMichael 1984; Friedmann and McMichael 1989). The sale of agricultural produce abroad was crucial to the development of the Basin and to the wider economy. Income from agriculture was taxed and helped to finance irrigation as well as state infrastructural expenditures (including railways, dams, irrigation schemes) and agricultural research.

On most of the Basin's typical family farm properties, capital was invested in machinery and equipment and for the purchase of other manufactured inputs. Labour displacement was one outcome; another was continued increases in output, which, in buoyant economic times, allowed for the further expansion of agriculture. The state supported this expansion through a plethora of policies which both subsidised production and assisted producers to market output. Broadacre production — and later irrigation — became the basis for economic prosperity in the Basin, with expansion based largely on overseas demand for primary products.

In more recent times — and as a result of global problems associated with world recession, international debt problems, world overproduction of agricultural products, and protection and subsidisation of overseas producers (see Goodman and Redclift 1989) — Australian farmers have faced falling prices for their commodities.

As a semi-peripheral economy Australia has been neither economically able to continue to subsidise its own producers, nor politically capable of preventing competing nations from

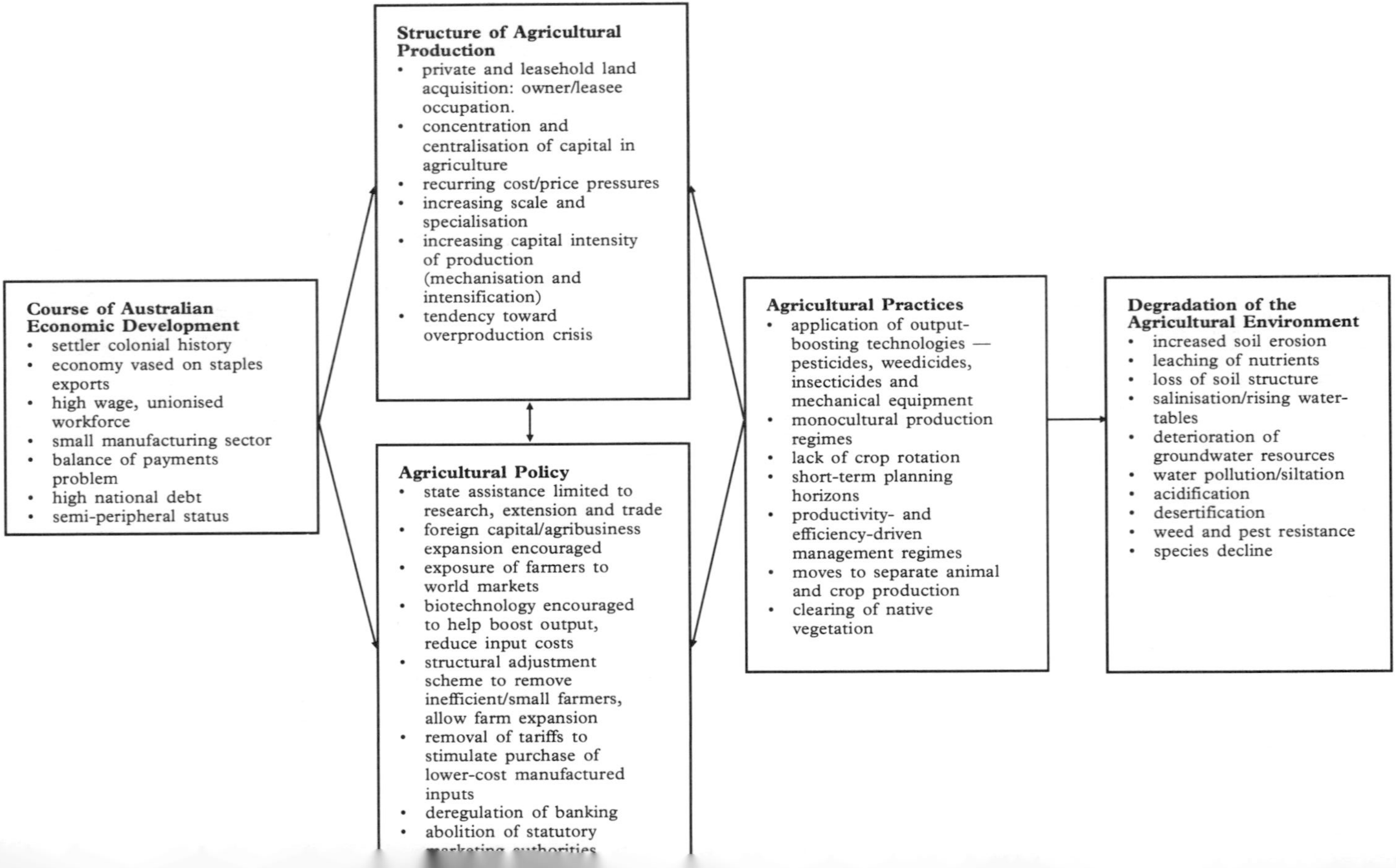

Course of Australian Economic Development
- settler colonial history
- economy vased on staples exports
- high wage, unionised workforce
- small manufacturing sector
- balance of payments problem
- high national debt
- semi-peripheral status

Structure of Agricultural Production
- private and leasehold land acquisition: owner/leasee occupation.
- concentration and centralisation of capital in agriculture
- recurring cost/price pressures
- increasing scale and specialisation
- increasing capital intensity of production (mechanisation and intensification)
- tendency toward overproduction crisis

Agricultural Policy
- state assistance limited to research, extension and trade
- foreign capital/agribusiness expansion encouraged
- exposure of farmers to world markets
- biotechnology encouraged to help boost output, reduce input costs
- structural adjustment scheme to remove inefficient/small farmers, allow farm expansion
- removal of tariffs to stimulate purchase of lower-cost manufactured inputs
- deregulation of banking
- abolition of statutory marketing authorities

Agricultural Practices
- application of output-boosting technologies — pesticides, weedicides, insecticides and mechanical equipment
- monocultural production regimes
- lack of crop rotation
- short-term planning horizons
- productivity- and efficiency-driven management regimes
- moves to separate animal and crop production
- clearing of native vegetation

Degradation of the Agricultural Environment
- increased soil erosion
- leaching of nutrients
- loss of soil structure
- salinisation/rising water-tables
- deterioration of groundwater resources
- water pollution/siltation
- acidification
- desertification
- weed and pest resistance
- species decline

financially underwriting their own farmers. Since it does not have the intention or the resources to enter a trade war with the US or the EC, Australia is limited to providing economic arguments, as one of the Cairns group of 'free traders', for the dismantling of both protectionist barriers and of subsidisation abroad. To date, the result has been less than encouraging, with Australian producers facing a highly competitive world market — one in which the US, for balance of payments reasons, is seeking expansion of agricultural sales in overseas markets (Peterson 1990) and where the EC, for domestic political reasons, has been reluctant to alter the conditions giving rise to oversupply.

The effect is that Australia continues to purchase increasingly expensive industrial items from abroad while paying for these through the sale of increasingly cheaper primary products (see Crough, Wheelwright and Wilshire 1980). The government's 'solution' to this structural terms of trade/balance of payments problem has been to develop an agricultural policy which endorses agribusiness efficiency principles for agriculture. That is, it has assisted in the restructuring of agriculture where greater volumes of output will be produced by fewer, more technologically sophisticated farmers.

Agricultural Practices are viewed in the model as being the main determinant of environmental degradation but they are directly related to both the Structure of Agricultural Production and Agricultural Policy. Agricultural Policy today is defined by principles of economic rationalism, deregulation and 'user pays'; consequently, the structure of agricultural production is being required to 'adjust' in a manner which increases scale, specialisation, mechanisation and intensification (Buttel and Gertler 1982). Structural adjustment policies help to accelerate the move to larger, more capital-intensive farming and, concomitantly, to assist in the removal of the smaller and less 'efficient' farmers. The retraction of vestiges of state support for family farm agriculture places pressure on those without sufficient capital resources for expansion to leave agriculture.

Proponents of such trends consider a 'streamlined' agriculture will be the key to Australia's success. Austin (1989: 94) optimistically views the future:

> the rural boom of the 1990s will be based on new markets, increased vertical integration, processed commodities, new commodities, more corporate farming and rapidly increasing production in northern Australia . . . large areas of the inland will open up and agricultural production will expand substantially.

Austin does not, however, evaluate the environmental costs which society would inevitably face from the greater intensification of farming. Such intensification would not occur without the greater use of existing resources and the increased application of agrichemicals and other manufactured inputs. As Munton, Marsden and Whatmore (1990: 110) have explained:

> For most farmers, choice over technological strategy is constrained. With a long term tendency for the real market price of farm products to fall, producers can only maintain or increase their margins by lowering unit costs faster than prices fall . . . the most effective means of maintaining margins has been to apply ever-increasing quantities of industrial inputs.

This is not to argue that individual producers will be incapable of altering their production regimes — even in quite dramatic ways. What we do suggest is that in aggregate terms, individual changes will be insignificant to overall Basin agriculture and will do nothing to challenge the present system of capital accumulation in agriculture. The present regime of production is bound to continue in spite of the finding that the benefits of new technologies accrue primarily to off-farm interests such as banks and agribusiness interests, especially agricultural input manufacturers and food processors (Munton *et al.* 1990).

The model recognises that Agricultural Policy has both a direct and an indirect effect on Agricultural Practices, the indirect effect occurring through its effect on the Structure of Agricultural Production. What is significant here is the realisation that the state has a major influence on the form and pace of change in agriculture (see Buttel and Gertler 1982). State policy is based on an assumption that it is largely through mechanisation and specialisation that farmers will achieve economies of scale (see DPIE 1989; Bureau of Rural Resources 1991). What is usually overlooked is that while output and efficiency — where efficiency is narrowly defined as maximum output achieved through the use of a minimal level of inputs (thereby denying energy criteria or environmental costs) — might increase, mechanisation and specialisation have a direct bearing on environmental quality. This is explained by Buttel and Gertler (1982: 105–106):

> Prevalent mechanization patterns affect environmental quality in several ways through increased soil compaction [and] by preventing or rendering more difficult the utilization of soil conservation practices . . . specialization tends to result in degradation of agroecosystems because of monoculturing and abandonment of nutrient recycling. Monocultural production tends to lead to greater reliance on pesticides and inorganic fertilizers because of the reduction or elimination of crop rotations . . . Intensification has mainly taken the form of growth in the use of commercial fertilizers and pesticides . . . The use of heavy amounts of synthetic

> inorganic fertilizers has resulted in problems of run-off, eutrophication, pollution of water supplies [and other problems identified in Figure 3.2] . . . The utilization of pesticides has been associated with poisoning of humans, animals, soil, and water; with toxic residues in food; and with the increased risk of 'system failure' as pests become resistant or as predator populations decline.

While there is evidence that Australian farmers believe strongly in stewardship of the land — considering it is essential to take care of farmlands so as to benefit future generations (Vanclay 1986; and see Chapter 6) — it is the structural conditions under which farming takes place which limit the opportunities for farmers to undertake environmentally sound practices (Lawrence 1987; Cameron and Elix 1991).

Degradation, in other words, is not an accidental outcome of present-day activities but is, rather, a logical result of the combination of the Structure of Agricultural Production and Agricultural Policy — both of which are shaped by the present Course of Australian Economic Development (see Figure 3.2). In terms of improvements to the environment, it is unlikely that much will be achieved by tinkering at the edges of existing policies. In fact, as Agricultural Policy is reduced in importance as a result of conscious government efforts to limit support for agriculture, then the relative importance of the Course of Australian Economic Development increases. Overseas organisations and international trends begin to play an expanding role in determining (through the Structure of Agricultural Production) the forms which Agricultural Practices take.

In a real sense, Japanese or Korean-based corporations will be making decisions about land use on Australian farms in a much more direct manner than in the past. If consumers from those countries demand grain-fed beef then rural Australia will respond by developing feedlots — no matter what costs might be borne by the environment (see Lawrence and Vanclay 1992). If they demand cotton rather than wool, then cotton production will expand — despite its greater environmental impact. If increased rice sales are possible, this will be viewed by Australian producers as an opportunity for niche marketing — no matter what increased irrigation will do to salt levels in the Basin. There would seem to be no halting this process. The state, in abrogating what has been its traditional responsibility for influencing resource management (Dumsday, Edwards and Chisholm 1990), is, in effect, leaving decisions about the environment to foreign capital. Foreign capital is notoriously insensitive to the environment. If Basin farmers are told that the only means of making a profit is to separate animal and crop production, increase the use of agrichemicals, use heavy machinery, plant and graze the land more intensively or adopt short-term rather than long-term planning horizons, that is what — in relation to the needs of international agribusiness — they will be forced to do.

In the context of widespread debt the adoption of short-term production strategies provides opportunities for loan repayments to be met, allowing the farmer to remain in business in the hope of more prosperous times ahead. Unfortunately, such strategies reduce the likelihood of investment in on-farm conservation measures (Buttel and Gertler 1982). Yet, according to Strange (1988: 100), 'in an age when careful use of scarse and fragile natural resources is far more important than flooding the market with food surpluses, [the level of] resource consumption, not output, should be the measure of a farm. More conservation, not more production, is needed.' For Strange, it is essential that the market be rejected as the determining factor in agricultural production. The market generates overproduction through induced technological change and farm expansion. State intervention — to control production, improve farm income, to encourage sustainable practices and to regulate and monitor environmental practices — is seen to be crucial for redressing land and water degradation caused by farm production.

What this analysis has emphasised, however, is that the federal and state governments are structurally bound to continue with the present course of Basin agriculture. They are economically incapable of providing support for new forms of production and require, instead, intensification of existing forms as a way of ameliorating Australia's balance of payments problems. The essence of this problem is Australia's dependent, semi-peripheral place in the global economy.

According to this rather bleak scenario, environmental degradation will continue despite the good intentions of landcare groups, conservationists, environmentalists and others wishing to develop a more sustainable form of agriculture in the Basin. Confusion about the form of future economic development in the Basin is best demonstrated in a publication of the Riverina Regional Development Board (1991). In that report recommendations were made for the region to increase local processing of agricultural and pastoral products, cut and pulp more timber, to increase the quantity of crops grown in irrigation areas and to expand the intensive livestock industry in line with market forces. There was no assessment made of the likely environmental impacts of these proposals despite existing evidence suggesting that intensification of existing practices would result in higher levels of pollution (Beale and Fray 1990).

Instead, the Board endorsed, as its major recommendation, the promotion of the Riverina as 'the pure food centre of Australia'.

NOTE

The first two sections of this chapter formed part of 'Rural Restructuring in the Murray–Darling Basin, Australia: Economic Imperatives and Ecological Consequences' which was presented to the Rural Sociological Society's 'Food Systems and Agrarian Change in the Late Twentieth Century' Conference, Columbus, Ohio, 21–22 August 1991.

REFERENCES

Austin, N. (1989) 'Agriculture's Brave New World', *Australian Rural Times* 31 January, 7 February.

Austin, N. (1990a) 'The Stifled Heartland', *The Bulletin* 31 July 1990.

Austin, N. (1990b) 'The Growing Revolution', *The Bulletin* 16 July 1990.

Australian National Opinion Polls (ANOP) (1991) *Research Commissioned by the National Farmers Federation*, Canberra.

Baumgardt, B. and Martin, M. (eds) (1991) *Agricultural Biotechnology: Issues and Choices,* Purdue University Agricultural Experiment Station, Indiana.

Beale, B. and Fray, P. (1990) *The Vanishing Continent*, Hodder and Stoughton, Sydney.

Begg, J. and Peacock, J. (1990) 'Modern Genetic and Management Technologies in Australian Agriculture', in D. Williams (ed.), *Agriculture in the Australian Economy* (3rd edn), Sydney University Press, Melbourne.

Bell, J. and Pandey, U. (1986) 'Some Social Aspects of the Present Crisis in Australian Agriculture', Farmer Study Group Paper, University of New England, Armidale, New South Wales.

Boreham, P., Clegg, S., Emmison, J., Marks, G. and Western, J. (1989) 'Semi-peripheries or Particular Pathways: The Case for Australia, New Zealand and Canada as Class Formations', *International Sociology* 4 (1).

Bureau of Rural Resources (1991) *Strategic Technologies for Maximising the Competitiveness of Australia's Agriculture-Based Exports*, Bureau of Rural Resources, Canberra.

Busch, L., Lacy, W., Burkhardt, J. and Lacy, L. (1991) *Plants, Power and Profit*, Basil Blackwell, Oxford.

Buttel, F. and Gertler, M. (1982) 'Agricultural Structure, Agricultural Policy, and Environmental Quality: Some Observations on the Context of Agricultural Research in North America', *Agriculture and Environment* 7.

Buttel, F. and La Ramee, P. (1987) 'The "Disappearing Middle": A Sociological Perspective', Paper presented to the Annual Meeting of the Rural Sociological Society, Madison, Wisconsin, August.

Byman, W. (1990) 'New Technologies in the Agro-Food System and US–EC Trade Relations', in P. Lowe, T. Marsden and S. Whatmore (eds), *Technological Change and the Rural Environment*, Fulton, London.

Cameron, J. and Elix, J. (1991) *Recovering Ground: A Case Study Approach to Ecologically Sustainable Rural Land Management*, Australian Conservation Foundation, Melbourne.

Campbell, K. and Dumsday, R. (1990) 'Land Policy', in D. Williams (ed.), *Agriculture in the Australian Economy* (3rd edn), Sydney University Press, Melbourne.

Clegg, S. (1980) 'Restructuring the Semi-peripheral Labour Process: Corporatist Australia in the World Economy?', in P. Boreham and G. Dow (eds), *Work and Inequality*, Volume 1, Macmillan, Melbourne.

Collingridge, M. (1991) 'Just Surviving: Rural Social and Community Services', Paper presented to 'The State of Survival' Conference, University of Sydney, 22–23 July.

Committee to Examine Changes to Land and Water Controls in Southern NSW (1988) *Report of Interdepartmental Committee*, Sydney, September.

Commonwealth Scientific and Industrial Research Organisation (CSIRO) (1990) *Australia's Environment and its Natural Resources — An Outlook*, Institute of Natural Resources and Environment, CSIRO, Canberra.

Cook, P. (1989) 'Stewardship of Our Natural Resources: A Shared Responsibility', in Department of Primary Industries and Energy, *Proceedings of the First Community Conference of the Murray–Darling Basin Ministerial Council's Community Advisory Committee*, Australian Government Publishing Service, Canberra.

Crabb, P. (1988) *Managing Water and Land Use in Inter-State River Basins*, Research Papers No. 1, School of Earth Sciences, Macquarie University, Sydney.

Crofts, B., Harris, M. and O'Mara, P. (1988) 'Variations on Farm Output and Its Effects on the Non-Farm Sector', *Quarterly Review of the Rural Economy* 10 (3).

Crough, G., Wheelwright, T. and Wilshire, T. (eds) (1980) *Australia and World Capitalism*, Penguin, Melbourne.

Department of Primary Industries and Energy (DPIE) (1989) *International Agribusiness Trends and their Implications for Australia*, Australian Government Publishing Service, Canberra.

Dumsday, R., Edwards, G. and Chisholm, A. (1990) 'Resource Management', in D. Williams (ed.), *Agriculture in the Australian Economy* (3rd edn), Sydney University Press, Melbourne.

Dunkley, G. and Kulkarni, A. (1990) 'Structural Change and Industry Policy in Australia', *Regional Journal of Social Issues* 24.

Ecologically Sustainable Development Working Groups (1991) *Draft Report — Agriculture*, Australian Government Publishing Service, Canberra, August.

Edwards, K. (1988) 'How Much Soil Loss is Acceptable?', *Search* 19 (3).

Fagan, B. and Bryan, D. (1991) 'Australia and the Changing Global

Economy: Background to Social Inequality in the 1990s', in Social Justice Collective, *Inequality in Australia: Slicing the Cake*, Heinemann, Melbourne.

Farmfacts (1991) *Trends in On-Farm Investment*, Farmfacts NSW.

Fray, P. (1991) 'On Fertile Ground?', *Habitat Australia* 19 (2).

Friedmann, H. and McMichael, P. (1989) 'Agriculture and the State System: The Rise and Decline of National Agricultures, 1870 to the Present', *Sociologia Ruralis* 19 (2).

Gibson, B. (1989) 'Land Degradation: Agriculture and Environment', in Department of Primary Industries and Energy, *Proceedings of the First Community Conference of the Murray–Darling Basin Ministerial Council's Community Advisory Committee*, Australian Government Publishing Service, Canberra.

Goodman, D. and Redclift, M. (1989) 'Introduction: The International Farm Crisis', in D. Goodman and M. Redclift (eds), *The International Farm Crisis*, Macmillan, London.

Goodman, D., Sorj, B. and Wilkinson, J. (1987) *From Farming to Biotechnology*, Basil Blackwell, Oxford.

Jones, E. (1988) 'Managing Market Meltdown: Lessons from the Crash', *Journal of Australian Political Economy* 23.

Junor, R. (1989) 'Community Implications of a Natural Resources Management Strategy for the Murray–Darling Basin', in Department of Primary Industries and Energy, *Proceedings of the First Community Conference of the Murray–Darling Basin Ministerial Council's Community Advisory Committee*, Australian Government Publishing Service, Canberra.

Kloppenburg, J. (1988) *First the Seed*, Cambridge University Press, New York.

Knopke, P. and Harris, J. (1991) 'Changes in Input Use on Australian Farms', *Agriculture and Resources Quarterly* 3 (2).

Lacy, W., Lacy, L. and Busch, L. (1988) 'Agricultural Biotechnology Research: Practices, Consequences, and Policy Recommendations', *Agriculture and Human Values* 5 (3).

Lawrence, G. (1987) *Capitalism and the Countryside*, Pluto Press, Sydney.

Lawrence, G. (1989) 'The Rural Crisis Downunder: Australia's Declining Fortunes in the Global Farm Economy', in D. Goodman and M. Redclift (eds), *The International Farm Crisis*, Macmillan, London.

Lawrence, G. (1990) 'The Management of Change in Resource Use and Its Consequences in the Murray–Darling Basin', Paper presented at the Community Advisory Committee of the Murray–Darling Basin Ministerial Council Workshop on Management of Change in Resource Use, Shepparton, Victoria, 21 November.

Lawrence, G. and Vanclay, F. (1992) 'Agricultural Change and Environmental Degradation in the Semi-periphery: The Murray–Darling Basin, Australia', in P. McMichael (ed.), *Food Systems and Agrarian Change in the Late Twentieth Century*, Cornell University Press, Ithaca.

Lawrence, G. and Williams, C. (1990) 'The Dynamics of Decline', in T. Cullen, P. Dunn and G. Lawrence (eds), *Rural Health and Welfare in Australia*, Arena, Melbourne.

Lewis, P. (1990) 'Rural Population and Workforce', in D. Williams (ed.), *Agriculture in the Australian Economy* (3rd edn), Sydney University Press, Melbourne.

Lowe, P., Cox, G., Goodman, D., Munton, R. and Winter, M. (1990) 'Technological Change, Farm Management and Pollution Regulation: The Example of Britain', in P. Lowe, T. Marsden and S. Whatmore (eds), *Technological Change and the Rural Environment*, Fulton, London.

McMichael, P. (1984) *Settlers and the Agrarian Question*, Cambridge University Press, New York.

Mooney, P. (1988) *My Own Boss*, Westview, Boulder.

Munton, R., Marsden, T. and Whatmore, S. (1990) 'Technological Change in a Period of Agricultural Adjustment', in P. Lowe, T. Marsden and S. Whatmore (eds), *Technological Change and the Rural Environment*, Fulton, London.

Murray–Darling Basin Community Advisory Committee (1991) *Surviving Change — Chance or Choice?* Murray–Darling Basin Ministerial Council, Canberra, June.

Murray–Darling Basin Ministerial Council (1989) *Draft Murray–Darling Basin Natural Resources Management Strategy: Getting it Together*, Murray–Darling Basin Ministerial Council, Canberra.

Murray–Darling Basin Ministerial Council (1990) *Natural Resources Management Strategy: Towards a Sustainable Future*, Murray–Darling Basin Ministerial Council, Canberra.

Musgrave, W. (1990) 'Rural Adjustment', in D. Williams (ed.), *Agriculture in the Australian Economy* (3rd edn), Sydney University Press, Melbourne.

O'Connor, J. (1990) 'The Second Contradiction of Capitalism: Causes and Consequences', Paper delivered at the Conference on New Economic Analysis, Barcelona, Spain, 30 November to 2 December.

O'Reilly, D. (1988) 'Save Our Land!', *The Bulletin* 2 August.

Ockwell, A. (1990) 'The Economic Structure of Australian Agriculture', in D. Williams (ed.), *Agriculture in the Australian Economy* (3rd edn), Sydney University Press, Melbourne.

Osborne, R. (1989) 'Local Government's Role', in Department of Primary Industries and Energy, *Proceedings of the First Community Conference of the Murray–Darling Basin Ministerial Council's Community Advisory Committee*, Australian Government Publishing Service, Canberra.

Otero, G. (1991) 'Biotechnology and Economic Restructuring: Toward a New Technological Paradigm in Agriculture?', Paper presented to the Annual Meeting of the American Sociological Association, Cincinnati, Ohio, 23–27 August.

Peterson, M. (1990) 'Paradigmatic Shifts in Agriculture: Global Effects and the Swedish Response', in T. Marsden, P. Lowe and S. Whatmore (eds), *Rural Restructuring*, Fulton, London.

Powell, R. and Milhouse, N. (1990) 'Capital Investment and

Finance', in D. Williams (ed.), *Agriculture in the Australian Economy* (3rd edn), Sydney University Press, Melbourne.

Redclift, M. (1987) *Sustainable Development: Exploring the Contradictions*, Methuen, London.

Redclift, M. (1990) 'The Role of Agricultural Technology in Sustainable Development', in P. Lowe, T. Marsden and S. Whatmore (eds), *Technological Change and the Rural Environment*, Fulton, London.

Reeve, I. and Stayner, R. (1990) *Uncoupling: Relationships Between Agriculture and the Local Economies of Rural Areas in NSW*, Rural Development Centre, Armidale.

Rickson, R., Saffigna, P., Vanclay, F. and McTainsh, G. (1987) 'Social Bases of Farmers' Responses to Land Degradation', in A. Chisholm and R. Dumsday (eds), *Land Degradation: Problems and Policies*, Cambridge University Press, Cambridge.

Riverina Regional Development Board (1991) *Regional Strategy 1991–2001: Vision and Strategy*, Riverina Regional Development Board, Wagga Wagga.

Strange, M. (1988) *Family Farming: A New Economic Vision*, University of Nebraska Press, Lincoln.

Tait, J. (1990) 'Environmental Risks and the Regulation of Biotechnology', in P. Lowe, T. Marsden and S. Whatmore (eds), *Technological Change and the Rural Environment*, Fulton, London.

Vanclay, F. (1986) 'Socio-economic Correlates of Adoption of Soil Conservation Technology', unpublished M.Soc.Sci. thesis, Department of Anthropology and Sociology, University of Queensland, St Lucia.

Vanclay, F. and Cary, J. (1989) *Farmers' Perceptions of Dryland Soil Salinity*, School of Agriculture and Forestry, University of Melbourne, Melbourne.

Wettenhall, G. (1991) 'Selling off the Farm: The Second Green Revolution', *Australian Society* June.

Wonder, B. and Fisher, B. (1990) 'Agriculture in the Economy', in D. Williams (ed.), *Agriculture in the Australian Economy* (3rd edn), Sydney University Press, Melbourne.

4 THE SEMANTICS OF 'FOREST' COVER: HOW GREEN WAS AUSTRALIA?

JOHN CARY and NEIL BARR

There is a public perception that most of Australia's land degradation is a consequence of the removal of trees. This view has gained currency through publicity of Australia's growing problem of land salinisation. Seepage salinisation in dryland areas occurs because historical land use patterns of land clearing and modern agricultural processes of cropping and grazing have resulted in less water being used by the ecosystem than was used by the pre-existing natural ecosystems. Unused water becomes the source of either increased water runoff or higher watertables in the soil. Dryland salinity is caused by saline seepage resulting from rising watertables. This phenomenon is particularly evident in the wheat belt in south-west Western Australia and to a lesser extent in western and northern Victoria.

The commonly proposed solution to the problem of dryland salinity is the wide-scale replanting of trees. The tree cover at the time of white settlement is often taken as the datum for the ecologically 'correct' level of tree cover for Australia. This has posed a challenge in historical research for biological scientists. There is not a rich store of information about Australia's early vegetative cover. Most of the information is located in obscure and diverse sources. The little information more readily accessible has sometimes been used in a cavalier fashion.

THE EVERETT MAP

In the late 1860s the Victorian government opened the way for free selection of the state's Crown land. To help the would-be selectors, the Lands Department prepared a map of the state showing where there was forest and where there was open plain. The map was compiled by Arthur Everett from records held in the office of the Surveyor General, together with other material such as the observations of the Government Botanist Von Mueller (Everett 1869). Everett's map was originally conceived to encourage the settlement of new farming land. Today it is used to encourage forest conservation. It has been one of the basic foundation stones on which have been built our estimates of the destruction of native forest by European settlement. The other major foundation has been satellite imagery. For Victoria, Everett's 1869 map has been used as a basis for comparison with satellite maps (Woodgate and Black 1988). For an Australian comparison, the CSIRO used a map of vegetation at European settlement compiled between 1968 and 1974 from a variety of more recent sources by Carnaham (1976).

The result of these comparisons has been spectacular, and somewhat exaggerated, cartographic images of a lost forest cover (Eckersley 1989). A recent, more sober but still inadequate assessment has been presented in the *Atlas of Australian Resources: Vegetation* (1990: 7) which is based on Carnaham's map of Australia's vegetation in the 1780s and in which it is observed that 'reconstruction of parts of the natural vegetation, particularly in south-eastern Australia, remains largely speculative because of inadequate historical information'.

THE SEMANTICS OF 'FOREST'

The title of the Everett map implies that all woody vegetation, excluding only the sparsest woodland, was 'forest'. This was a culturally determined definition which led to the mapping of any vegetation impeding easy settlement. Everett nominated 11 wooded categories and a remaining category of 'open country'. In the aggregation of such categories, a variable landscape with patches of woodland and patches of open grass becomes encoded into one simple classification of forest. This is particularly the case with modern technology, such as digitally based, computerised Geographical Information Systems (GISs), where vegetational diversity is encoded into simpler map formats. This embracing definition of 'forest' is crucial to the understanding of how Everett's map has been subsequently used and abused. In Australia there was, and still is, an extensive gradation in tree cover. There has always been only small areas of closed forest where the upper tree canopy is closed. There was extensive open forest along the south-eastern coast of Australia. This is now considerably diminished; but today's 'open' forest cover is generally considerably denser than at the time of white settlement. This is not recorded on comparative vegetation maps such as those in the *Atlas of Australian Resources: Vegetation* (1990). Tree cover in a large area of Australia — in the margin beyond the coastal fringe — was woodland. The density of tree cover was often low with an open, park-like aspect. The localities of greatest tree loss are in the land which has been used for cropping in the 'wheat zones' — land which formerly was often very open woodland or scrub and heathland.

Anyone attempting to express the complexity of the extent and diversity of Australia's tree cover in a two-dimensional map faces considerable difficulty in describing what currently exists. There are much greater difficulties in attempting this task for the state of affairs at the time of white settlement. The problem is compounded by the dynamics of the changing forest cover. Forest cover increased after white settlement and then later decreased. So, in grappling with the descriptions of the past, our first problem is to decide which particular past is relevant.

THE FIRST FARMERS

Australia has been farmed for thousands of years. The Aboriginal people used fire to control game and increase the productive capacity of the land. Early Europeans, such as Abel Tasman, Joseph Banks, James Cook and Arthur Phillip, commented on the Aborigines' use of fire. The land which the first Europeans explored had been skilfully managed and shaped by continuous and creative use of fire. Aboriginal firestick farming was reported in south-eastern Australia, in Queensland (Leichhardt 1847: 354), in central and northern New South Wales (Oxley 1820: 5, 174) and in south-west Western Australia (Stokes 1846: 400). Burning the bush encouraged lush young growth from the grasses. This attracted kangaroos, wallabies and other game which the Aborigines hunted. Burning stimulated the production of manna by the manna gum (Curr 1965: 171). Burning cleared tracks and herded game. The early seaborne explorers recorded observations of Aborigines' constant use of fire. Tasman saw Aborigines setting fire to rainforest (Sharp 1968: 44, 111). Cook saw fires all along the eastern seaboard and noted the Aborigines' efficiency at firing large tracts of country (Cook 1969: 638–639).

It was the same inland. Hume and Hovell made frequent references to the Aborigines' use of fire. Many times they described the land all around as lit up by the natives' grass fires. They fed their horses by leap frogging from one area of fresh, unburnt grass to the next (Hovell and Hume 1831: 17–64). Thomas Mitchell (1848: 412–413) described the importance of fire in his journals of his inland travels:

> Fire, grass, kangaroos, and human inhabitants, seem all dependent on each other for existence in Australia; for any one of these being

> wanting, the others could no longer continue. Fire is necessary to burn the grass, and form the open forests, in which we find the large forest-kangaroo . . . the omission of the annual periodic burning of the grass and the young saplings, has already produced in the open forests near Sydney thick forests of young trees, where, formerly, a man might gallop without impediment, and see whole miles before him. Kangaroos are no longer to be seen there; the grass is choked by underwood; neither are there natives to burn grass.

The work of generations of these farmers changed the landscape (Jones 1968). The Aborigines helped create and maintain eucalypt-dominated forests. Analysis of pollen grains trapped in the mud sediments of Lake George near Canberra indicates that in the wetter areas of Australia the Aborigines arrived to a land covered with a diverse forest of casuarina, native cypress, pine, eucalypt and rainforest species (Singh, Kershaw and Clark 1981). Rainforest trees, casuarinas, cypress and pine were not able to compete with the fire-resistant eucalypts in the new fire regime.

It is attractive to assume that the land's first human inhabitants shared our much more recent conservation ethic (Anderson 1989). The Aborigines reshaped the land within the limits of their technology. Their fires decreased tree cover and reduced the genetic diversity of the forest. They changed the soil structure with the constant regime of fire, increasing erosion rates (Hughes and Sullivan 1981). The Aboriginal farming system did not conserve the landscape of Australia. It created a new landscape which was more productive than the landscape they found. The ecological system produced by the Aborigines was not in a timeless state. The Aboriginal society was in a continual flux, with its cultures changing through time and with considerable regional diversity (Mulvaney 1975). If the Europeans had arrived in a different era they may have found a different landscape. In the end, the Aborigines' grassland was the undoing of their society. The landscape they produced was particularly inviting to the European colonisers.

When the first Europeans arrived in eastern Australia they commonly used the phrase 'park-like' to describe the landscape that was the legacy of the Aborigines' burning regime. William Westgarth (1853: 32) described the Australian countryside:

> The open forest, free from underwood, with its grassy carpet beneath and its park-like aspect, is common to all latitudes of the country . . . to the pastoral facilities which this kind of country offers, ready-made as it were at the hand of Nature, is to be traced the rapid progress of the Australian colonies.

The Aborigines had managed their game 'parks' in much the same way as the American Indians (see Jennings 1976; Cronon 1983) and the tribespeople of the African savanna.

Other observers in the 1840s and 1850s had descriptions of the landscape similar to those of Westgarth. In New South Wales, William Wentworth (1820, cited in Bromby 1986) wrote that the greatest asset of the Australian land was its freedom from timber. The squatters occupied the lightly timbered woodland and open grassland. Occasionally they cleared some forest but the essence of their business was using as little hired labour as possible. While their runs were unfenced the squatters could burn their pastures and unintentionally maintain the open forests in the same manner as the Aborigines.

FALL AND RISE OF THE EUCALYPT

The early Selection Acts, from the 1860s, gave squatters the opportunity to secure their land tenure. The main impact of the squatters on the forests was less from direct than from indirect destruction. In 1861 Neil Black, a squatter in western Victoria, observed that the eucalypts around Terang were dying in great numbers. The same observations were made elsewhere. Security of tenure allowed squatters to build fences. Fences meant the end of intentional grassfires. This sounded a death knell for many of the older trees. The regular fire regime under the Aborigines had acted as a break on the soil insect population. When the fires stopped, the insect population increased and mature trees, weakened by water stress due to soil compaction, lacked the strength to withstand the increased numbers of leaf eaters. Squatters like Black blamed an explosion in the possum population. He wondered if when the 'Aborigines die away the opossums on which they live chiefly increase and destroy the gum trees' (Kiddle 1961). Alfred Howitt (1890), a keen anthropologist and natural scientist, blamed a combination of soil compaction and insect population explosions.

The end of the burning meant new life for re-emerging eucalypt saplings across the countryside. The eucalypts' regrowth had been controlled by a combination of regular burning by the Aborigines and grazing by the small marsupials which lived on the native grasslands. The destruction of the clumpy native grasses destroyed the habitat of these animals and their numbers declined. Investment in fencing drastically reduced the frequency of burning. The human-made ecosystem which kept the eucalypts in check was shattered.

In 1890, Alfred Howitt wrote an article for the Royal Society of Victoria in which he concluded that the forests in Gippsland in south-eastern Victoria were more widely spread and denser than at the time of white settlement.

> After some years of occupation, the whole country [of the Snowy River] became covered with forests of saplings . . . and at the present time has so much increased . . . that it is difficult to ride over parts which one can see by the few scattered old giants [were] at one time open grassy country . . . in the Omeo district, young forests of eucalypts of various kinds are growing where a quarter of a century ago the hills were open and park like (Howitt 1890: 118–119).

Whole new forests appeared in other districts. In northern New South Wales the creation of the Pilliga forest has been described by Eric Rolls as a direct result of the European settlement (Rolls 1981). On the New South Wales coast, country which today is thickly timbered was once open grassland with few trees (Murray 1990).

The poet Les Murray used some telling prose to describe the loss of the management system which the Aborigines had maintained for centuries: 'the wilderness we now value and try to protect came with us, the invading Europeans. It came in our heads, and it gradually rose out of the ground to meet us' (Murray 1984). Europeans created a literal wilderness with guns and smallpox and then, as a consequence of this displacement, an illusory wilderness of thickening forests (Griffiths 1991).

WHY THE HENTYS CHOSE PORTLAND

Everett's map and the early vegetation maps in the *Atlas of Australian Resources* can be compared with the more qualitative information available from the diaries of early settlers and explorers. These descriptions show that some of the recent assessments of vegetative change give an impression that the original forest was more dense and, more importantly, more extensive than was actually the case.

To show the variety of the original land cover in Australia we will explore the land surrounding the first permanent European settlement in Victoria. Its vegetative cover depicted in the Everett map is shown in Figure 4.1.

The Portland region is a fair case to consider as it had a more extensive and more varied forest cover than was typical for most other regions of Australia. The forest cover of the Portland area in 1869 has been claimed to have been 89 percent of the total land area; in 1972 (on the basis of landsat images) there was 17 percent forest cover; and in 1987 there

Figure 4.1 Portion of a 1869 map known as the Everett map: Portland detail (Map Collection, State Library of Victoria)

was 16 percent cover (Woodgate and Black 1988). From the 1869 Everett map, most of the land claimed to be unforested existed west of the Glenelg River and adjacent to the border with South Australia. Such claims appear to provide persuasive support for arguments dramatically to reafforest farm land. The accuracy of such claims warrants investigation.

The Portland coast was one of the last stretches of the Australian coast to be charted by Europeans. The first to do so was Lieutenant James Grant in the *Lady Nelson* in 1800. He observed the land around Cape Bridgewater to be 'flat land covered with brushes and large woods inland' (Learmonth 1934: 2). The next sighting was an 1802 French scientific expedition under the control of Nicholas Baudin. The French made their way into Portland Bay from the east. Baudin found the coastline of Portland Bay to be formed of dunes and the abutting land to be 'high and a little wooded'. Standing off Bridgewater Bay, Baudin described the coast as 'high and well wooded towards the interior of the country'. The adjacent Discovery Bay was 'followed by sandy and arid land' (Learmonth 1934: 14).

The Henty family, the first squatters in Victoria, settled at Portland in 1834. Edward Henty had previously come by boat from Van Diemen's Land to Bridgewater Bay, immediately to the west of Portland Bay, in 1833. Edward Henty kept a daily journal in which he described a three-day journey 24 kilometres inland from the Surry River at the head of Portland Bay, made a fortnight after his arrival at Portland in 1834 (Learmonth 1934):

> Dec. 1 — Arrived at 6 p.m., made boat fast in middle of river and started three days' walk in the bush . . . We walked two miles, made a hut, made supper and turned in, crossing a beautiful ridge of land, several marshes and plains with grass up to our knees, of this year's growth.
>
> Dec. 2nd. — Started 4 a.m., walked four miles over beautiful steep hills well sprinkled with wattle and covered with kangaroo grass, trefoil, and a silky grass, description unknown to us . . . We walked north for two hours over beautiful land. We then came to a stringybark forest and it took us two hours to get through it . . . After passing the forest we got into more splendid country, well watered, timbered with wattle, blackwood, and a few small gums . . . We walked a mile and a-half and came to a river called by us Clark's River . . . The land to the north side of the river is generally rocky, but the grass and trefoil is as thick as it can grow and no other timber but blackwood and wattle, which abounds. This would make a beautiful sheep run, for every mile you have a marsh of two or three hundred acres well covered with grass and seldom if ever flooded . . . You can see three-quarters of a mile

> either way, so this cannot be called thickly timbered. We walked six miles and took a fresh cut back to the river, which we crossed and walked down four miles — land of the finest description, grass as thick as it can grow (Learmonth 1934: 85–86).

On the following days they found more land with plenty of grass. At Mount Clay, Henty climbed a large gum tree and observed 'as far as the eye could extend I saw nothing but open hills and very extensive marshes' (Learmonth 1934: 87).

One hundred years later, in 1934, this country was less open than when explored by Henty (Learmonth 1934: 87). Today, Mount Clay is covered with dense forest. On forest maps and vegetation maps the land adjacent to Mount Clay is referred to as dense scrub or low forest; and the surrounding land as very open woodland.

The government surveyor Thomas Mitchell came to Portland in 1836 and was surprised to find the Henty settlement. Mitchell came via the Glenelg River to the west. He found the hinterland to the east of the Glenelg and north of Cape Bridgewater much more substantially vegetated. He recorded a ride of 32 kilometres through forest swamp, swamp, hilly ridge and grassland which brought him to Portland. Mitchell described densely timbered land to the west of Portland Bay (Mitchell 1836, cited in Learmonth 1934). The thickness of the trees impeded the progress of his drays because the fallen timber covered so much of the surface. Many of the trees were 1.8 metres, and some as much as 2.4 metres, in diameter.

Mitchell then travelled south to approach Portland, He made for a 'fine green hill', which he named Mount Eckersley. It was 'covered with a luxuriant crop of kangaroo grass'. Like Henty, he was 24 kilometres north of Portland Bay but about 11 kilometres to the west of where Henty had explored. Mitchell (1839: 236–238) continued: 'the intervening country seemed so low, and swamps, entirely clear of timber, appeared in so many places, that I could scarcely hope to get through . . . I was agreeably surprised to find also, on descending, that the rich grass extended among the trees on the lower country.' Mitchell continued to pursue a course through the woods until he reached the Surry River and camped on the rich grassy land beyond it. It was the river from which Henty commenced the journey we have already described.

In 1844, the writer Rolf Boldrewood took up a pastoral run, Squattlesea Mere, on the lower Eumeralla River, 16 kilometres inland from Portland Bay and just east of where Henty travelled. Boldrewood (1896: 98) lyrically described his 20 000 hectare run as 'wood and wold, mere and marshland, hill and dale'.

> On the south lay open slopes and low hills with flats between . . . At the back were again large marshes, with healthy flats and more thickly timbered forests. Overall was a wonderful sward of grass, luxuriant and green . . . All the land I looked on was deeply-swarded, thickly-verdured as an English meadow (Boldrewood 1896: 42–44).

The land at the head of Portland Bay and north of Mount Clay alternated in swamps, heaths and belts of trees. All the early settlers commented on the amount of grass available on the marshy flats (Bonwick 1970: 91).

Mitchell had told the Hentys of the beautiful downs, 70 kilometres to the north, that he had passed beyond the dense forest that provided a northern land barrier to the Portland settlement (Bride 1898: 263). A year later the Hentys cut a track through the forest to Mount Eckersley and drove sheep through to a run at Heywood and then to runs at Merino Downs on the Wannon, and later, Muntham and Sandford. The Hentys' run at Merino Downs was reputed to be the finest run in Australia, carrying approximately three sheep to every hectare, even in an unimproved state (Roberts 1970: 162).

As Mitchell had made his way towards Portland he entered the Portland region 60 kilometres west of the Grampian Mountains and about 130 kilometres north of Henty's settlement. Beyond Pigeon Ponds Creek he had seen 'hills of the finest forms, all clothed with grass to their summits, and many entirely clear of timber' (Mitchell 1839: 204). Near present-day Casterton, Mitchell (1839: 207) observed:

> A river winding amongst meadows which were fully a mile broad, and green as an emerald. Above them rose swelling hills of fantastic shapes, but all smooth and thickly covered with rich verdure. Behind these were higher hills, all having grass on their sides and trees on their summits, and extending east and west, throughout the landscape, as far as I could see.

This was the land on which the Hentys established their runs. Granville Stapylton, second in command of the Mitchell expedition, enthused about this country: 'Here we have undulating ground clear of timber except occasional picturesque clumps of trees' (Stapylton 1836).

North of Casterton, in the Nangella and Wando Vale country, Mitchell had observed 'an open grassy country, extending as far as we could see, — hills round and smooth as carpet — meadows broad, and either green as emerald' (Mitchell 1839: 211). To the east of the Grampian Mountains the land was open grassland. To the west of Portland the land was scrub and heath with little forest cover (Bonwick 1970).

Figure 4.2 View of the Hentys' Muntham in south-west Victoria about 1860: Thomas Clark (City of Hamilton Art Gallery)

Figure 4.3 Koonongwootong Valley north of Casterton in about 1860 (to the east of where Thomas Mitchell travelled): Thomas Clark (private collection)

The Hentys had seen some areas of Portland's pastoral hinterland that were not the ones Mitchell described. In 1835, 13 months before Mitchell's visit, in a letter from Van Diemen's land, Thomas Henty reported '200,000 acres of good land may be found not a very long way from the Bay' (Learmonth 1934: 67). The likely locality was open country inland from Yambuk and Port Fairy (Learmonth 1934: 69–70). The downs that Mitchell discovered, the tussock grassland and open savanna, which represented another 200 000 acres (see Gibbons and Downes 1964), were additional to this. In total, the grasslands of which the Hentys were aware represented 8 percent of the land in the Portland region. The open country elsewhere, to the north and to the west, suggests that the amount of 'unforested' land at the time of European settlement was considerably more than the 11 percent extrapolated from the Everett map by Woodgate and Black (1988).

CONCLUDING COMMENTS

What can we learn from this consideration of historical accounts of the nature of the original vegetative cover? The qualitative information of a first-hand verbal or written description is heavily distilled and generalised once it is reduced by the constraints of presentation in a map. The accounts of the first Europeans to walk across the Portland region indicate large changes in the vegetative cover within relatively short distances. For the most part, south-western Victoria was a mix of open woodland, bands of thicker forest, heathlands and tussock grasslands; it was not generally densely wooded. The forests were mostly open in nature. The explorers were able to travel through most of the forest either on foot or with horses and drays, without the need to clear tracks. Grasslands suitable for pastoral activity comprised a significant area of the region. Within 10 years of European settlement all the so-called 'forest' in the south-west, apart from some land west of the Glenelg River, had been taken up for grazing under pastoral licences (Peel 1971).

When maps of original vegetation cover are compared with modern digitised computer-generated maps we are often mesmerised by the sophistication of the measurement technique and ignore the crudeness of the assumptions that have to be made when the maps are simplified for comparison purposes. Digitising early maps for comparison with current maps produces an impression of precision not present on the original maps. When we describe vegetative change by taking two maps where the vegetative diversity is markedly different, with each map only inadequately describing vegetative diversity, and then make a map of the differences between them, we are presented with a misleading picture: vegetation change has been mapped but not adequately described. There are several unacknowledged difficulties.

One difficulty is a matter of timing. By the time of this first attempt at vegetation mapping the periodic burning had ceased and the density of cover in both unsettled and some settled areas would have increased. This was a matter which would not have concerned Everett. Another difficulty lies in the simplification of complex information. It was not possible in 1869, and still difficult today, to produce a single map depicting areas of vegetative cover the nature of the cover and an accurate measure of the density of the cover. In order to make informed judgements about estimates of the changes in tree cover in Australia we need to consider the density of trees and the nature or type of vegetative cover, which might be considered as forest. The detailed maps in the *Atlas of Australian Resources: Vegetation* (1990) are an attempt to produce such information, using a complex coding system of eight categories of growth form, six categories of foliage cover, and identification of the dominant floristic type for the locality. All this information is contained in two large maps of Australia, which do not adequately depict the locality variation we have discussed. When such maps are reduced to the size used for public communication such simplification produces the exaggerated cartographic images of lost forest cover to which we referred at the start of this chapter.

We have devoted this chapter to a consideration of the semantics of forest: the meaning of forest in its historical and current use in Australia and the nature of forest at the time of European settlement. Such a consideration challenges claims that Australia's public forests should be closed in order to preserve them. Most of Australia's current public forests have always been 'managed'. Alfred Howitt's comments about the earlier nature of the East Gippsland forest contradicts the 'green' debate over logging practices in the south-east forests. We are not arguing that the current level of forest cover is appropriate, but rather that earlier estimates of a pre-European arcadia are often seen through misty eyes. The tree loss has been considerable, but not as extensive as some wish to portray. The loss of the trees has become a symbol of the cause of land degradation. Symbolic beliefs are often necessary for community consensus so that agreed community activity may occur. Symbolic beliefs should not intrude upon the task of establishing the causes of land degradation problems and establishing rational and feasible solutions. It is necessary to see beyond the symbols.

There are many areas of land degradation in Australia where the removal of trees has been an important or major contributor to land degradation. It is not axiomatic that the removal of trees results in deteriorating land quality. There are other areas of Australia where extensive removal of trees has had little influence on land quality. For some of the areas of worst degradation — Australia's salinity-affected irrigation areas — the removal of trees has been a minor contributor in a complex hydrogeological process. We should also not assume that all saline land is a reflection of European farming. Salt lakes were not uncommon at the time of European settlement in the less well drained parts of Australia. Learmonth, one of the first selectors in the Western District of Victoria, found Lake Burrumbeet to be intensely salty (Bride 1898: 95, 98); and Thomas Mitchell observed salty lakes on the south-west Victorian plains (Mitchell 1839: 266).

Finally, we need to consider economic and social matters in prescribing tree planting as a solution to salinity. Solutions involving the replacement of shallow-rooted annual pastures with deep-rooted improved perennial pasture are likely to be more economically attractive to the dryland farming community than proposals for extensive tree planting.

REFERENCES

Anderson, C. (1989) 'Aborigines and Conservationism: The Daintree–Bloomfield Road', *Australian Journal of Social Issues* 24 (3): 214-227.

Boldrewood, R. (1896) *Old Melbourne Memories,* Macmillan, London.

Bonwick, J. (1970) *Western Victoria: Its Geography, Geology, and Social Condition* (the narrative of an educational tour in 1857), Heinemann, Melbourne.

Bride, T. (1898) *Letters from Victorian Pioneers,* Trustees of the Public Library Victoria, Melbourne.

Bromby, R. (1986) *The Farming of Australia,* Doubleday, Sydney.

Carnaham, J. (1976) 'Natural Vegetation', in Department of National Development and Energy, *Atlas of Australian Resources: Natural Vegetation,* second series, Division of National Mapping, Australian Government Publishing Service, Canberra.

Cook, J. (1969) *An Account of a Voyage Round the World with a Full Account of the Voyage of the Endeavour in the year MDCCLXX Along the East Coast of Australia,* Smith and Paterson, Brisbane.

Cronon, W. (1983) *Changes in the Land: Indians, Colonists and the Ecology of New England,* Hill and Wang, New York.

Curr, E. (1965) *Recollections of Squatting in Victoria* (first published 1883), Melbourne University Press, Melbourne.

Department of National Development and Energy (1976) *Atlas of Australian Resources*, Volume 6, *Vegetation*, second series, Division of National Mapping, Australian Government Publishing Service, Canberra.

Department of National Development and Energy (1990) *Atlas of Australian Resources*, Volume 6, *Vegetation*, third series, Division of National Mapping, Australian Government Publishing Service, Canberra.

Eckersley, R. (1989) *Regreening Australia: The Environmental, Economic and Social Benefits of Reforestation*, Occasional Paper No. 3, CSIRO, Canberra.

Everett, A. (1869) *Map of the Distribution of Forest Trees in Victoria*, compiled under the direction of R. Brough Smyth, La Trobe Library, Melbourne.

Gibbons, F. and Downes, R. (1964) *A Study of the Land in South-western Victoria*, Soil Conservation Authority of Victoria, Melbourne.

Griffiths, T. (1991) 'History and Natural History: Conservation Movements in Conflict?', in D. Mulvaney (ed.), *The Humanities and the Australian Environment*, Australian Academy of Humanities, Canberra.

Hovell, W. and Hume, H. (1831) *Journey of Discovery to Port Phillip New South Wales by Messrs W.H. Hovell and Hamilton Hume in 1824 and 1825*, A. Hill, Sydney.

Howitt, A. (1890) 'The Eucalypts of Gippsland', *Transactions of the Royal Society of Victoria* 2 (1): 116–120.

Hughes, P. and Sullivan, M. (1981) 'Aboriginal Burning and Late Holocene Geomorphic Events in Eastern New South Wales', *Search* 12: 277–278.

Jennings, F. (1976) *The Invasion of America: Indians, Colonialism and the Cant of Conquest*, Norton, New York.

Jones, R. (1968) 'Geographical Background to the Arrival of Man in Australia', *Archaeology and Physical Anthropology in Oceania* 3: 186–215.

Kiddle, M. (1961) *Men of Yesterday: A Social History of the Western District of Victoria 1834–1890*, Melbourne University Press, Melbourne.

Learmonth, N. (1934) *The Portland Bay Settlement*, McCarron Bird, Melbourne.

Leichhardt, F. (1847) *Journal of an Overland Expedition*, Boone, London.

Mitchell, T. (1839) *Three Expeditions to the Interior of Eastern Australia*, Volume 2, Boone, London (facsimile edn, 1965).

Mitchell, T. (1848) *Journal of an Expedition into the Interior of Tropical Australia*, Longman, Brown, Green and Longmans, London.

Mulvaney, D. (1975) *The Prehistory of Australia*, Penguin, Melbourne.

Murray, L. (1984) 'Eric Rolls and the Golden Disobedience', in *Persistence in Folly*, Angus and Robertson, Sydney.

Murray, L. (1990) 'In a Working Forest', in R. McDonald (ed.), *Gone Bush*, Bantam, Sydney.

Oxley, J. (1820) *Journals of Two Expeditions into the Interior of New South Wales,* Murray, London.

Peel, L. (1971) 'The First Hundred Years of Agricultural Development in Western Victoria', in M. Douglas and L. O'Brien (eds), *The Natural History of Western Victoria,* Australian Institute of Agricultural Science, Horsham.

Roberts, S. (1970) *The Squatting Age in Australia,* Melbourne University Press, Melbourne.

Rolls, E. (1981) *A Million Wild Acres: 200 Years of Man and an Australian Forest,* Thomas Nelson, Sydney.

Sharp, A. (1968) *The Voyages of Abel Janzoon Tasman,* Clare, Oxford.

Singh, G., Kershaw, A. and Clark, R. (1981) 'Quaternary Vegetation and Fire History in Australia', in A. Gill, R. Groves and I. Noble (eds), *Fire and the Australian Biota,* Australian Academy of Science, Canberra.

Stapylton, G. (1836) *The Journal of Granville William Chetwynd Stapylton,* Mitchell Library, Sydney.

Stokes, J. (1846) *Discoveries in Australia,* Boone, London.

Westgarth, W. (1853) *Victoria: Late Australia Felix, or Port Phillip District of New South Wales,* Oliver and Boyd, Edinburgh.

Woodgate, P. and Black, P. (1988) *Forest Cover Changes in Victoria: 1869–1987,* Department of Conservation, Forests and Lands, Melbourne.

5 ENVIRONMENTAL MANAGEMENT AND CAPITALIST AGRICULTURE

BRIAN FURZE

In recent years concerns relating to the relationship between notions of development and environment have become increasingly apparent. Scholars, activists and others have begun to speak of a crisis in development and an environmental crisis, though all do not necessarily see the two as being related or share common ideas on how to overcome them. There is, however, a growing recognition that the process of 'development' as it is currently practised and the assumptions which go with it are in need of reconceptualisation.

The question of environmental management is implicitly linked to the notion of development as a social process. As Redclift (1987: 3) explains, 'the environment is transformed by economic growth in a material sense but is also continually transformed existentially, though we — the environment users — often remain unconscious of the fact'. Redclift is pointing to the nexus between social relations and nature. The environment is transformed by economic growth (a particular model of development) as are social relations. Consequently, within social relations there are implicit and explicit notions regarding nature and its use. This can be termed 'environmental perception' and its relevance is found both in an examination of the results of agriculture within capitalism and in the search for alternative agrarian social relations.

The point about environmental perception is important if one considers the structural contradictions which social groups (in this case, farmers) encounter in the face of the dominant model of development based on capitalist accumulation. The economic relations of capital agriculture which farmers experience may force them into an agrarian practice which is based on the appropriation of nature as a factor of production. Yet the social relations they experience may provide a perception of nature (land, for example) in terms of a resource which must be conserved for both the immediate and future generations. While this view may be instrumentalist (that is, to ensure intergenerational farming) it does include a conservation ethic and, importantly, a quite different view of land use from that found within the dominant economic system.

In any exploration of the relationship between environmental management and capitalist agriculture it is important to begin with the conceptualisation of development. While it can be argued that all societies 'develop' and experience social changes in some form, the term's use — in the western-centric, progressionist sense to imply a movement along a simplistic continuum to industrial capitalism — was determined by precise historical circumstances.

Larrain (1989: 1) locates the emergence of the notion of development with the emergence of capitalism, for 'it was capitalism that for the first time allowed productive forces to make a spectacular advance, thus making it possible for the idea of material progress and development to arise'. Development, in this way, became equated with industrial capitalism.

The emergence of what can loosely be grouped as modernisation theories coupled with an era of capitalist expansion represented a forceful and influential paradigm which suggested that the path to economic and social prosperity was the path of capitalist industrialisation. This process would break the shackles of tradition, thereby opening up an era of economic prosperity and scientific rationality. The accompanying social changes, it was argued, allowed for greater individualism including greater individual freedom.

The influence of this paradigm can be seen by its attempted replication in a variety of social, economic and political systems. While the modernisation theory has as its basis a simplistic linear view of development and social change based on dominant Western notions of the 'good life', its emergence as a global process has had a profound effect. Importantly, many of the assumptions it carried, particularly those relating to industrialisation and the creation of surplus, have crossed ideological boundaries. The industrialisation of the production process has resulted in a truly dominant notion of development.

Implicit in this paradigm has been a particular relationship between the individual, society and nature. The rise of individualism as a way of life and the emergence (and later dominance) of a mode of production which has been characterised by rapid industrialisation, accumulation and technological rationality resulted in the development of social relations based on the estrangement of the individual from society and the reification of nature as a factor of production. Thus, the basis for a relationship between society and nature has been anthropocentric, with nature harnessed for the 'benefit' of humanity, a benefit defined within the parameters of capitalist accumulation.

The connection between the capitalist mode of production in its monopoly form, especially the development of the capitalist world economy (see Wallerstein 1980; Amin 1974), and the environmental crisis has been ably explored by writers such as Stretton (1977), Bahro (1982, 1986) and Redclift (1984) and provides the framework for this chapter. Acknowledging that the environmental crisis cannot be divorced from the economic and social relations of the dominant mode of development, this chapter seeks to consider some implications for agrarian practice.

Food is a basic human need and a basic human right. However, a food production system and the social relations of agriculture are socially constructed. A food system can be defined as the totality of tangible and intangible means employed by a particular human community for the production, conservation, distribution and consumption of food (George (1985) more fully explores this definition). The food supply system is shaped by cultural perceptions, economic forces and relations of production and political influence — all the realms of human agency. These also shape environmental perception and the use of the environment (that is, the nexus between agrarian social relations and nature).

Susan George (1980, 1985) has suggested that prolonged or intense interaction with external food supply systems may accelerate change within the indigenous food system to such a point where the community may lose control over its own environment. It is fruitful, therefore, to analyse the nature of the dominant food supply system and its effect on agrarian social relations. Furthermore, both the environment and agrarian social relations must be interpreted within a framework of dominant notions of development. The suggestion that natural resources are distributed according to rules which are not created or influenced by economic and social forces neglects the form of the international economy, the class structure of individual countries and the nature of the relationships between countries. As Redclift (1984: 1–2) suggests, so many of the causes of the environmental crisis are structural, and anything other than an analysis which includes social, economic and political factors lacks credibility.

THE NEED FOR A NEW PARADIGM: CAPITALIST AGRICULTURE AND ENVIRONMENTAL CONSIDERATIONS

Monopoly capitalism has inspired a fundamental alteration to the world agricultural system. A domination of technological and scientific thought has emerged and this, combined with the rationality of the capitalist marketplace, is central to its analysis. Ultimately, the economic and social spheres of monopoly capitalism have become embodied with formal rationality, value efficiency and the rational organisation of both commodities and people (Hearn 1985).

One outcome of this new rationality is a further alteration to the status of the natural world. Value efficiency equates with a perception of the natural world as a resource with technology determining how that resource might be best utilised. This results in the transformation of agrarian production and rural society as they are co-opted into the global market. It is the market which becomes the institutionalisation of individualism, rationality and, in Schumacher's (1983: 36) words, 'non-responsibility'. This has a profound effect on the social relations of agrarian production and traditional community development processes.

The transformation from tradition to modernity results in an altered relationship to nature. People's older, more intuitive ways of interacting with the land and each other are labelled 'superstitious' (Devall and Sessions 1985). There is little place in the rationalised world for folk law or what is often construed to be non-scientific thinking.

Agriculture and the mass food industry it serves are altered to conform to the new ideology. The end result of this process is an industrialised food complex with a high degree of specialisation and interdependence. Economic and technological rationality results in a concentration of ownership of land centred on humankind's domination of the natural world. Vertically integrated agribusiness companies gain increasing control of the marketplace and, ultimately, rural life is fundamentally altered. For Schumacher (1983), the main danger to agriculture lies in the application of inappropriate and amoral principles of business which treat natural resources (and people) as no more than inputs to a production process.

Two important points emerge from this. First, under the economic arrangements of monopoly capital, a contradiction occurs between the public purpose of agrarian production (the right to healthy food as a fundamental human right) and the private accumulation of profit. This point is taken up by others in this book. Second, a contradiction exists between the rationality of the capitalist marketplace (both locally and, importantly, globally) and the reliance on nature to provide the foundation from which farmers can compete. On the one hand there is the tendency for the market to be the institutionalisation of non-responsibility (Schumacher 1983). On the other hand there is an urgent need for social responsibility.

Technology is a useful case in point. Redclift (1984: 101) defines technology as 'the means adopted by [humankind] to change or influence the environment . . . [Technology] is the application of scientific ideas to the environment, providing us with the knowledge by which . . . we may be able to make ourselves masters and possessors of nature.' Technology is viewed as having the potential to transform people's relationship with nature. The existence of a form of technology and its use implies certain choices which are found in economic and social processes. Consequently, the form of technology may or may not act as a liberating agent (*vis-à-vis*, for instance, food production and distribution) and this is dependent on the social and economic positions of those using or adopting it (Redclift 1984).

Monopoly capital can and does influence the liberatory potential of technology. The form of the appropriation of nature can be altered, most recently through the development of biotechnology (see, for example, Chapter 16). According to Redclift (1987: 2), 'we are, literally, "producing" nature for the first time, while we are busy destroying it for the last time'.

Farmers thus tend to face a complex and formidable set of economic and social relations arising from their position within the marketplace and the production process. Given the expansion-driven rationality of the capitalist marketplace and the incursion of monopoly capital in the form of, for example, agribusiness corporations, farming families can face a process of proletarianisation (such as through contract farming) and/or displacement (see Chapter 5). With a rural ideology that can and does allow for a belief in a form of land stewardship and the particular relationship with nature that this belief implies, farmers trying to follow the dictates of capitalist economic rationality are faced with tough decisions relating to the use of their land. (The contributors in Part II of this volume explore the results of this.) The problem, though, is how to escape these contradictions.

ALTERNATIVE AGRARIAN PRACTICE AND ENVIRONMENTAL MANAGEMENT PARADIGMS

Attempts at implementing new agrarian social formations face structural obstacles which occur because of the social and economic formations already in existence. This raises an important question: how can alternative forms of agrarian social relations (and alternative models of rural development) be implemented? The answer lies in part with those who are labelled 'social change agents'. The analysis of the dialectic between the dominant development paradigm and attempts by human agents to instigate alternative social relations can inform us of the obstacles to their creation as well as uncovering structural processes operating within the dominant model. The question of the educative role of social change agents is of crucial importance.

People have long been involved in the search for alternatives — and the role of Utopian vision in directing social change is of importance here. In particular, if one is to avoid model building for its own sake (and if the issues relating to social change and alternative development are to be linked to the practice of achieving it) what role does Utopian vision play in directing change? In other words, how significant is it for those attempting to instigate an alternative agrarian practice to have a Utopian vision?

Lewis Mumford (1962: 1) has suggested that Utopia can mean the ultimate in human hope or in human folly. It is an ambiguous term which can suggest 'what ought to be' as well as that which can never be, that is, the action to achieve alternative social and economic formations, or the futility of trying to achieve the impossible. These contradictory ways of looking at Utopia produce a tension between what Max Weber (1963: 144) describes as 'the actually existent and the ideal' and it is this tension which produces the importance of Utopian thought in the search for alternatives. As Weber (in Gerth and Mills 1958: 128) suggests, humankind 'would not have attained the possible unless time and again [it] reached out for the impossible'. Mannheim (1952: 236) has suggested, too, that 'with the relinquishment of utopias, [humankind] would lose [its] will to shape history and therewith [its] ability to understand it.'

Utopian thought is thus important in the search for alternatives. However, in the rational world of monopoly capitalism, it is frequently dismissed as irrelevant to social life: those who seek alternatives to dominant notions of development may be marginalised on the basis of the perceived usefulness of their thought, the assumptions of their thought and/or the potential for achieving that which they seek. For farmers who are squeezed between the structural contradictions of capitalist rationality and an alternative relationship to nature, the decisions are indeed difficult.

Environmental management paradigms which are based on alternatives to dominant agrarian social relations can be located within a Utopian tradition to the extent that they represent possible alternative futures for Australian agriculture. The understanding of the 'actually existent' has provided a foundation for what constitutes an alternative agrarian practice.

Merrill (1976a: 286–287) lists six important considerations for alternative agriculture. It:

1. needs to revise the total impact of capitalist agriculture on resources;
2. should analyse the long-term effects of pesticide and chemical usage;
3. must ask if the displacement of rural culture by the impact of agribusiness and rationality is a positive thing;
4. needs to analyse the consequences of an agricultural system dependent on non-renewable fuels;
5. should acknowledge the effects of monopoly capital in agriculture;
6. must ask to what degree the polluting techniques of capitalist agriculture can be replaced by renewable and self-sustaining systems.

Routley and Routley (1980) suggest that a self-sustaining, ecologically sound future for agriculture depends on renouncing the relationship between the social character of production and its private purpose. In other words, there is a need to alter the previously held world-view. A dominant world-view must be in place which looks to an agriculture based on self-determination and which strives to achieve its public purpose — human health and survival.

One thing is certain. It is not merely enough to alter existing social and economic relations. There needs to be a reconceptualisation of the nexus between these and nature.

The following sections briefly canvass some of the paradigms put forward for an alternative agrarian practice and, through that, environmental management within agriculture. The development of a simplistic schema within which to canvass the notion of 'alternative' is itself fraught with danger. What is included and what is left out, what is put into one category and not another, is subject to the whims and theoretical precepts of authors as well as their interpretations of historical circumstances.

The following schema represents what can be considered to lie within the tradition of alternatives to capitalist rural development. It consists of capitalism's historical alternatives, those of agrarian production under state socialism and the tradition of rural populism as an alternative means to centralised agrarian production, and considers the emerging ecological or 'green' critique, leading to consideration of the potential for a red–green alliance.

STATE SOCIALISM

The traditional alternative to the exploitative capitalist relations of production has been state socialism. However, within the parameters of agrarian production, state socialism may hardly be appealing. The advantages offered to society as a whole by the transfer of power to the state may not extend to aspects of the real causes of the crisis in capitalist agriculture — the assumptions of the dominant development paradigm. In fact, the quest for freedom may result in a stronger commitment to the assumptions of the dominant development paradigm. That is, the industrialisation of agriculture may occur within the framework of state rather than private accumulation. While this alters the relationship between the individual and the production process, it does not really address the nexus between agrarian social relations and nature.

Critics of the state socialist system agree that the interests of the individual are better served under socialism than under capitalism. However, these interests are still limited (by the state apparatus) and so a fully holistic system is not achieved. Thus, in order for state socialism to work it must go beyond industry, technology and the assumptions of the dominant development paradigm. If it fails to do this, there is no guarantee of preventing the crisis in agriculture or making any material alteration to the existing relationship between agrarian social relations and nature. The crucial point is that a form of state socialism must go beyond reconstituting agrarian social relations. It must also reconstitute the nexus between social relations and nature.

Lawrence (1987), in assessing the prospect for agrarian change in Australia, warns that those who wish to continue farming may have little choice but to accept the evolution of profoundly state regulated agriculture. There is little doubt that, if Routley and Routley's (1980) point about renouncing the social character of production and its private purpose is accepted, then a decentralised, democratic socialist agriculture, albeit one regulated in a manner which ensures sustainability, provides an alternative pathway.

Alterations to dominant agrarian social and economic relations are necessary, and it could be argued that these will flow on to an altered relationship between society and nature, as increased yield is not equated with efficiency under the capitalist inspired 'get big or get out' rubric. Yet there is no guarantee that this will happen and the notion of efficiency could still be equated with production and output rather than with ecologically sustainable agrarian practices. The prospect for state socialism providing an alternative model of agrarian social and economic relations is very much dependent on the conceptualisation of nature and its relationship to production.

RURAL POPULISM

Populism as an alternative development strategy can be traced to the early nineteenth century and to Owen (Britain), Proudhon (France) and Herzen (Russia). Originally developed as a critique of capitalist industrialisation, it has, as a body of thought, been characterised by a quest for small-scale co-operative villages within which industry and agriculture could develop symbiotically under a system of decentralised control and management.

Populism has had a rather checkered history, being seen as both a liberatory theory and an example of the institutionalisation of thought which concentrates on a romantic vision of the past rather than being useful for the present and the future. In its recent form, much populist thought has been equated with the notion of communality.

The doyen of communality, E.F. Schumacher (1983: 54) suggests that 'when it comes to action we obviously need small units because action is a highly personal affair'. Communal social forms, however, offer not only the advantages of action. They can offer an alternative where direct democracy, self-sufficiency (to varying degrees) and self-empowerment can be achieved (Bookchin 1982). Local food production may also have more lasting survival value (Merrill 1976b). The cooperative venture would encourage the trading of fresh, highly nutritional food in an atmosphere of self-help and mutual freedom.

The cooperative alternative also offers the opportunity to create a cultural identity which is separate from the prevailing social and economic system (McLeod 1976: 205). It is therefore easier to bring about some form of change at the local level through individual and cooperative action. Lastly, as a social movement, the value of the cooperative may be to develop and implement alternative forms of organisation and technology, production and distribution (McLeod 1976: 202).

In the Australian context some difficulties exist. Since change is attempted within a dominant system, some form of linking remains for marketing and distribution. As an alternative lifestyle, it may be possible to maintain identity but problems exist in bringing about structural change. Whilst the philosophy of change by example may well be useful, the fact remains that alternative lifestyle cooperatives are marginalised and/or stereotyped within Australia (see for example, Munro-Clark 1986; Sommerlad, Dawson and Altman 1985).

Operating within a dominant social and economic system has obvious limitations. While some of these centre on social legitimacy in the dominant society, they also include such things as a focus on individual rather than structural change. This is not meant to imply that change by example is not possible or that the individual cannot bring about structural change. It does, however, recognise the difficulty attached to this form of change given the extent and degree of reification of the assumptions of the dominant development paradigm (see, for example, Cock 1985; Furze 1989).

THE ECOLOGICAL CRITIQUE

The rapidly emerging ecological critique of dominant notions of development have gone beyond populism to recognise the need for reconstituted social relations and a renewed, holistic relationship with nature. However, it would be incorrect to suggest that the ecological critique represents an homogeneous school of thought.

One strand, which can be called environmentalism, has been offered as an alternative. What is advocated here is a change in awareness so that actions taken in spheres of, for example, production, consumption and politics are taken with an awareness of the impact on the environment. Within this alternative, social movements are the vehicles for action. Agitation for social change occurs through attempts at the alteration of the nature of that which is seen to be the cause. Action is therefore widespread but often issue specific.

This may not offer a valid alternative. First, environmentalism may fail to uncover the real reasons for the crisis in capitalist agriculture. Although the environmental ethic certainly aims for self-sustaining agriculture, it may fail to take into account the structural contradiction between the capitalist marketplace and the social expectations of the agricultural producer. Issues are reduced to environmental ones rather than economic/rational ones, of which environmental issues form only a part.

Second, environmentalism, because it is issue specific, fails to address the systemic causes of the crisis. Social movements propose action strategies to specific issues — the manifestation of the crisis rather than the causes. Environmentalism obviously offers some advantages but it tends to be single-issue based and fails to attack the real causes of the problem.

Another strand in environmental thought is that of self-management. For those advocating this alternative, a future for agriculture must be premised on two things — an holistic understanding of the dominant development paradigm and an acknowledgement of the benefits of local level management and distribution. The self-management alternative offers some of the benefits of the options already discussed as well as some of its own. Self-management can avoid the compulsive growth of our present society. It would bring food production to that which is required at the local level whether this be for consumption, trade or both (Routley and Routley 1980). In this, it draws heavily on rural populism.

The production of food for needs at the local level reduces the dependence on science and technology to that which is beneficial for the bioregion. It would thus influence, for example, the role of technology by affecting its design and its social character. The inclusion of the bioregion in such a model thus distinguishes it from populist notions.

Finally, self-management should lead to self-regulation, as the gemeinschaft-like qualities of devolved social, economic and political relations interact to establish stability within the bioregion. The centrality of the bioregion and the relationship of people to nature thus would ensure appropriate agrarian social relations which reconceptualise the dominant development paradigm.

In Australia, the ecological critique has seen the emergence of alternative agrarian practices such as those based on 'sustainable agriculture'. This term hides a myriad of assumptions, beliefs and practices, but the essential underpining is an attempt to reconstitute the agrarian practice/nature relationship. The question still remaining though, is how to instigate alternatives within the framework of agrarian social and economic relations based on capitalist accumulation.

A RED–GREEN ALLIANCE?

One of the difficulties in instigating alternative agrarian social relations is that of understanding the structural nature of those social relations presently existing. Without this understanding, the development of an alternative agrarian practice is problematic because real difficulties arise in the relationship between dominant and alternative practices. Yet it is not enough merely to know the processes and outcomes of the dominant logic of capitalist agriculture. There is a real need to have a developed model of alternative practices to be put into place and it is in this context that a case can be mounted for an environmental management model with a red–green alliance as its base.

Potentially, such an alliance would provide theoretical and

conceptual foundations for both the critique of existing agrarian social relations (a task undertaken by Marxist and non-Marxist critical theorists) and an alternative model of agrarian practice (a task taken up by environmentalists). The light shed on these areas by the respective camps should illuminate both perspectives.

Practically, however, there are real challenges for a red–green alliance. For example, Pepper (1986) takes up the call for a closer alignment between labour and the environmental movement, based on a socialist tradition. Defining socialism as 'a social system based on common ownership of the means of production and distribution and which displays an attachment to ethical and democratic values as well as an emphasis on the distinction between common and state ownership' (1986: 120) he locates both the environmental and the labour movements within this tradition. *Ipso facto*, commonality exists which should be drawn upon to instigate socialist organisation. Yet, where is the common ground? Although both labour and nature may benefit from alternative forms of social organisation the common goal is by no means clear to the respective parties.

In the context of agrarian social relations and environmental perception in Australia the likelihood of a red–green alliance appears, at least at this stage, remote. Despite an alteration to agrarian social relations with the proletarianisation of farmers and the incursion of monopoly capital with its very damaging social, economic and environmental consequences, the record of red–green alliances in agriculture is not particularly impressive. There is not only the historical conservatism of the rural sector (see, for example, Lawrence 1987; Costar and Woodward 1985) but also the bankrupt state of institutionalised labour, socialist and communist parties (Frankel 1987).

In practical terms, therefore, the finding of common ground on which to form a red–green alliance for an alternative agriculture in Australia may not be easy. While there might be a great deal to be gained, the schism arises from different problematics emanating from divergent world-views (Bell 1987). At this stage, the socialist project does not necessarily take the environment into account (Redclift 1984) and the environmental project does not necessarily take socialism into account. This will be difficult to overcome because the traditional conservatism of the rural sector and the more general tendency of political parties to reinforce the status quo would appear to offer little hope in the near future for the widespread adoption of alternative forms of agrarian practice.

Perhaps it is pertinent here to contextualise the above through a brief look at an alternative model of agrarian practice, that of permaculture. The term 'permaculture' comes from the words 'permanent' and 'agriculture', and has been coined to mean a 'conscious design and maintenance of agriculturally productive ecosystems' through the 'harmonious integration of landscape and people providing their food, energy, shelter and other material and non-material needs in a sustainable way'. Further, 'without permanent agriculture there is no possibility of a stable social order' (Mollison 1988: ix).

There are three broad principles which guide the permaculture system:

1. Care of the earth (provision for all life systems to continue and multiply).
2. Care of people (provision for people to have access to those resources necessary for their existence).
3. Setting limits on population and consumption. By governing needs, resources can be set aside to further the above principles (Mollison 1988).

Integral to the philosophy is the notion of 'permanence', or a mature ethic which sees (according to a general rule of nature) cooperative species and associations of self-supporting species making healthy communities. The design is not only one of non-human relationships but one of human and non-human relationships. In other words, it seeks to establish a specific relationship between the individual, society and nature.

'Communal permanence' is an ethic which seeks to establish a specific relationship between the individual, society and nature based on the notion of sustainability and using the ethical basis of permaculture as a guideline for social relations. Mollison (1988: 3–6), quite specifically states that there is more than one way to achieve both permanence and stability of land use and society, though the ideal of communal permanence is integral to all possible paths.

Because of the notion of communal permanence and because the basis of permacultural systems can be applied equally to urban and rural dwellers, it is believed possible to establish an ethical basis for permacultural communities from the local level to the nation-state. This basis is made up of the following:

1. The emphasis of the duties and responsibilities of people to nature and people to people.
2. The adoption of an ethic of 'right livelihood', which emphasises the use of labour and skills in ethical pursuits.
3. Consumer societies (which are the basis for consumption

in the permaculture nation-state) which are a mosaic of small, well-managed and effective systems.
4. The meaning of life is realised through action toward common ideals, in serving the whole.
5. Security is found in the renunciation of ownership over people, money and real assets.
6. Leisure time becomes time to express individual (as opposed to group or communal) capacities and is a plentiful resource (Mollison 1988).

An important aspect to any ethic which seeks to introduce alternative social and economic formations is the question of change. It is not simply enough to develop an alternative ethic; there must be an enunciation of pathways to achieve social change. In other words, there needs to be a statement concerning the linking of theory (or model building) with practice.

Mollison (1988: 1) suggests that:

> There is no longer time to waste nor any need to accumulate more evidence of disasters; the time for action is here. I deeply believe that people are the only critical resource needed by people. We ourselves, if we organise our talents, are sufficient to each other. What is more, we will survive together, or none of us will survive . . . A person of courage today is a person of peace. The courage we need is to refuse authority and to accept only personally responsible decisions.

Further, he suggests that

> It has become evident that unity in people comes from a common adherence to a set of ethical principles, each of us perhaps going our own way, at our own pace, and within the limits of our own resources, yet all leading to the same goals, in our case that of a living, complex, sustainable earth. Those who agree on such ethics, philosophies and goals form a global nation (Mollison 1988: 3).

The paths to achieving such a society are varied. Because there is no well-defined model (or little discussion) of what change strategies are to be employed, the question is left unanswered. Mollison and Holmgren (1990: 95) advise that 'we are shaping a tool and an idea, how the application of either of these is made is for each of us to decide, and to refine'.

The permaculture model provides a series of what could be called 'first principles' which govern the relationship between the individual, society and nature within a sustainable agrarian system. Although the model is somewhat skeletal, it does provide a basis for social action via a vision of what 'should be'. The pathways taken for the creation of a permaculture system, by being left flexible and in the hands of those who seek to establish such a system, recognise the need for change from the bottom up.

Potential exists, but to what extent this can be realised is dependent on the change strategies of those who follow the differing paths, including their awareness of the dynamics of existing social, economic and political formations. Here, they could well encounter the same difficulties as those who have attempted rural land sharing as an alternative model — the forms of the structural linkages with the dominant social and economic system, social legitimacy and an understanding of the ways in which the dominant society structures and replicates itself.

ENVIRONMENTAL MANAGEMENT AND SUSTAINABLE RURAL DEVELOPMENT?

Within the alternative agrarian models discussed above, there is not one which stands out as 'the option'. All have attractions and all have limitations. Perhaps this is a reflection of the undeveloped stage of social theory as it applies to models of alternative agrarian practice. Perhaps also it is a reflection of the difficulties faced by those who seek to instigate such a practice in the face of a dominant paradigm which has gained mainstream credibility and which, because of its characteristics, pushes alternative model building to the side, labelling it Utopian with all the negative connotations that accompany that term.

It would seem though that there is a need to establish an alternative, one which reconceptualises the nexus between the individual, society and nature. The schism between the public purpose of agrarian production and the private accumulation of profit must be overcome. Perhaps then, what is needed is a red–green alliance which draws upon the various left-inspired trenchant criticisms of capitalism but incorporates as its base a profound environmental ethic. Out of this fusion will come a vision of the alternative forms of social organisation which can and should develop. Perhaps such an alliance offers the advantages of centralisation (distribution mechanisms, for example) integrated with devolved political, economic and social relations based on the bioregion. Indeed, a form of centralisation may well be a necessary first step in any case if local-level initiatives continue to be co-opted into monopoly capitalism or marginalised, thereby being robbed of important social legitimacy.

Much work needs to done before an environmentally sound, alternative agrarian practice can be achieved. Its development is the task in front of us; it is becoming, not surprisingly, *the* task of the 1990s.

REFERENCES

Amin, S. (1974) *Accumulation on a World Scale: A Critique of the Theory of Development*, Monthly Review Press, Hassocks.

Bahro, R. (1982) 'Capitalism's Global Crisis', *New Statesman* 17 December.

Bahro, R. (1986) *Building the Green Movement*, GMP, London.

Bell, S. (1987) 'Socialism and Ecology: Will Ever the Twain Meet?', *Social Alternatives* 6 (3).

Bookchin, M. (1982) *The Ecology of Freedom*, Cheshire, New York.

Cock, P. (1985) 'Sustaining the Alternative Culture? The Drift Towards Rural Suburbia', *Social Alternatives* 4 (4).

Costar, B. and Woodward, D. (eds) (1985) *Country to National: Australian Rural Politics and Beyond*, Allen and Unwin, Sydney.

Devall, B. and Sessions, G. (1985) *Deep Ecology: Living as if Nature Mattered*, Peregrine, Salt Lake City.

Frankel, B. (1987) *The Post-Industrial Utopians*, Polity Press, Oxford.

Furze, B. (1989) 'The Distance Between "Utopian" Vision and Individual Action: Some Perspectives Through the Study of a Rural Landsharing Collective', *Regional Journal of Social Issues*, Summer.

George, S. (1980) *How the Other Half Dies: The Real Reasons for World Hunger*, Penguin, Harmondsworth.

George, S. (1985) *Ill Fares the Land: Essays on Food, Hunger and Power*, Writers and Readers, London.

Gerth, H. and Mills, C. (eds) (1958) *From Max Weber: Essays in Sociology*, Oxford University Press, New York.

Hearn, F. (1985) *Reason and Freedom in Sociological Thought*, Allen and Unwin, Boston.

Larrain, J. (1989) *Theories of Development: Capitalism, Colonialism and Dependency*, Polity Press, Cambridge.

Lawrence, G. (1987) *Capitalism and the Countryside*, Pluto Press, Sydney.

McLeod, D. (1976) 'Urban–Rural Food Alliances: A Perspective on Recent Community Food Organizing', in R. Merrill (ed.), *Radical Agriculture*, Harper Colophon, New York.

Mannheim, K. (1952) *Ideology and Utopia*, Harcourt Brace, New York.

Merrill, R. (1976a) 'Toward a Self-sustaining Agriculture', in R. Merrill (ed.), *Radical Agriculture*, Harper Colophon, New York.

Merrill, R. (1976b) 'Introduction', in R. Merrill (ed.), *Radical Agriculture*, Harper Colophon, New York.

Mollison, B. (1988) *Permaculture: A Designer's Manual*, Tagari Publications, Tyalgum.

Mollison, B. and Holmgren, D. (1990) *Permaculture One* (3rd edn), Tagari Publications, Tyalgum.

Mumford, L. (1962) *The Story of Utopias*, Viking Press, New York.

Munro-Clark, M. (1986) *Communes in Rural Australia: The Movement Since 1970*, Hale and Iremonger, Sydney.

Pepper, D. (1986) 'Radical Environmentalism and the Labour

Movement', in J. Western (ed.), *Red and Green: The New Politics of the Environment,* Pluto Press, London.

Redclift, M. (1984) *Development and the Environmental Crisis: Red or Green Alternatives?,* Methuen, London.

Redclift, M. (1987) *Sustainable Development. Exploring the Contradictions,* Routledge, London.

Routley, V. and Routley, R. (1980) 'Social Theories, Self Management and Environmental Problems', in D. Mannison, M. McRobbie and R. Routley (eds), *Environmental Philosophy,* Monograph Series No. 2, Department of Philosophy, Australian National University, Canberra.

Schumacher, E. (1983) *Small is Beautiful: A Study of Economics as if People Mattered,* Abacus, London.

Sommerlad, E., Dawson, P. and Altman, J. (1985) *Rural Land Sharing Communities: An Alternative Economic Model?,* Australian Government Publishing Service, Canberra.

Stretton, H. (1977) *Capitalism, Socialism and the Environment,* Cambridge University Press, London.

Wallerstein, I. (1980) *The Capitalist World Economy,* Cambridge University Press, Cambridge.

Weber, M. (1963) *The Sociology of Religion,* Beacon Press, Boston.

6 THE SOCIAL CONTEXT OF FARMERS' ADOPTION OF ENVIRONMENTALLY SOUND FARMING PRACTICES

FRANK VANCLAY

Land degradation is Australia's most serious environmental problem and has been recognised as a problem since the 1930s (Messer 1987). Many techniques for preventing land degradation exist, yet these have not been widely adopted by farmers. There are many soil conservation practices, some of which have been around for 40 years or more, which, according to the soil scientists, would greatly reduce the long-term land degradation currently experienced on Australian farms. At first glance, the lack of adoption appears to be surprising, since many of these techniques would require little or no change to overall farming practice and many could be implemented without significant cost to the individual farmer (Donald 1982; Chamala, Keith and Quinn 1982; Pampel and Van Es 1977). Some techniques, especially the establishment of deep-rooted perennial pasture species such as lucerne, which are recommended salinity mitigation strategies, are considered to be profitable for farmers (Oram 1987; Thorne 1991). Why, then, do Australian farmers not adopt the soil conservation technology that is available?

The lack of adoption of conservation farming practices by farmers indicates that land degradation is not primarily a technical problem. The issue is not the lack of techniques of soil conservation or of sound land management practices but the social, structural, perceptual and financial situations and processes that act to prevent farmers from adopting those techniques. It is, therefore, important to examine the situation of farmers to determine from their point of view reasons for the non-adoption of practices that technical experts clearly believe would solve, or at least reduce, the soil degradation problem.

THE CALL FOR A CHANGE IN FARMERS' ATTITUDES

Politicians and conservationists often comment in off-the-record statements that farmers' attitudes to the environment are not conducive to effective land management and that the solution to the environmental problems experienced on Australia's farms would most likely require the changing of farmers' attitudes. Many urban people also consider that farmers are not the stewards of the land that they would have us believe they are and that farmers have little intrinsic concern for the land. Of course, it is very unlikely that any person in the public arena would wish to go on public record stating this, even if this is a personal viewpoint. Nevertheless, former Minister for Primary Industries Kerin (1984) and former Prime Minister Hawke (1989: 44) have made remarks indicating that a change of attitude would be important in dealing with the land degradation problem.

The adoption of a land ethic is actively promoted by Dr Brian Roberts, a previous chairman of the Soil Conservation Association of Australia, and a member of the Commonwealth Soil Conservation Advisory Committee (see Roberts 1990), and it is also the platform of many (soil) conservation organisations. The Soil and Water Conservation Association of Australia considers that the problems of inadequate land management will remain 'until such time as the whole community accepts the need for stewardship and adopts a land conservation ethic' (Standing Committee on Environment, Recreation and the Arts (SCERA) 1989: 105). As a witness to the Inquiry into the Effectiveness of Land Degradation Policies and Programs, Dr Smiles, Chief of the CSIRO Division of Soils, also endorsed the notion that 'there needs to be a public reappraisal of attitudes to land management' (SCERA 1989: 61). Furthermore, the National Soil Conservation Program has five goals, the fifth being 'that the whole community adopt a land conservation ethic' (SCERA 1989: 66).

These calls have a political element. By placing the failure of soil conservation adoption on farmers, governments can claim that the responsibility for the problem lies with farmers, not with government. Indeed, education campaigns to increase farmers' awareness of the problem are likely to be far less costly than other potential action. From a political perspective, the call for attitude enhancement is understandable. The land degradation issue is complex and costly, with actual payoff generally outside political time-frames. In many cases success would not be obvious and it is quite likely that small efforts by governments would have very little effect. But all this rests on the premises that farmers' attitudes are actually environmentally negative and that attitudes adequately predict behaviour.

FARMERS' ATTITUDES TO CONSERVATION

Although a fundamentally important concept in psychology, 'attitude' does not enjoy a uniformly accepted definition. However, as a general approximation, attitudes are some form of 'learned predisposition to respond in a consistently favorable or unfavorable manner with respect to a given object' (Fishbein and Ajzen 1975: 6). Methodologically speaking, an attitude is an intervening hypothetical construct mediating the influence of an external stimulus on an individual's response to that stimulus (Figure 6.1).

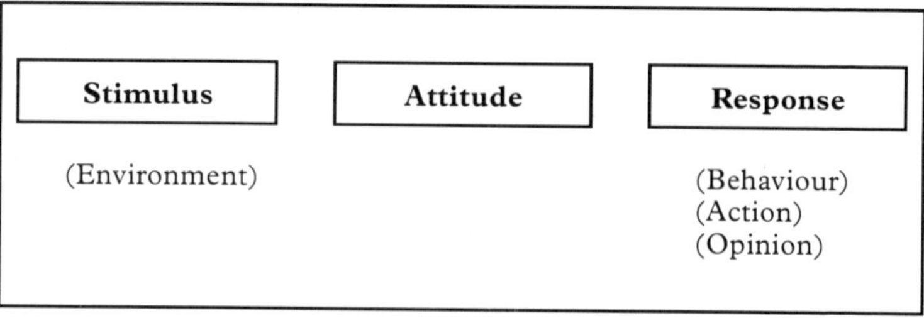

Figure 6.1 Simplistic attitude model

It is generally accepted that farmers will adopt soil conservation technology when they consider themselves to be at risk (that is, perceive land degradation on their land) (Rickson *et al.* 1987). The attitude then mediates the relationship between the perception of the environment (stimulus) — in this case the recognition that land degradation is occurring — and the appropriate response, which is the adoption of soil conservation technology. According to this model, farmers with the right attitude (those who are conservation minded) will adopt soil conservation technology, while those with environmentally destructive attitudes, or less positive attitudes, will continue to let land degradation occur unabated.

Vanclay (1986, 1992) tested this model by developing five attitude scales to measure the different aspects of farmers' attitudes to the environment: stewardship, conservation is economic, the importance of conservation, seriousness of off-site damage, and no erosion problem. Each scale consisted of several items scored on a five point Likert scale (1 = strongly disagree; 5 = strongly agree) with the total scale scores averaged to allow scoring on the original measurement. The scales were developed from a bank of attitude items completed by a sample of 92 Darling Downs farmers for a major study of farmers' responses to soil erosion.

Darling Downs farmers had high scores (on average agreed) for the scales 'stewardship' (mean = 3.8), 'importance of conservation' (3.9) and 'conservation is economic' (4.0), with lower scores (on average were undecided) for the scales 'no erosion problem' (3.1) and 'seriousness of off-site damage' (2.8). There was very little variance in the scores, indicating consensus among farmers on these issues (Table 6.1).

Stewardship refers to the notion that farmers are stewards of the land and that farming is a way of life that places implicit responsibility on farmers to look after the land for future generations. The stewardship concept recognises that farmers may have to make uneconomical decisions in order to protect the land. It embraces the notion that there is more to farming than economic management. Ninety-six percent of Darling Downs farmers had an attitude favourable to the notion of stewardship (i.e. had scores > 3.0 on the stewardship scale).

Almost all the Darling Downs farmers believed in the importance of conservation. This was measured by a scale covering issues of the importance of conservation to farmers and to the community in general, especially as far as the future was concerned. All farmers also believed that soil conservation was economic. Given these attitudes of Darling Downs farmers, this should suggest that the adoption of soil conservation technology would not be a problem on the Darling Downs.

Table 6.1 Attitude scales — descriptive statistics

	No. of items	Cronbach alpha*	Min. score	Max. score	Mean score	Std dev.	% who agree
Stewardship	9	.74	2.1	4.9	3.8	0.48	96
Importance of conservation	8	.68	2.4	4.9	3.9	0.48	94
Conservation is economic	11	.73	3.2	5.0	4.0	0.41	100
No erosion problem	10	.63	2.2	4.2	3.1	0.45	60
Seriousness of off-site damage	5	.53	1.4	4.2	2.8	0.64	30

Potential range from 1.0 to 5.0; n = 90.
*Reliability coefficient.
Source: Vanclay (1992).

Not only did these farmers have appropriate attitudes, they also believed that soil conservation is economic. There is no reason to believe that farmers on the Darling Downs should be any different, at least to any major degree, than farmers anywhere else in Australia. It is very likely that similar results would be obtained from any representative sample of Australian farmers in any geographical location. Although attitudes are learnt, and therefore changeable, they tend to be stable over time since they are the result of years of socialisation and of internalisation of experiences. It is unlikely that these attitudes would change from year to year or season to season. However, despite the attitudes remaining stable, other influences affecting resultant behaviour may change, so behaviour may change even though the underlying attitudes remain the same.

Unfortunately, the existence of these attitudes is not associated with the adoption of soil conservation technology and the absence of land degradation on the Darling Downs. Soil scientists argue that large proportions of Darling Downs (and other Australian) farms are not adequately protected against land degradation. Combining the figures of the Queensland Department of Primary Industries (Soil Conservation Services Branch) with categories developed by the NSW Soil Conservation Service, it was determined that 45 percent of the Darling Downs farmers in the study had not adequately protected their farms against soil erosion (Vanclay 1986). Clearly, the fact that farmers have appropriate environmental attitudes does not guarantee that they will adopt the necessary practices. It also suggests that educational campaigns aimed at improving the attitudes of farmers are likely to fail. What then does explain the contradiction between farmers' attitudes and their lack of adoption of necessary practices?

THE CONTRADICTION BETWEEN ATTITUDES AND BEHAVIOUR

The first obvious explanation is that the measurement of farmers' attitudes is affected (biased) by the potentially enormous influence of social desirability in this sort of attitude measurement. As a corollary, it might be claimed that farmers are actually hostile to the environment, despite the results of these attitude scales. Farmers are astute enough to be aware of the politics of conservation and the socially desirable answer is obvious to anyone responding to the questionnaire. It would be difficult to rule out the social desirability argument, except perhaps by repeating the research including some sort of social desirability measure. Alternatively, it could be assumed that social desirability affects farmers equally; that the mean scale score for each farmer is elevated, but that differences between farmers are still meaningful. If this were to be the case, those with higher conservation scores would be expected to have higher levels of adoption of technology and be more likely to protect their farm from land degradation. Vanclay (1986) undertook discriminant and regression analysis to identify the socioeconomic correlates of adoption of soil conservation technology, using 'protection', a dichotomous measure indicating whether the farmer had adopted sufficient and appropriate soil conservation techniques to protect the land adequately against soil erosion, as the dependent variable. In that study, there was no indication that stewardship or conservationism (as measured by any of the five scales) was positively associated with protection. Furthermore, there was some evidence that the non-protectors actually had stronger conservation attitudes. This finding supports the general conclusion that, even if the attitude scales are contaminated by social desirability response bias, farmers' attitudes to conservation do not predispose them to adopt soil conservation technology.

The finding that attitudes are not generally predictive of behaviour is not unique to this study and is well recognised in psychology. It is only under certain conditions that attitudes are expected to have any strong effect on behaviour. Myers (1989: 558) lists these conditions as:

1. when other influences that affect our attitudes and are actions are minimised;
2. when the attitude is specifically relevant to the behaviour; and
3. when we are keenly aware of our attitudes.

If we consider farmers responding to the environment and determining whether or not to do anything (Figure 6.2), the connection between the stimulus (the environment) and the response (adoption) is not only, if at all, affected by their attitudes. It is also affected by their perception of the environment (the perception screen) and by their personal and financial situation or context (the context/situation screen).

Farmers are unlikely to adopt soil conservation technology to the satisfaction of soil scientists if they have differing perceptions about the nature and extent of land degradation on their land. Furthermore, even where they do perceive land degradation, they are unlikely to adopt soil conservation technology if they lack appropriate information and/or have other demands on their capital and time. In the context of Myers' conditions:

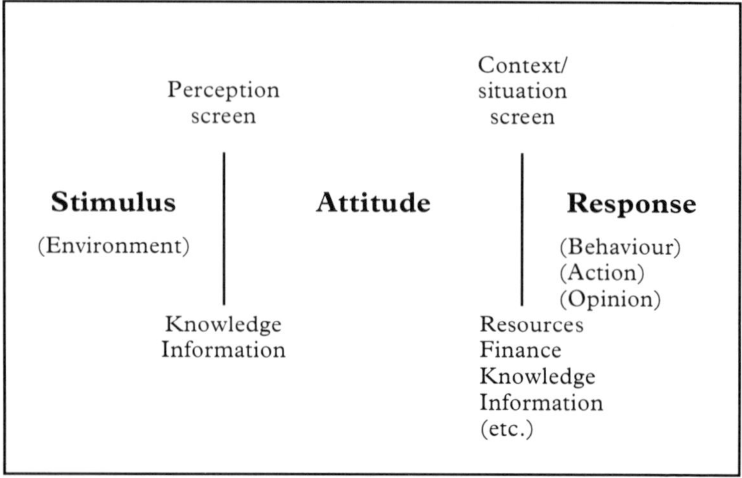

Figure 6.2 Modified attitude model

1. The situation of farmers' adoption of soil conservation technology is heavily affected by a wide range of other influences. The range of possible actions the farmer may undertake in response to the situation is considerable and there are many influences on farmers' attitudes, including processes that may lead to the denial of the problem (cognitive dissonance).
2. Attitudes to the environment (such as stewardship and conservationism) are very general attitudes consisting largely of 'motherhood' statements which also experience a high degree of social desirability. These attitudes do not determine specific behaviour. In the situation of land degradation, even where farmers recognise it occurring on their farms, there is a wide range of appropriate behaviours and there is much conflicting technical information.
3. Farmers are not generally aware of their attitudes to the environment. They think that surveys like this are a waste of time and do not conceive of or intellectualise their responses in the way that urban professional people might. Farmers are much more likely to respond according to notions of good farm management which exist in the farming subculture and the local peer group.

Recognising the wider situation of farmers and the general nature of conservation attitudes, it might be expected that there would be strong attitudinal support but low adoption.

In the Darling Downs there is a further explanation for the contradiction. Darling Downs farmers not only had high scores on the 'stewardship', 'conservation is important', and 'conservation is economic' scales, they also believed that there was no (real) erosion problem and that the loss of soil through soil erosion is exaggerated by people who are not farmers. They also tended not to believe (were undecided) in the 'seriousness of off-site damage'. With their beliefs that soil conservation practices are economic and the small nature of erosion problems generally, farmers considered that most of the work required to be done to protect farms against erosion was already done (or soon would be) and that no major changes to agricultural management practices or technology was required. This suggests that the farmers may be less concerned about land degradation than soil scientists consider they should be and that they may not appreciate the full implications or seriousness of the erosion problem.

FARMERS' CONCERN ABOUT LAND DEGRADATION

Farmers are concerned about land degradation as a general community problem but tend to consider that it is not a problem that will affect them personally (see Tables 6.2, 6.3 and 6.4). Farmers consistently understate and misperceive the extent to which their farms are affected by land degradation (Cameron and Elix 1991; Chamala *et al.* 1982; Rickson and Stabler 1985). Research by Rickson *et al.* (1987) indicated that were farmers to appreciate the full extent of land degradation that does occur on their farms, they would quite likely act to prevent it. Rickson *et al.* found that farmers' estimates of anticipated yield losses for a nominated hypothetical erosion rate of 5 mm per annum (the currently accepted, estimated average erosion rate for unprotected properties on the Darling

Table 6.2 Farmers' concern about erosion (Darling Downs)

	Darling Downs	Local area	Own farm
Not a problem	0	0	15
A small problem	1	17	37
A medium problem	11	37	37
A major problem	88	47	11
Total (*n* = 90)	100	100	100

Modified from Rickson *et al.* (1987).

Table 6.3 Farmers' concern about current salinity (Victoria)

	Central Highlands	Neighbourhood	Own farm
Not a problem	4	23	53
A small problem	20	30	32
A medium problem	30	23	10
A major problem	45	24	5
Don't know	2	0	0
Total (n = 131)	100	100	100

Vanclay and Cary (1989).

Downs) exceeded the actual yield losses in field experiments at a nearby site on the Darling Downs simulating that erosion rate (Figure 6.3). The median anticipated yield loss by farmers was 50 percent of current yield after 20 years of erosion at 5 mm per annum, while the maximum yield loss obtained in any of the three experimental plots was actually only 30 percent after removal of 100 mm of top soil (Rickson *et al.* 1987).

The actual yield loss obtained will vary enormously according to local conditions, especially on the Darling Downs where there are very deep soils and very shallow soils in very close proximity (the gilgai phenomenon). Consequently, there may be locations on the Darling Downs, and elsewhere in Australia, where actual yield losses will be much higher than the figures reported here. Nevertheless, the research demonstrates that farmers do recognise the relationship between erosion and yield loss, and on average, anticipate large yield reductions at the currently estimated levels of soil erosion.

Discussions with farmers revealed that farmers could not accept that there was an erosion rate of 5 mm per annum on

Table 6.4 Farmers' concern about future salinity (20 years from now) (Victoria)

	Central Highlands	Neighbourhood	Own farm
Not a problem	3	12	28
A small problem	4	11	24
A medium problem	14	35	24
A major problem	75	39	21
Don't know	5	3	2
Total (n = 130)	100	100	100

Vanclay and Cary (1989).

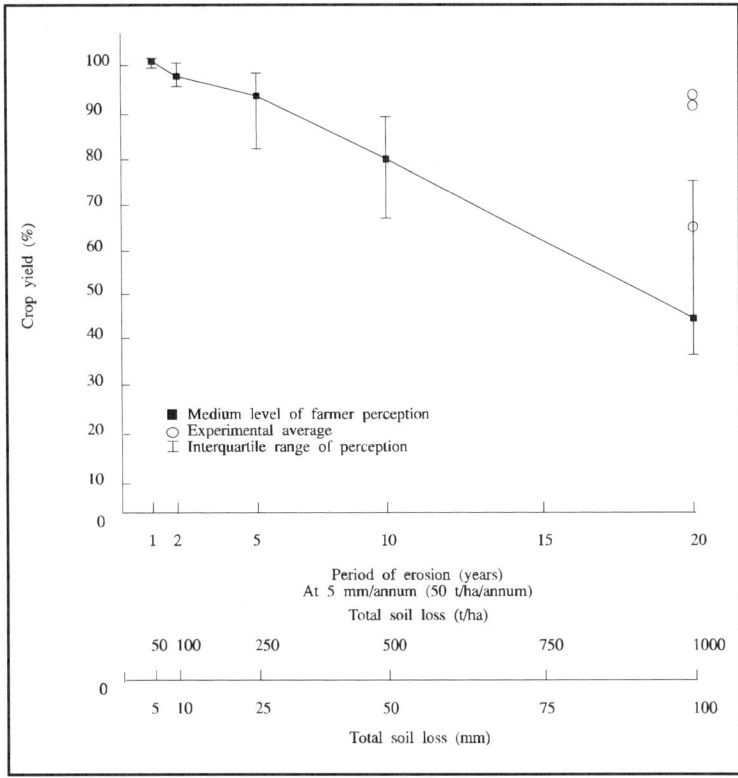

Figure 6.3 Soil erosion and crop productivity (adapted from Rickson *et al.* 1987)

their farms. The anticipated yield losses are high because they consider 5 mm per annum erosion to be an enormous and greatly exaggerated erosion rate.

Nevertheless, these results indicate that if farmers can be brought to accept that high rates of erosion are actually occurring, they may be likely to adopt soil conservation technology. However, it is also likely that their estimates of yield losses will be moderated as they come to accept higher erosion rates as being normal.

THE NATURE OF LAND DEGRADATION AND OF FARMERS' RESPONSES

Land degradation has pervasive and intensive forms (Barr and Cary 1984). While farmers generally respond positively to the more obvious forms of erosion, extreme intensive forms of accelerated erosion may evoke a fatalistic acceptance response

(Chamala, Rickson and Singh 1984; Williams 1979). However, the insidious nature of pervasive forms of land degradation, especially in their early stages, are such that the effects are slight, not obvious, and even if noticed are easily dismissed as being due to other factors. Roberts (1991) relates a story about an old South African farmer who, when asked whether he had seen any changes on his farm over his lifetime, replied, upon serious reflection, 'I think the rocks are growing'.

The principal forms of land degradation occurring on the Darling Downs are sheet and rill erosion, the subtle processes which together are responsible for eroding an estimated 5 mm of soil per annum (50 tonnes per hectare per annum) on inadequately protected cultivated land. Sheet erosion is not obvious at the point of erosion, while rill erosion is recognisable by the very small gullies (rills) that form as water runs across the paddock. Although in aggregate, these forms of land degradation erode an enormous amount of soil, at the individual level, on a day to day basis, the effects of this soil loss are not obvious. However, the ways to prevent soil erosion are well understood and soil erosion processes are relatively easy for farmers to understand.

Although soil conservation technology is costly and may require some change to farm management, there are clear benefits to the individual farmers who adopt it, probably within the farming lifetime of that farmer. Nevertheless, while there may be substantial, perhaps almost uniform, technical agreement that various forms of management practices (such as contour cultivation and stubble mulching) and structural practices (such as the use of contour banks, grassed waterways and diversion banks) are practical, appropriate and required on most farms, technical disagreement exists over the appropriateness of so-called conservation cropping (at least when defined as zero or minimum tillage) (see Chapter 14). While farmers may, in principle, agree with extension officers about the use of some of the recommended practices, they tend to disagree over the extent of their use. For example, there is very little disagreement that steep slopes should not be cultivated and should be left as permanent pasture. Disagreement occurs in determining what constitutes a steep slope with, typically, extension officers recommending permanent pasture on slopes which farmers might normally cultivate (for example, slopes just over 8 percent).

Soil salting is more insidious in nature, in that it is not only a slow process but also temporally and spatially distorted. According to the currently accepted model of dryland salinity, farmers who experience salting are not generally those farmers whose farming practices will prevent soil salting. The process of water infiltration into the watertable has occurred since the land was first cleared and salinity control techniques implemented now are unlikely to have an effect on watertable levels for a very long time. The model of the salting process is complex — although not beyond farmers' understanding — but is not necessarily generally accepted by them. Salinity control measures are costly, have almost no short-term or even medium-term effect on the severity of salting and may require major changes to farm management. There is also considerable disagreement about the technical and plausible solutions to the problem. This is recognised in the extension agencies in the development of what are referred to as 'best-bet' strategies. Consequently, there is little incentive for farmers to participate in soil salting control (Vanclay and Cary 1989), and farmers' participation is likely to require a great deal of faith on the part of the farmers (Barr and Cary 1992). Importantly, there is some suggestion that farmers are being asked to grow trees where trees are unlikely to have grown before (see Chapter 4; and Barr and Cary 1992).

FARMERS' KNOWLEDGE OF LAND DEGRADATION

The research concern of Vanclay and Cary (1989) in the Central Highlands of Victoria was the extent of farmers' knowledge about soil salting. Seventy-nine percent of farmers knew that dryland salting was a consequence of tree loss and/or watertable changes. Seventy-six percent knew that the solution to salinity required tree planting and/or the growing of deep-rooted species. However, in an area where all farmers were in close proximity to discharge areas, only 57 percent were aware that salt-tolerant species were early indicators of soil salting. Clearly, of these three issues, farmers' knowledge of the early warning signs was the limiting factor in their overall picture of the salinity process.

The above figures relating to farmers' knowledge levels are overstated since very lax criteria were used. In terms of the early warning signs, any salt-tolerant species was accepted. In reality, many salt-tolerant species, such as salt bush, appear only at the very late stages of salting, and even the recognised salt-indicator species such as barley grass and spiney rush occur at such a stage that it may be impossible to reverse the salting process.

For farmers to be environmentally conscious, it is important that they recognise the subtle changes in pasture composition (for example, subterranean clover giving way to strawberry clover) and the lack of prolific growth among plants as being indicative of potential salting. When farmers recognise potential salting at this early stage they will begin to see that they themselves are personally at risk and will be more motivated to participate in salinity control measures, or in community programs aimed at controlling salinity.

The same argument holds true for soil erosion. If farmers become aware of the early warning signs, they will begin to recognise these signs on their own and other farmers' properties. The recognition of the farm to be at risk from land degradation is one of the factors that is predictive of soil conservation technology adoption (Vanclay 1986). Furthermore, if farmers become aware of the early warning signs, they have an opportunity to respond to those visual cues while there is still time. The potential to dismiss the need for adoption because of a fatalism associated with dramatic intensive events is reduced. It is most desirable that farmers be made aware of the early warning signs.

The problem here is that the early warning signs are general, not specific, indicators of a 'problem'. In all situations of land degradation, the early warning signs could be attributed to other causes. For salinity, soil erosion, acidification and some other forms of land degradation, the early warning signs include poor seed germination rates, change in species composition in pasture, reduced proliferation and lack of vigour in plants. Farmers could easily and logically explain many of these changes as being due to the lack of fertiliser, too much fertiliser, the wrong sort of fertiliser, poor quality fertiliser, poor quality seed, the lack of rainfall, too much rainfall at the wrong time, too high temperatures, too low temperatures, pests, weeds, the influence of the neighbours' activities, residual influences from previous crops and sprays, and many other possible causes. Furthermore, for most of these early warning signs, it would also be impossible for an extension officer or other expert to determine precisely what was the real cause.

For soil erosion by water, the warning signs could also include coloured runoff, turbid creeks and dams, build-up or loss of soil around fences, silted-up creeks and soil on the roadways. However, most farmers are so used to seeing these signs that they regard this as the usual situation.

The early warning signs for salinity include salt-indicator species such as barley grass and spiney rush. However, these plant species are not restricted to salty environments and are frequently associated with general waterlogging, not necessarily being due to rising (salty) watertables. Nevertheless, the prolific establishment of these species is usually associated with high groundwater salinity (Jenkin and Morris 1982), although farmers may not believe this to be the case.

Given that in most cases farmers have not experienced reduced yields that are not seasonal fluctuations, farmers would be very reluctant to interpret early warning signs as indicators of land degradation, that is, as evidence that they would need to change their farm practices; rather they would be far more likely to accept one of the many other possible explanations.

Since the visual cues to land degradation appear after significant degradation has already occurred and since these visual cues are easily dismissed, there is enormous potential for soil testing kits in promoting adoption of conservation farming techniques. Soil testing kits, when used on a regular basis, can identify soil salting, acidification and other forms of land degradation long before significant degradation happens and well before the visual cues become apparent. Since these kits give concrete indicators (that is, digital readout) of rising salt levels or of increasing acidification, they provide indisputable evidence to the farmer that there is an increasing land degradation problem (see Powell and Pratley 1991).

MEDIA IMAGES OF LAND DEGRADATION

Farmers have 'conservation-oriented' attitudes and are aware that land degradation is an important environmental issue, at least as a general issue. Yet they fail to perceive themselves to be at risk from land degradation. Rickson *et al.* (1987) have demonstrated that farmers do accept that land degradation has serious economic implications and that significant yield losses may occur if they were to experience land degradation. The reason why farmers do not consider themselves to be at risk is because they do not know the early warning signs of land degradation and because of the presentation of land degradation in its most severe form in media images of land degradation. Any examination of land degradation in the media, be it the popular press (see *The Australian Magazine* 29–30 April 1989), the conservation press (see *Habitat Australia* February 1988), and much, if not all, of the extension literature all rely on dramatic visual images of severe forms of land degradation (dry, salt-encrusted bare patches; 3 metre deep gullies).

Very few farmers actually experience land degradation in the severe forms depicted in the media and extension literature. Consequently, while farmers' awareness and attitudes are being heightened by the images, the images themselves are telling farmers that because they do not have land degradation like those images they must not have a problem. We can surmise that the media and extension literature have been counter-productive in the message they portray.

Since farmers are unlikely to adopt soil conservation technology when they do not believe themselves to be at risk (Bultena *et al.* 1981; Chamala *et al.* 1982; Pampel and Van Es 1977; Rickson and Stabler 1985), the promotion of dramatic

Figure 6.4 Our land needs help (photograph by Rick Davies, from *Habitat Australia* 16 (1), published by the Australian Conservation Foundation)

DRYLAND SALTING

The Problem

Soil salting is the single greatest threat facing Victoria's environment. It occurs in both dryland (non-irrigated) and irrigated areas. In this leaflet we look at dryland salting which is now widespread throughout the northern and western parts of Victoria, and is starting to show up in other parts of the state as well.

The problem affects not only farms and farmers but rivers, roads, farm and town water supplies, and, wildlife habitat. In other words it is a problem of great concern to the community at large.

Figure 6.5 Dryland salting (photo courtesy of the Department of Conservation and Environment, Victoria)

images and the consequent ignorance of the early warning signs by farmers are major barriers to adoption of soil conservation techniques. It is obviously desirable that farmers recognise the early warning signs and that extension literature reduce the emphasis on dramatic images.

This may be difficult. Soil conservation departments employ publicity/publications officers who usually have an advertising, public relations or journalism background. They seem to highlight the more dramatic images (perhaps an important technique in advertising) and seem incapable of harnessing images that are less dramatic. The presentation of the less dramatic early warning signs may require a radical rethinking of how extension information is communicated to farmers.

FARMING SUBCULTURE

Society is not homogeneous. There are many groups within society and each group tends to develop a particular subculture, a set of behaviours and attitudes expected of people in that group. While still remaining part of the wider culture, each group develops a unique identity. The farming community is no different from any other group in this respect. Subcultures are not prescriptive and there can be diversity within the subculture. Furthermore, many people may belong to more than one subculture and may be placed in situations of having contradictory role expectations. Not all farmers will subscribe to all aspects of the farming subculture. Because we are socialised into our culture and subcultures, in a continuous and subtle process, we may not be aware that these subcultures actually exist and we may not be conscious of all the aspects of the subculture. However, this lack of recognition or cognition of the subculture does not prevent our subcultures from being an important part of our life or a significant factor in determining our behaviours and attitudes. Peer pressure is part of the enforcement of the subcultural expectations: but peer pressure is intense and readily experienced by the individual. Peer pressure explains why individuals do things (and do not do things) which might not be in accordance with more widely accepted behaviour. Socialisation into a subculture accounts for people's normal everyday behaviour, is not intense and not necessarily experienced or perceived by each individual.

The obvious manifestation of the farming subculture is in farmers' dress: their checked long sleeve shirts, moleskin trousers, elastic sided riding boots and akubra hats. When it looks like rain, driza-bones are an essential part of the farmer's attire. A preference for country and western music is also expected.

While this may be something of an overgeneralisation, the farming subculture does emphasise certain attitudes. There are three major components that can generally be distinguished: stewardship; farming as a way of life; and a unique form of (rural) political and social conservatism. These are sometimes grouped together as agrarianism, although agrarianism is not a unidimensional phenomenon (Flinn and Johnson 1974) and tends to concentrate on the conservatism and non-economic orientation to farming, often ignoring the stewardship aspect (see Flinn and Johnson 1974; Buttel and Flinn 1975, 1977; Carlson and McLeod 1978; Buttel *et al.* 1981; Craig and Phillips 1983; Singer and Freire-de-Sousa 1983; Molnar and Wu 1989). Australian farmers exhibit high levels of agrarianism (Craig and Phillips 1983).

The subculture works at two levels: at promoting certain attitudes, ideas and beliefs that members of the subculture are supposed to have; and at specifying certain behaviours and practices. This means that much of the behaviour of individuals may be in response to subcultural expectations (the sociological model), rather than as a result of the individual's own attitudinal response mechanisms (the psychological model). Part of the problem in relation to the adoption of new techniques or management practices is that these techniques and practices are not generally accepted within the subculture. Consequently, only the innovators within the community are likely to adopt them (see Rogers 1983). However, once these techniques have gained wider acceptance within the subculture, the majority of farmers can be expected to adopt them. The task, therefore, is to promote the acceptance of techniques and management practices within the farming subculture.

Some years ago there was a subcultural insistence that farms have perfectly straight furrows. In fact, rural show days often had competitions to determine who could produce the straightest furrows. This has now been replaced with an acceptance of contour cultivation. Contour cultivation is not universally adopted, of course, with some farmers finding the management of their farm too difficult with contour cultivation. Nevertheless, there is no longer a subcultural obstacle to the adoption of contour cultivation. Gradually, other aspects of conservation farming will need to become accepted parts of the farming subculture.

'Recreational ploughing' or 'recreational tillage' also needs to be subdued. Farmers have a strong work ethic, yet farming is an activity that may mean at certain times of the year there is not a lot of work that needs immediate attention. Some farmers feel that unless they are doing something productive, like driving tractors, they are not working and many farmers are not used to the idea of leisure. There is clear evidence that many Australian farmers overcultivate their land, and that they use ploughing as a therapy to avoid some of the stresses of life. Ideally, the importance of the tractor in farming life needs to be de-emphasised and farmers need to be better trained to deal with the sometimes long periods of leisure time they are likely to enjoy.

SITUATIONAL CONSTRAINTS ON FARMERS' ADOPTION

Land degradation processes are complex, some control measures are costly and there is limited short-term return for investment in soil conservation technology. In fact, in situations of high interest rates, with future discounting entering into the calculations, it is likely that many soil conservation techniques and management practices are not economic (Quiggin 1987). Except where land degradation reduces the capital value of the land or yields, land degradation is an externality to the economic situation of farmers. The off-site consequences of erosion and, in the case of soil salting, the temporal as well as spatial separation of consequences from management practices, makes land degradation external to the immediate concerns of the farmer. Consequently, in situations of conflicting technical information, unpredictable markets and uncertain return for investment in soil conservation, it may be economically rational for farmers to avoid widespread adoption of soil conservation technology.

In recent years, farm incomes have been considerably diminished owing to reductions in the world market prices for the major Australian agricultural commodities, the removal of the floor price for Australian wool and general deregulation of tariffs. It is estimated that the average net farm income for Australian farmers in 1991–92 will be around $2000, with further reductions expected in future years (Lawrence and Vanclay 1991). Land values have also fallen, reducing farmers' equity in their properties and functioning to prevent further borrowing for capital improvement as well as preventing the transition of the smaller farmers out of agriculture (Lawrence and Vanclay 1991).

Many Australian farmers are in a situation of 'agricultural involution' (Geertz 1963), not being able to afford to undertake any capital improvement and not being able to change management strategies that involve any risk, or a perception of risk (Lawrence and Vanclay 1991). No matter how environmentally aware these farmers may be, if they don't have the capital to outlay, or are prevented from borrowing further, they can't undertake any adoption of soil conservation strategies that involve additional spending. Their economic situation means that their primary concern must be their immediate economic survival. The economic barrier is a major barrier to change, but it is not the only barrier. Adoption of conservation farming techniques has also been lower than desirable during times when commodity prices and farm incomes were high. When farmers' income is severely limited they have no flexibility (at least when they have adequate incomes they have the possibility of adoption of new techniques and practices).

Having farmers adopt the recommended soil conservation technology and management techniques will often require their going against economic self-interest; their putting aside other priorities for capital; their rejecting some of their own ideas and knowledge about their local environment; and their accepting the models and knowledge of the extension agencies. As Barr and Cary (1992) suggest, much more than a leap of faith on the part of the farmer is often required.

Farmers tend to be older than the extension officers they deal with. Too often, extension officers and agencies fail to appreciate the experiences and knowledge of farmers. Farmers may have had a lifetime in dealing with their land. It is true that land degradation continues to exist on farms, but it continues to exist on many of the farms that may have complied with previous extension agency dictates about desirable land management practices. In many cases, advice given by extension agencies in the past has been wrong, or at least has not lived up to expectations, and may have caused more problems than it has provided solutions for farmers (see Chapter 7). Some farmers are tired of hearing from yet another extension officer that all they have to do to protect their land from land degradation is to adopt a particular practice or technique. Often, information provided by extension agencies has been ill-timed, making farmers consider extension to be irrelevant. Farmers also complain that they are treated as 'idiots' by extension agencies (Woodhill 1991). The language used by the extension agencies and their staff is often patronising and prescriptive. We do not need to 'educate farmers' — rather we need to learn from them, or at least to understand their situation. The subculture of extension officers also tends to promote a patronising attitude towards farmers. Extension officers generally perceive farmers' beliefs to be different from their own, although in fact they tend to share the same beliefs as farmers (Earle, Brownlea and Rose 1981).

Farmers are also placed in the situation of receiving contradictory advice. Soil conservation agencies are not the only agencies in the business of extension. In some states, the same agency that promotes soil conservation may have other extension officers whose job it is to promote commercial innovations. In other states, government agencies are structured so that commodity-based extension is the responsibility of one agency, while conservation extension is the responsibility of another agency. In both situations, there may be very little communication between (and within) agencies, and farmers are given different — and often conflicting — information. In addition to state government agencies, many other groups are in the business of extension. Various commodity bodies such as the Australian Meat and Livestock Corporation as well as agribusiness interests, particularly agricultural chemical manufacturers and distributors, also actively provide advice to farmers. In addition, farmers seek information from rural publications and from each other.

Even where farmers may accept the information presented to them, they are not always in a situation where they can comply. They have conflicting goals concerning the use of their time, the use of their capital, and the ideal ways to manage their farms. In times of unpredictable markets, farmers may wish to maintain flexibility. This limits the use of the deep-rooted perennial pasture species desirable for salinity control, because it locks farmers into grazing. Obviously, where achievement of sound land management requires considerable capital investment on behalf of the farmer, in terms of investment in new equipment, structural practices, seed and agricultural chemicals, the economic situation of the farmer is important. Where farmers have low equity levels, when interest rates are high, market prices for produce low and farmers have competing priorities for capital expenditure, investment in soil conservation is likely to be minimal.

The stage of life of the farm family may be important in determining goals for time and capital. There may be conflicting goals of improved housing or education of children. Many farms suffer from poor quality housing because farms are inherited, and farm housing tends to date from the time the land was first settled. Furthermore, owing to patterns of inheritance and the fact that farm families have tended to be relatively large, farms may have numerous shareholders, with the non-farming shareholders having different expectations and demands for the use of capital than the farming shareholders, usually preferring dividends to reinvestment of capital. The large size of farm families and the resulting inheritance 'battles' have produced considerable tension over land use and property disbursement.

In other situations, the problem is exacerbated because the land degradation may be an off-site and non-point problem. That is, the farmers largely responsible for the problem may be different from the ones experiencing the negative effects of the problem. This is particularly the case with dryland salinity. Accessions to the watertable occur in areas of preferential recharge, such as the rocky tops of hills which have vertical or oblique sedimentary strata and on the slopes often used for cropping and grazing. These recharge areas may be a considerable distance from the discharge areas, usually the low lying areas, where the effects of salting are noticed. Consequently, where the farmers experiencing soil salting are convinced that they need to do something, the accepted model of salinity requires them to convince other farmers in the recharge areas — who may not be experiencing any ill effects from salting and are not likely to — to participate in salinity control programs. Farmers who engage in such programs as the planting of trees (particularly on rocky hilltops) and the growing of deep-rooted pasture species on hill slopes cannot restrict any benefit that adoption might have on watertable levels to themselves. Furthermore, because of the very slow rate of movement of subterranean water, it is likely that the farmers who suffer most will not benefit until a very long time has passed.

Although governments provide some financial incentives to encourage adoption of soil conservation measures, these seldom cover the full costs borne by farmers. All this suggests that farmers are unlikely to participate in adoption for conservation reasons alone. However, farmers tend not to act in an economically 'rational' way; they respond, instead, to farming culture and the notions of good farm management that exist within their community. In terms of farm management, farmers generally do what they regard is required, often consciously knowing that such an activity may not be economically rational. Such activities are justified as being part of the farming way of life (stewardship) and necessary in order to improve the farm for their children. This has meant that at times farmers' expenditure patterns have been inappropriate — something which has resulted in financial trouble with increasing interest rates and declining product prices and land values.

The constraints upon adoption, therefore, are not necessarily economic, but opinion-related. Farmers fail to adopt soil conservation technology because they are not satisfactorily convinced that it is necessary. In terms of salinity control measures, Vanclay and Cary (1989) found that many farmers failed to participate in adoption for very practical reasons. Some of the reasons they gave were wrong but, based on the logic and premises of the farmer's explanation, the decision not to adopt was sensible. The task for extension agencies in Victoria was to change farmers' opinions of the particular deep-rooted species being recommended (phalaris and lucerne) — something which would necessitate increasing the farmers' level of knowledge of these species.

THE VALUE OF LANDCARE

The cause of the problem of land degradation is social, not technical, in nature. However, in terms of the economic and social situation of farmers, the lack of adoption is understandable. Extension agencies have alienated farmers and have failed to appreciate the extent of farmers' knowledge and experience in dealing with land management issues. The problem is also social in that many of the consequences of land degradation are off-site or outside the farming lifetime of the farmers. The solutions to land degradation therefore require community concern for the economic situation of farmers and community support for adoption of conservation farming strategies. Landcare, as a strategy for group extension, is likely to provide a suitable model for overcoming many of the problems that have been described.

In particular, Landcare is likely to empower farmers by making farmers themselves responsible for setting the agenda of the land management issue they address and the strategies they employ. It legitimises their indigenous local knowledge. Landcare will also potentially create a public acceptance of new ideas and land management strategies. Because farmers are meeting other farmers in a forum specifically to discuss land management strategies, new ideas are likely to gain legitimacy within the farming subculture much more readily than they would by diffuse innovation processes that occurred with individual extension strategies. As a local initiative, individual Landcare groups can respond to the particular needs of the members of that group. Landcare is also likely to be a suitable organisation to assist farmers in dealing with the off-site consequences of land degradation.

However, in order to be effective, Landcare groups will need to be satisfactorily served by well-informed extension agencies in order to ensure that their information and other resource needs are met. State governments that believe Landcare will be a strategy for reducing their commitment to extension ought to reconsider. The enthusiasm and energy of Landcare groups will quickly be lost if they are not properly serviced. Extension agencies and individual extension officers who see Landcare as part of their domain, rather than as a mechanism for empowering farmers, are likely to jeopardise the effectiveness of Landcare and alienate farmers again. The concept of 'ownership' that farmers attach to their Landcare group and the activities of their group is likely to be very important in the success of that group and of those activities.

Landcare groups also need to ensure that they develop appropriate community organisational structures. Burnout of group leaders, loss of enthusiasm and eventual decline in the group will occur unless appropriate safeguards are implemented. Landcare is similar to any community organisation in most respects, and Landcare groups and the extension agencies sponsoring Landcare can learn a great deal from community organisational structures (see Chamala and Mortiss 1990).

Extension agencies also need to be aware that while Landcare is about empowering farmers, only some farmers are actually empowered. Landcare becomes another organisation in which the politically and socially astute elite of the local community can dominate others. Far from Landcare empowering all farmers, it is quite likely that certain individuals will have their interests served by Landcare at the expense of other individuals (see Chapter 8).

Funding for Landcare needs to be carefully considered. Already there is the perception amongst some farmers that Landcare is just another way of getting money (Woodhill 1991). It is important not to let Landcare degenerate into just another social club, and it is likely that funding should reward positive action undertaken by each group.

There also needs to be some consideration given to the effectiveness of Landcare both in terms of implementation of conservation farming strategies (however this may be defined in each region) by members of Landcare groups and in terms of the coverage of Landcare. If large numbers of farmers are not being adequately serviced by Landcare, and continue to use environmentally unsound farming practices, consideration will need to be given to encourage their participation in Landcare, or to find other mechanisms which will encourage their adoption of conservation farming strategies.

CONCLUSION

Land degradation is primarily a social, rather than a technical, problem. A disproportionate amount of research has been spent on physical research, with insufficient attention being placed on the social aspects of land degradation. Land degradation is social in nature for the very reason that solutions to land degradation exist but are not adopted by farmers for a wide variety of social, economic, cultural, perceptual and situational reasons.

Farmers do not have environmentally hostile attitudes. Rather, they endorse concepts of stewardship and conservation. It is highly unlikely that attempts to improve farmers' attitudes will increase the adoption of soil conservation practices. Although adoption of soil conservation technology may not be economically rational for the individual in the short term, farmers do see conservation as having wider economic rewards and do appreciate that land degradation does significantly affect future yields. Farmers are also sufficiently concerned about the issue of land degradation. However, while most see their local area to be at risk, only few consider that their own farms are at risk from land degradation. This is due to a misperception and underestimation of the land degradation processes. Farmers fail to recognise the early warning signs of land degradation because most media and extension literature usually present dramatic images of severe forms of land degradation. The protection of Australia's farmland does not require promotion of changes in farmers' attitudes, but does require an increase in farmers' knowledge of the land degradation processes and symptoms and the acceptance of conservation farming techniques within the farming subculture. Regular use by farmers of soil testing kits is likely to lead to the early detection of land degradation and to the adoption of soil conservation technology and more appropriate land management practices.

The encouragement of conservation farming strategies through Landcare groups is likely to lead to greater acceptance of these ideas within the farming subculture, and will lead to greater adoption. However, there are barriers to adoption which also need to be addressed.

REFERENCES

Barr, N. and Cary, J. (1984) *Farmer Perceptions of Soil Salting: Appraisal of an Insidious Hazard*, School of Agriculture and Forestry, University of Melbourne, Melbourne.

Barr, N. and Cary, J. (1992) *Greening a Brown Land: The Search for Sustainable Land Use in Australia*, Macmillan, Melbourne.

Bultena, G., Nowak, P., Hoiberg, E. and Albrecht, D. (1981) 'Farmers' Attitudes Toward Land Use Planning', *Journal of Soil and Water Conservation* 36 (4): 37–41.

Buttel, F. *et al.* (1981) 'The Social Bases of Agrarian Environmentalism: A Comparative Analysis of New York and Michigan Farm Operators', *Rural Sociology* 46 (3): 391–410.

Buttel, F. and Flinn, W. (1975) 'Sources and Consequences of Agrarian Values in American Society', *Rural Sociology* 40 (2): 134–151.

Buttel, F. and Flinn, W. (1977) 'Conceptions of Rural Life and Environmental Concern', *Rural Sociology* 42 (4): 544–555.

Cameron, J. and Elix, J. (1991) *Recovering Ground: A Case Study Approach to Ecologically Sustainable Rural Land Management,* Australian Conservation Foundation, Melbourne.

Carlson, J. and McLeod, M. (1978) 'A Comparison of Agrarianism in Washington, Idaho, and Wisconsin', *Rural Sociology* 43 (1): 17–30.

Chamala, S., Keith, K. and Quinn, P. (1982) *Adoption of Commercial and Soil Conservation Innovations in Queensland: Information Exposure, Attitudes, Decisions and Actions,* Department of Agriculture, University of Queensland, St Lucia.

Chamala, S. and Mortiss P. (1990) *Working Together for Land Care: Group Management Skills and Strategies,* Australian Academic Press, Brisbane.

Chamala, S., Rickson, R. and Singh, D. (1984) *Annotated Bibliography of Socio-economic Studies on Adoption of Soil and Water Conservation Methods in Australia,* Department of Agriculture, University of Queensland, St Lucia.

Craig, R. and Phillips, K. (1983) 'Agrarian Ideology in Australia and the United States', *Rural Sociology* 48 (3): 409–420.

Donald, C. (1982) 'Innovation in Australian Agriculture', in D. Williams (ed.), *Agriculture in the Australian Economy* (2nd edn), Sydney University Press, Sydney.

Earle, T., Brownlea, A. and Rose, C. (1981) 'Beliefs of a Community with Respect to Environmental Management: A Case Study of Soil Conservation Beliefs on the Darling Downs', *Journal of Environmental Management* 12: 197–219.

Fishbein, M. and Ajzen, I. (1975) *Belief, Attitude Intention and Behaviour,* Addison Wesley, Reading.

Flinn, W. and Johnson, D. (1974) 'Agrarianism Amongst Wisconsin Farmers', *Rural Sociology* 39 (2): 187–204.

Geertz, C. (1963) *Agricultural Involution,* University of California Press, Berkeley.

Hawke, R. (1989) *Our Country Our Future: Statement on the Environment,* Australian Government Publishing Service, Canberra.

Jenkin, J. and Morris, J. (1982) 'Salinity Problems Outside the Irrigation Areas', in *Conference Proceedings of the AIAS and AAES, Salinity in Victoria,* La Trobe University, October.

Kerin, J. (1984) 'Restoring the Balance of Conservation and Agriculture', *Habitat Australia* 12 (6): 12–14.

Lawrence, G. and Vanclay, F. (1991) 'Rural Restructuring in the Murray–Darling Basin, Australia: Economic Imperatives and Ecological Consequences, Paper presented to the Rural Sociological Society 'Food Systems and Agrarian Change in the Late Twentieth Century' Conference, Columbus, Ohio.

Messer, J. (1987) 'The Sociology and Politics of Land Degradation in Australia', in P. Blaikie and H. Brookfield (eds), *Land Degradation and Society*, Methuen, London.

Molnar, J. and Wu, L. (1989) 'Agrarianism, Family Farming, and Support for State Intervention in Agriculture', *Rural Sociology* 54 (2): 227–245.

Myers, D. (1989) *Psychology* (2nd edn), Worth, New York.

Oram, D. (1987) *The Economics of Dryland Salinity and its Control in the Murray River Basin of Victoria: A Farm Level Approach*, Occasional Paper No. 11, School of Agriculture, La Trobe University, Melbourne.

Pampel, F. and Van Es, J. (1977) 'Environmental Quality and Issues of Adoption Research', *Rural Sociology* 42 (1): 57–71.

Powell, D. and Pratley, J. (1991) *Sustainability Kit Manual*, Centre for Conservation Farming, Charles Sturt University, Wagga Wagga.

Quiggin, J. (1987) 'Land Degradation: Behavioural Causes', in A. Chisholm and R. Dumsday (eds), *Land Degradation: Problems and Policies*, Cambridge University Press, Cambridge.

Rickson, R., Saffigna, P., Vanclay, F. and McTainsh, G. (1987) 'Social Bases of Farmers' Responses to Land Degradation', in A. Chisholm and R. Dumsday (eds), *Land Degradation: Problems and Policies*, Cambridge University Press, Cambridge.

Rickson, R. and Stabler, P. (1985) 'Community Responses to Non Point Pollution From Agriculture', *Journal of Environmental Management* 20 (3): 281–294.

Roberts, B. (1990) *The Birth of Landcare*, University College of Southern Queensland Press, Toowoomba.

Roberts, B. (1991) 'The Big Shift: From Me Now to Them Later', Paper presented to the Hawkesbury Centenary Conference on Agriculture, the Environment and Human Values, University of Western Sydney — Hawkesbury, Richmond, October.

Rogers, E. (1983) *Diffusion of Innovations* (3rd edn), Free Press, New York.

Standing Committee on Environment, Recreation and the Arts (SCERA) (1989) *The Effectiveness of Land Degradation Policies and Programs*, Report of the House of Representatives Standing Committee on Environment, Recreation and the Arts, Australian Government Publishing Service, Canberra.

Singer, E. and Freire-de-Sousa, I. (1983) 'The Sociopolitical Consequences of Agrarianism Reconsidered', *Rural Sociology* 48 (2): 291–307.

Thorne, P. (1991) 'Perennial Pastures that Pay', *Landcare News* 8: 1–3

Vanclay, F. (1986) 'Socio-economic Correlates of Adoption of Soil Conservation Technology', unpublished M.Soc.Sci. thesis, Department of Anthropology and Sociology, University of Queensland, St Lucia.

Vanclay, F. (1992) 'Farmer Attitudes or Media Depiction of Land Degradation: Which is the Barrier to Adoption?', *Regional Journal of Social Issues* 26.

Vanclay, F. and Cary, J. (1989) *Farmer' Perceptions of Dryland Soil*

Salinity, School of Agriculture and Forestry, University of Melbourne, Melbourne.

Williams, M. (1979) 'The Perception of the Hazard of Soil Degradation in South Australia: A Review', in R. Heathcote and B. Thom (eds), *Natural Hazards in Australia*, Australian Academy of Science, Canberra.

Woodhill, J. (1991) 'Landcare — Who Cares? Current Issues and Future Directions for Landcare in NSW', Discussion Paper from the 1990 Review of Landcare in NSW, Landcare and Environment Program, Centre for Rural Development, University of Western Sydney — Hawkesbury, Richmond.

7 EFFECTIVENESS OF EXTENSION STRATEGIES

BRUCE FRANK and SHANKARIAH CHAMALA

The contribution of agricultural extension to rural development has been changing gradually, to a point where critical questions are being asked of the discipline's capacity to demonstrate an effective return to investment by governments. Improvement policies and associated agricultural extension programs have rested firmly on the modernisation view of change, strongly influenced by the North American work of the more conservatively trained rural sociologists (see discussion in Buttel, Larson and Gillespie 1990). Such policies stressed the importance of diffusion to the traditional sector, which appeared to lack the motivation and opportunities for economic development (Hardeman 1978). Diffusion theory had assumed that development occurred as technological information 'trickled down' from modern to traditional social systems (Saint and Coward 1977), making the core and periphery homogeneous. However, there were peripheries into which growth did not filter (Berry 1972) and inequities tended to increase rather than decrease.

This chapter will trace the development of the extension discipline from the period of rapid economic growth in the 1960s, through the tightening economic climate of the 1970s and 1980s, to the present. It will then suggest ways in which the extension discipline needs to move in order to be an effective force in agricultural development.

ECONOMIC GROWTH IN THE 1960s — DIFFUSION OF INNOVATIONS

An 'individual adopter' perspective dominated early sociological research in extension. The process of innovation adoption by individuals within a social system was first described by the diffusion of innovations model (Griliches 1957; Rogers 1962; Jones 1967). Warner (1974) noted that the diffusion model's S-shape defined the nature of learning, with knowledge being accumulated during the adoption process. As relevant information reduced the level of uncertainty about an innovation

(Hiebert 1974), the efficiency of its use increased. That is, the adopter learnt from experience.

The 'individual adopter' perspective focused much of the earlier research on communication and motivation issues. Information flow was presumed to be the primary factor limiting modernisation — and consequently development (Rogers and Shoemaker 1971; Brown, Malecki and Spector 1976; Saint and Coward 1977). The ultimate goal sought by individuals is a 'state of well-being which minimises tension' (Rogers 1962: 301; Harvey 1973: 120). This desired state of satisfaction can be viewed as an expression of perceived reward (Coughenour 1976; Frank 1988), which must outweigh the efforts and costs that an individual expects to incur. Satisfaction encourages an individual to continue the rewarded behaviour, while dissatisfaction stimulates avoidance (Homans 1974), or active rejection of an innovation (Coughenour 1976; Yapa and Mayfield 1978; Thenes 1979; Lahren 1979; Frank 1988).

Innovativeness or entrepreneurship was thought to be a critical precursor for the adoption of technological innovations, their diffusion and consequent large-scale economic development (Hardeman 1978; Brown *et al.* 1979). Early agricultural extension policies were based on 'individual modernism' defined by Benvenuti (1962: 41) as 'a predisposition to seek solutions to developments in personal life by means of scientific knowledge and the help of available modern technology'. Consequently, the extension discipline developed as a means of encouraging innovativeness, in order to improve the economic productivity of national agricultural sectors. Implicitly, it assumed that scientific knowledge was desirable and could be used to resolve problems limiting productivity — something demonstrated in the politically stable nations of the Western world (Moore 1963, in Long 1977).

The 'adopter perspective' predominated until Brown *et al.* (1976, 1979) showed that marketing strategies of the innovation's propagators (public or private), existing infrastructure and the distribution of resources such as information, access to capital, transport services and processing also influenced adoption. Although emphasis shifted from the individual, research had shown that one of the most important attributes of an individual innovator was the capacity for abstract thinking, expressed as conceptual skill or managerial aptitude. Conceptual skill influences an individual's ability to perceive relationships, form mental images, consider them in abstract terms and translate them into positive action on the farm

(Emery and Oeser 1958; Crouch 1970, 1972; Bembridge 1975, 1977).

THE TIGHTENING ECONOMIES OF THE 1970s

The agricultural extension strategies of the 1960s and 1970s were determined by the research, development and diffusion model, with emphasis on individual adoption behaviour. This was consistent with the prevailing modernisation concept. Extension implied an 'extension' of the research product to the farmer. However, as Kuhn (1970) demonstrated, the conceptual approach selected within an existing dominant paradigm had implications for the assumptions made, questions asked, role played and policies recommended in any research or development project.

During the late 1970s the desirability of progression along the modernity continuum was challenged as being too deterministic (Long 1977), suggesting that changes in the organisation and activities of local populations were responses to external stimuli. Diffusion principles were highly compatible with the national growth strategies of the 1960s, and assumptions based on that model were now seen to be invalid (Goss 1979; Nachmias and Sadan 1976; Long 1977; Newby 1980; Bunce 1982).

An alternative perspective of modernisation described by Olshan (1981: 300) was that the 'essence of modernity is the perception of choice', and that it is not associated with particular technologies, types of social organisation, or beliefs and values. 'Individual modernity' therefore can be interpreted as the degree of freedom that an individual has to exercise choice towards an intrinsic goal of personal satisfaction; and the 'modernity of the culture' is the extent to which an individual's freedom of choice is collectively expressed by the community to which he or she belongs. An implicit assumption in this definition is the individual's right to choose *not* to behave in a manner consistent with what others perceive as rational economic development.

Not only was the modernity concept challenged, but the classical paradigm also was challenged on the basis of its North American ethnocentricity (Holden 1972), its list of unrecognised assumptions (Holden 1972; Saint and Coward 1977; Goss 1979; Brown *et al.* 1979), the imperfect definition or clarification of concepts (Warner 1974; Brown *et al.* 1979) and the classification of adopter categories (Lahren 1979; Chamala, van den Ban and Roling 1980). Further, biased emphasis on personal traits of potential adopters, social relationships and the importance of communication was noted by Warner (1974) and Goss (1979) and the corresponding lack of studies on the interactive effects of economic and sociological variables was noted by Brown (1968a,b), Warner (1974) and Brown *et al.* (1979).

THE 1980s — A SYSTEMS APPROACH

More recently, researchers have adopted systems methodologies (Checkland 1981; Conway 1985; Chamala 1987a; Chamala and Coughenour 1987; Byerlee and Tripp 1988; Frank 1988; Roling 1988; Chamala and Mortiss 1990) in order to examine holistic sets of interacting elements. Using a systems perspective, the influence of environmental, technical, social and economic factors on the adoption of innovations in the extensive cattle industry of north Queensland were studied by Frank (1988). He showed that an individual's adoption behaviour varied with socioeconomic need expressed as declining economic return and as an inverse function of resources per production unit.

Two implicit assumptions have influenced research and extension strategies. First, scientists have assumed that farmers make rational economic decisions about the adoption of new technology. Farmers manage agricultural systems within larger, more complex socioeconomic systems, but technological research has quantified a limited number of these relationships, while assuming implicitly that producer assessment is based exclusively upon a profit motive. However, some value characteristic of the innovation often substitutes for monetary profit. Second, an implicit assumption that new practices are both suitable for adoption and desirable was often based on proven productivity increases in a research environment rather than in a field setting (Saint and Coward 1977).

Although research aims to generate economic advantages, new practices may be unnecessary and even potentially harmful (Holden 1972; Brown *et al.* 1979). Scientists often evaluate success by the extent of use of new technologies (Saint and Coward 1977), with minimal attention to the distributional impact of potentially adverse outcomes on a social community (Goss 1979). In north Queensland, individual technological innovations which were introduced to reduce cattle losses have had adverse consequences when operating together. The introduction of stylos, nitrogen-based supplements and *Bos indicus* genotypes was promoted actively by research and extension agents to reduce cattle losses in the breeding herd. However, the nitrogen supplements and Brahman genotypes contributed to a doubling of stocking rates over 15 years — something which occurred at a time of depressed market conditions (Frank 1988). During the 1980s, the effect of several drought years on native pasture species compounded the problem, leading to denudation and associated soil erosion in the Burdekin dam catchment.

Although the outcome of the rapid increase in cattle numbers had been predicted by cattle producers and scientists 16 years earlier, there was little evidence that research policies anticipated the influence of adverse consequences in north Queensland. No land use or market research was conducted, and no constraints were applied to land use practices.

It is the perception of profitability or expected utility maximisation in an adopter's own environment which contributes to the rate of adoption, rather than profitability *per se* (Anderson 1974; Newbery 1975; Van Es and Pampel 1976; Frank 1988). Whereas technological research aims to improve financial profitability of an industry, there has been a failure to market innovations using systematic and socioeconomic dimensions. For example, cattle producers in north Queensland wanted to achieve and/or maintain a satisfying lifestyle (Frank 1988) so their property development was continuous and systematically directed towards that end. However, their progress was limited by different constraints at each level of management and each technological innovation was perceived in a unique manner for each situation.

At first, property development was dependent on the manager's level of cattle control and instrumental knowledge (or informally learned experiential skills). If satisfied at this stage, the manager did not see any socioeconomic advantage or utility to be gained by adopting new technology. If not satisfied, the first reaction to declining returns was to reduce costs such as labour. Often, some critical event prompted the manager to learn about an innovation by comparing its benefits and costs in assessing its utility (that is, its perceived usefulness, satisfaction and profitability). This learning process contributed to a core of experiential knowledge. Where managers perceived an advantage, they decided whether the necessary resources were available. Where such an advantage was apparent, relevant and applicable, an innovation diffused rapidly, as in the case of urea-molasses roller drums during the 1965–66 drought. However, where no advantage was apparent, the decision not to adopt was an active and rational act, under the prevailing environmental circumstances.

Cattle producers chose to adopt or reject an innovation by using their experiential knowledge and managerial ability to assess its benefits and costs for them. The technological adoption behaviour of north Queensland cattle producers was not consistent with profit maximisation or the perception of profitability, but varied with their social and economic need to establish and/or maintain a desired lifestyle. This was consistent with Olshan's (1981) concept of modernity.

To illustrate the importance of the producers' perception of the innovation's relevance to a situation, the example of *Stylosanthes* spp. in north Queensland will be used. Legume introduction in north Australia since 1962 was largely based on the observed success of *Stylosanthes humilis*, commonly known as Townsville Stylo. Scientists had assumed that a legume-based pasture in the dry tropics was desirable, an idea which attracted a large research and extension investment. However, very little adoption of new cultivars occurred in areas below 600 mm rainfall because cattle managers did not see a sufficiently favourable economic advantage over native pastures. In contrast, scientists conducted very little research or extension to improve the range management of native pastures, although these pastures had formed the basis of the cattle industry for over 100 years.

Until recently, scientists paid little attention to the significant but informal levels of indigenous knowledge held by local communities. In the 1960s, older rural residents of north Queensland perceived risks associated with stylos and land subdivision, and predicted overgrazing of native pastures along rivers and alluvial flats. Their attitude to free-ranging cattle appeared to be sensible, allowing herds to move freely, graze 'green-pick' areas and select from a range of feed sources for optimal nutrition. They erected fences only to block cattle while mustering — otherwise, gates were left open during the year. In many instances, botulism emerged as a problem after fencing limited the cattle range, increased stock densities, reduced opportunities for diet selection and forced growing cattle onto phosphorus-deficient country.

Primary producers have sound reasons for their behaviour, which often are not understood by scientists. Two examples may help to illustrate this. The insidious presence of vibrio (*Campylobacter foetus*) in a herd causes heifers to 'slip' two or three calves before developing resistance. These heifers mature at a higher body weight than those treated for the disease and have a lower level of calving mortalities.

Through their experience over space and time, primary producers are often aware of subtle variations in landscape not recognised by scientists. Although the brigalow-dominant dark clay soils in southern Queensland were classified uniformly, farmers in the Tara–Meandarra district recognised different types. They associated these variants with different forms of land use, reflected by ease of cultivation and sheep carrying capacity. By discussing land use practices with the farmers, Van Dijk (1979) was able to relate the boundaries of these variants to geomorphic development patterns of the clays. The farmers' local knowledge was a necessary but not sufficient condition for the explanation of several problems occurring throughout the district. These problems included copper deficiency, salinity and intractable areas of cultivation (Van Dijk 1980). At the time, Van Dijk's close association with the farmers, his acknowledgement of and dependence on their knowledge attracted criticism from his scientific peers.

The importance of indigenous knowledge sources is also reflected by the distinct contrast between the two groups classified as medium and high adopters in north Queensland (Frank 1988). Attributes of the medium adoption behaviour group described a state of stability. This was associated with time for experiential learning: both time of tenure and time to assess and continue with 'proven' management practices. Financial stability was indicated by a relatively high proportion of labour and financial resources per production unit (the breeding cow). This was associated with larger areas and proportions of fertile soils managed for that unit. The negligible effect of kangaroos perceived by medium — as compared to high — adopters indicated a level of ecological stability. The long, relatively isolated tenure indicated social stability, and a 'satisfying lifestyle'.

Conversely, the high adopters were shorter-term residents striving to achieve a balance by operating their system at a higher risk level, at least temporarily, in order to survive in a climate of low beef prices, high interest and inflation. They were adopting practices recommended by extension, based on research, to reduce their rate of economic decline. The higher risk level was stressful, not only for the individual managers and their families, but also (and perhaps more importantly) for the agroecological system in terms of excessive grazing pressure and consequent long-term land degradation. Many managers were very conscious of this high risk, but could not see an alternative option.

Although the managerial skill levels of both adopter groups were equivalent and the innovations may have offered favourable utility, medium adopters did not wish to change a managerial style which was clearly satisfying to them — one which had been in harmony with the environment for 13 to 40 years. The experience of the medium-level adopters suggests that where there is a stable ecological balance, non-adoption of apparently relevant technology is an intelligent response. This contrasts with the attitude of biological scientists who perceived specific advantages in their recommendations for the grazing industry, and who failed to appreciate the reasons behind the apparent low level of adoption.

As Crouch and Payne (1983) have suggested, the skill to live in harmony with the natural environment is developed from experience 'after many years of concerted effort'. That is, medium adopters had developed a knowledge base from their experiential learning to reduce uncertainty and improve their managerial efficiency. However, this skill was not recognised by scientists concentrating on technological research. Scientists believed implicitly that high levels of adoption were desirable, but collectively these practices had caused damage by increasing productivity to non-sustainable levels.

Both the medium and high adopter groups had a financial planning skill, which was a necessary but not sufficient condition for the adoption of higher-order technology levels. In the medium adopter group, this skill was associated with the decision not to adopt technology to levels recommended by research and extension authorities.

Initially, cattle producers reacted to economic decline by cost-cutting, but a pro-active skill became necessary if the economic decline continued. This skill required an abstract ability to develop forward plans, as reflected by conceptual skill or managerial aptitude. Pro-active responses of cattle producers aimed to increase herd productivity by using sets of practices. The use of records for financial planning offered a more abstract, longer-term improvement in herd productivity as property development reached a higher level. Business planning skills enabled an individual to adapt to critical events, so that adoption behaviour became a compensatory mechanism to re-establish stability.

If not pressed by declining returns, individuals with sufficient intrinsic ability for pro-active management maintained a 'medium' level of adoption behaviour, suggesting that rational economic development was a means of achieving a suitable lifestyle. In terms of modernity or 'collective self awareness' (Olshan 1981: 300), cattle producers exercised their freedom to choose from a wide range of management options as a means of achieving their own goals. Collectively, their rural community was attempting to control and direct its own evolution. It did not accept technology and material culture for its own sake but as a means of maintaining a desired way of living.

These examples from north Queensland show that social and

economic issues can override technological issues in the decision-making environment of the landholder. We can conclude that traditional extension strategies associated with the transfer of technology from researchers to farmers are not necessarily valid. Individuals can, and do, make sound decisions as they learn from experience within their social system. Their learning contributes to an indigenous knowledge base which is informal but most valuable. Innovations generated by research are not necessarily desirable, and it is important that the collective wisdom of communities be respected and actively involved as part of a knowledge system. This knowledge system will now be examined as part of a global system exerting major influences on agricultural extension policies.

THE INFLUENCE OF GLOBAL FORCES

Australian agriculture is being influenced by rapid changes occurring in international agricultural policies with agricultural extension strategies currently reacting to these international changes. These changes originate from surplus production, environmental concerns and changing habits of food consumption, as well as from responses to improvements in production and information technologies (see Chamala 1988; and Chapter 3). Political policies in Europe and the Middle East and marketing policies of Japan, the USA and developing countries influence the export competitiveness of Australian primary produce. Some examples expected to affect Australia include: the major share in exports and rebuilding obtained by the USA and the UK following the war in the Middle East; the trade surplus of Japan; the US deficit; changes in the European Economic Community; and the bilateral trade pacts of Canada and South America.

There is considerable evidence of an agricultural crisis in rural Australia (see Lawrence 1987, 1989). Structural responses to this crisis include a decreasing number of commercial family farms and agribusiness firms, and an increasing number of hobby and corporate farms (Chamala 1988). The cost-price squeeze resulting from low sugar, wheat and wool prices has reduced the average net farm income and equity in response to increasing farm debt. As a result, extension has become more involved in farm financial counselling services. Although federal and state governments have taken some initiatives to help local groups employ financial counsellors, help is often too late. Farmers need to learn how to manage the economic reality of high debt, rising interest and falling commodity prices and land values.

As land values fall, reflecting reduced economic margins on the family farm, it is likely that alternative, non-agricultural forms of land use will emerge, with the traditional rural emphasis of extension redirected to involve whole communities. Changes in the consumption of agricultural products will reflect a conscious shift towards health-promoting foods, with consumers exercising an increasing influence (Chamala 1988), with the primary extension audience likely to shift away from agriculture, and with rural people, rural problems and rural economic development to continue as the main focus (Dillman 1986). There will also be an increasing environmental concern, with a growing voice of non-agricultural 'green voters' being reflected in agricultural programs and policies (Chamala 1988). This has been evidenced in the recent cooperation between the Australian Conservation Foundation and the National Farmers Federation.

THE DEVELOPMENT OF INFORMATION SYSTEMS

As world trends occur from centralisation to decentralisation, from representative to participatory democracy, and from hierarchies to networking (Naisbitt 1982), the extension agents' relationship with clientele will change (Dillman 1986). New strategies are needed to adapt agricultural extension services to the increasingly complex and uncertain situations facing rural and urban communities today. If the role of the extension professional is a 'facilitator of wise decisions in the real world' (Piccone and Schoorl 1987), the main emphasis needs to be changed from one of transferring technical information content from research sources to that of managing the process of change within a social system. That is, the role of an extension agent needs to shift from technological problem-solving to one of managing knowledge systems.

In order to cope with community development problems, both old and emerging institutions will need to interact and create 'mutual interest networks that transcend geographical localities'. Rolls, Jones and Garforth (1986) predicted that the role of extension personnel will be as guides and counsellors to local sources of valid information, using rapidly changing and developing information technologies. They showed that along with the long-established role of disseminating research results, new dimensions for extension included information on marketing, credit, farmer organisations and rural service facilities as well as public involvement in many aspects of the rural environment.

Such a change has implications for extension training, staff requirements and their effective employment. It would encourage a self-help program of instrumental learning by resource managers, allowing extension organisations to concentrate on effective forms of management skills and information retrieval for the whole population within the agricultural system, including technological and socioeconomic perspectives. Many rural managers have sufficient instrumental skill to seek information from sources that they perceive as credible (Frank 1988), but further training will be required. Training programs for rural managers in resource management and information retrieval processes has been minimal, while almost total emphasis has been directed to the introduction of innovations which are technologically efficient.

Many of the predictions of Naisbitt (1982), Dillman (1986), Rolls *et al.* (1986) and Chamala (1988) are now clearly evident, as we observe the shifts from an industrial society to a post-industrial information society whose components are interdependent in space and time. Dillman (1986) noted that extension was expected to become the principal access point for information, so clients could avoid 'information overload' by accessing data banks for precisely fitting needs, using interactive TV computer transmission and video telephones. Dillman considered that non-geographic principles would be used for assembling audiences, as specific topics to meet unique needs would permit every targeted party to 'be reached simultaneously and even brought into contact with one another using the appropriate technology'. These concepts are being applied in remote areas of Australia using satellite technology.

Other strategies for overcoming communication constraints include the electronic dissemination of information about technology and marketing (Delaney and Chamala 1985; Chamala 1987a,b). In Queensland, expert systems such as Beefman and Wheatman programs help disseminate information in a new form, while Australia's first computerised crop management system (SIRATAC) continues to expand, now managing more than 30 percent of the Australian cotton crop. 'Elders Farm-Link' and 'Information Express' provide agricultural information on their videotex systems involving satellites, which also enable remote sensing to study land use patterns. The 'Computer Aided Livestock Marketing System' in the cattle industry, tele-auctions and tele-shopping are already being used.

In response to the changing structure of information systems, agricultural organisations are broadening their organisational missions and increasingly tending towards networking among institutions (Chamala 1988). Dillman (1986) considered that the mode of internal organisation for information flow would be less hierarchical, less concerned with clearance through several layers, and more interactive between the source and user. Chamala (1987a,b) noted that more industry involvement and better communication between research organisations are essential at the stages of priority setting, and research planning and development. The indigenous knowledge resources of individual farmers and their associations are now being recognised by authorities funding research and development, leading to 'producer-driven' rather than 'researcher-driven' programs. This is leading to control of 'intellectual property' by funding bodies, with significant implications for the traditional extension model. Research and extension organisations will need to assess their work for possible adverse consequences of technological innovations in social systems, and demonstrate a tangible socioeconomic benefit to the potential users of that innovation. This will be an integral component of the innovation package (Frank 1988).

Research and development organisations are moving with the trend from institutional help to self-help (Naisbitt 1982); for example, information centres supported by the agribusiness industry are being established. Australian farmers have always believed in self-help. In the US, 15 million Americans now belong to some 500 000 self-help groups. In many European countries, extension services are provided by grower organisations, while government services focus on regulatory functions and environmental aspects (Rolls *et al.* 1986). Consistent with this trend, Chamala (1987a,b) reported a movement towards the 'user pays' principle for extension services in Australia, where some state agricultural departments, notably in Tasmania, are charging for information and diagnostic services. While there is some scope for the 'user pays' principle, its impact on environment degradation, social equity and agricultural research funding needs to be examined in relation to natural resources and consumer interests.

There is an increasing role for commercial advisers and private consultants as agricultural extension services withdraw from traditional advisory fields (Chamala 1987a,b). There are several new developments in agribusiness services ranging from private consultants in farm management and integrated pest management to company-paid advisers. At the same time, there is an expanding use of volunteers in extension programs. Community participation by various professional associations, environment-conscious groups, producer organisations and other non-government organisations provide immense scope for using the services of volunteers in policy formulation, program planning, implementation and monitoring of projects.

Rural women are the most neglected of the human resources in agriculture. In Queensland, according to the 1990 census data, 28 800 women are currently employed in farming, fisheries and forestry as compared to only 12 300 in 1974. James (1989) stated that 'almost two and a half million Australian women live outside urban centres' and 'one third of Australian farmers and farm workers are women', with roles ranging from informal helper to independent operator. Increasing numbers of farm women are employed off the farm to supplement the farm-derived income, especially during rural crises (Lawrence 1987). During these critical transition periods, rural women are confronted with problems associated particularly with personal development issues (Alston 1990). If the problems are not addressed through counselling and the provision of services, the stability of the farm family could be threatened. Rural women's networks are developing to counteract this problem (Kingston and Brough 1990; Brenda McLachlan pers. comm.). Extension services need to provide personal development programs so that farm women can enhance their skills for managing the farm, producer organisations or Landcare groups. These enhanced skills will assist in the structural adjustment process of rural industry.

A SYSTEMS APPROACH: THE PAM MODEL FOR SUSTAINABLE AGRICULTURE

Systems approaches range from hard reductionist models to holistic soft systems methodologies. Chamala (1990) developed a general model called Participative Action Management (PAM) as a way to encourage participation by government, agribusiness, and non-government agencies in effective integrated natural resources management. The PAM model is not an extension education or communication method but an organisational system by which various interests in the community and other outside agencies are organised into groups at various levels which offer true partnership to all stakeholders in sustainable agriculture.

The Landcare movement is one of the most exciting and significant developments in land and water management in Australia. Landcare groups are being formed with some using the PAM model. The number of Landcare groups, which doubled in 1990, is expected to stabilise between 1000 and 1500 by 1995. These groups (Landcare and other development groups) will help define problems, design programs and implement and monitor a range of programs throughout the country. They will demand innovative approaches to provide services for rural development. Empowered groups of people are expected to take the lead collectively to solve problems and create new opportunities for sustainable development.

The application of additional methodologies such as the holistic systems approach derived from Conway (1985) and Marten (1988), the 'soft systems' work first published by Checkland (1981) and recently discussed by Checkland (1989) and Checkland and Scholes (1990), and action learning and action research principles (Bawden 1990; Zuber-Skerritt 1990, 1991) together offer exciting opportunities for extension workers. Agricultural systems are not simply technological but are a complex of biological, economic and social systems which are managed for the collective benefit of communities. Such management requires a global perspective (Atkinson 1991) with agriculture being only one part of the social system involved.

Limiting factors which are likely to affect system properties such as sustainability, equitability and stability need to be recognised and used to monitor the productivity in terms of time, space, flow and decision patterns. Policies which encourage the attainment of short-term goals at the expense of sustainability are perceived as counter-productive by the more pro-active managers. Conversely, policies which encourage responsible, sustainable land management need to be developed in conjunction with other relevant groups and organisations to ensure their desirable application in the long term.

A SYSTEMS APPROACH TO HUMAN RESOURCE DEVELOPMENT

Human resource development provides opportunities for formal and non-formal learning experiences over a definite time period to improve job performance relevant to particular roles.

Agricultural extension and management personnel in Australia are not widely encouraged to undertake postgraduate or short-term training, whereas it has become the norm for research personnel to have postgraduate degrees. After the Commonwealth Extension Services Grants were discontinued, many state governments stopped sending extension staff for advanced training.

The Commonwealth government's 'Industry Guarantee Training Scheme' was recently introduced to encourage investment in human resource development. As well as government bodies, it is essential that community leaders, agribusiness personnel and other non-government organisation (NGO) staff be encouraged to undertake training programs which are tailored to the emergent needs of social systems. Without increasing human ability across roles and disciplines, the opportunities to achieve sustainable development are limited.

CONCLUSION

Agricultural extension strategies have shifted from the 'transfer of technology' model with its emphasis on optimal productivity to the management of knowledge systems for sustainable and equitable productivity. This requires a multidimensional perspective involving multiple organisations as partners. These partners need to learn to work together so that each can benefit from the outcome of their activities. This process of working together, as described by the Participative Action Model (PAM), is based on the concept of individual empowerment and associated freedom of choice. It requires the devolution of power from hierarchical authorities so that common, sustainable benefits may be distributed throughout the community. Extension therefore is concerned with managing change in complex societies which are dynamic and uncertain. It has grown from agriculture to service the needs of all stakeholders within the community. In order to be effective, it needs to select from a wide range of methodologies generated by new ways of thinking, spread across traditional paradigm boundaries.

REFERENCES

Alston, M. (1990) 'Farm Women and Work', in M. Alston (ed.), *Rural Women*, Centre for Rural Social Research, Charles Sturt University, Wagga Wagga.

Anderson, J. (1974) 'Risk Efficiency in the Interpretation of Agricultural Production Research', *Review of Marketing and Agricultural Economics* 42 (3): 131–184.

Atkinson, B. (1991) 'The Global Management Approach: A Summary of the Useful Lessons as They Apply to 1. Using G.M.A. in General Situations, 2. Reporting on the Trends in Extension, 3. Identifying the Problems in our Extension Services' (a summary of the workshop run by FAO consultants Agnes Gannon and Maria Nejez, Townsville, 1991), *QDPI Workshop: System Methodologies in Queensland Agriculture,* Brisbane, August.

Bawden, R. (1990) *Systems Agriculture: Learning to Deal with Complexity*, Kentucky University Press, Kentucky.

Bembridge, T. (1975) 'The Communication and Adoption of Beef Cattle Production Practices in the Matabeleland and Midlands Provinces of Rhodesia', unpublished thesis (D. Inst. Agrar.), Faculty of Agriculture, University of Pretoria, May.

Bembridge, T. (1977) 'Organisation, Problems and Strategy of the

Rhodesian Extension Service', *South African Journal of Agricultural Extension* 6: 58–67.

Benvenuti, B. (1962) *Farming in Cultural Change*, van Gorcum, Assen.

Berry, B. (1972) 'Social Change as a Spatial Process', *International Social Development Review* 4: 11–19.

Brown, L. (1968a) 'Diffusion Dynamics — A Review and Revision of the Quantitative Theory of the Spatial Diffusion of Innovation', *Lund Studies in Geography*, Series B, *Human Geography* 29.

Brown, L. (1968b) *Diffusion Processes and Location: A Conceptual Framework And Bibliography*, Regional Science Research Institute, Philadelphia.

Brown, L., Malecki, E. and Spector, A. (1976) 'Adopter Categories in a Spatial Context: Alternative Explanations for an Empirical Regularity', *Rural Sociology* 41: 99–118.

Brown, L., Schneider, R., Harvey, M. and Riddell, J. (1979) 'Innovation Diffusion and Development in a Third World Setting: The Cooperative Movement in Sierra Leone', *Social Science Quarterly* 60 (2): 249–268.

Bunce, M. (1982) *Rural Settlement in an Urban World,* Croom Helm, London.

Buttel, F., Larson, O. and Gillespie, G. (1990) *The Sociology of Agriculture*, Greenwood, Westport.

Byerlee, D. and Tripp, R. (1988) 'Strengthening Linkages in Agricultural Research Through a Farming Systems Perspective: The Role of Social Scientists', *Experimental Agriculture* 24: 137–151.

Chamala, S. (1987a) 'Agricultural Extension — New Developments', Paper presented at Australian Association of Agricultural Faculties Twenty-eighth Annual Conference on Tropical Agriculture, held at the University of Queensland, 9–13 February.

Chamala, S. (1987b) 'Strategies to Overcome Communication Deficiencies in Achieving Sustainable Farming Systems', *Journal of the Australian Institute of Agricultural Science* 53: 164–169.

Chamala, S. (1988) 'Megatrends Affecting Australian Agriculture — Restructuring Extension to Meet the Challenge', *Proceedings of the Australasian Agricultural Extension Conference,* Brisbane, October: 811–828.

Chamala, S. (1990) 'Establishing a Group — A Participative Action Model', in S. Chamala and P. Mortiss, *Working Together for Land Care: Group Management Skills and Strategies,* Australian Academic Press, Brisbane.

Chamala, S. and Coughenour, M. (1987) 'Model for Innovation Development, Diffusion and Adoption (MIDDA)', *Journal of Extension Systems* 3 (1): 47–55.

Chamala, S. and Mortiss, P. (1990) *Working Together for Land Care: Group Management Skills and Strategies,* Australian Academic Press, Brisbane.

Chamala, S., van den Ban, A. and Roling, N. (1980) 'A New Look at Adopter Categories and an Alternative Proposal for Target Grouping of Farming Community', *Indian Journal of Extension Education* 16: 1–16.

Checkland, P. (1981) *Systems Thinking, Systems Practice*, John Wiley and Sons, New York.

Checkland, P. (1989) 'Soft Systems Methodology', *Human Systems Management* 8: 273–289.

Checkland, P. and Scholes, J. (1990) *Soft Systems in Action*, John Wiley and Sons, Chichester.

Conway, G. (1985) 'Agroecosystem Analysis', *Agricultural Administration* 20: 31–55

Coughenour, M. (1976) 'A Theory of Instrumental Activity and Farm Enterprise Commitment Applied to Woolgrowing in Australia', *Rural Sociology* 41 (1): 76–98.

Crouch, B. (1970) 'Today, Tomorrow, Never — A Sociological Study of the Factors Determining the Adoption of Agricultural Innovations by Woolgrowers in the Yass River Valley, New South Wales', unpublished PhD thesis, Australian National University, Canberra.

Crouch, B.R. (1972) 'Innovation and Farm Development: A Multidimensional Model', *Sociologia Ruralis* 12 (3/4): 431–449.

Crouch, B. and Payne, G. (1983) 'Value Orientation of Pastoralists in the Arid Zone of Queensland and its Relation to Adoption of Sheep Management Practices', *Proceedings of Social Science Symposium,* CSIRO Division of Wildlife and Range Research, Canberra.

Delaney, N. and Chamala, S. (1985) *Electronic Information Technology Relevant to Agricultural Extension in Australia — A Review and Discussion Paper,* Department of Agriculture, University of Queensland, December.

Dillman, D. (1986) 'Cooperative Extension at Beginning of 21st Century', *The Rural Sociologist* 6 (2): 102–116.

Emery, F. and Oeser, O. (1958) *Information, Decision and Action — A Study of the Psychological Determinants of Changes in Farming Technique,* University of Melbourne Press, Melbourne.

Frank, B. (1988) 'Factors Influencing the Adoption of Selected Innovations in the Extensive Beef Cattle Industry of the North Queensland Dry Tropics', unpublished PhD thesis, James Cook University of North Queensland, Townsville.

Goss, K. (1979) 'Consequences of Diffusion of Innovations', *Rural Sociology* 44 (4): 754–772.

Griliches, Z. (1957) 'Hybrid Corn: An Exploration in the Economics of Technological Change', *Econometrica* 25 (4): 501–522.

Hardeman, J. (1978) 'Innovation and Agrarian Structure: Government Versus Peasant', *Tijdschrift voor Econ. en Soc. Geografie* 69 (1/2): 27–35.

Harvey D. (1973) *Explanation in Geography*, Arnold, London.

Hiebert, L. (1974) 'Risk, Learning and the Adoption of Fertiliser Responsive Seed Varieties', *American Journal of Agricultural Economics* 56 (4): 764–768.

Holden, D. (1972) 'Some Unrecognised Assumptions in Research on the Diffusion of Innovations and Adoption of Practices', *Rural Sociology* 37 (3): 463–469.

Homans, G. (1974) *Social Behaviour: Its Elementary Forms* (rev. edn), Harcourt Brace Jovanovich, New York.

James, K. (ed.) (1989) *Women in Rural Australia,* University of Queensland Press, St Lucia.

Jones, G. (1967) 'The Adoption and Diffusion of Agricultural Practices', *World Agricultural Economics and Rural Sociological Abstracts* 6 (9): 3.

Kingston, C. and Brough, E. (1990) *The Regional Women's Network Pilot Project — A Discussion Paper*, Queensland Bureau of Regional Development and Queensland Department of Primary Industries, Brisbane.

Kuhn, T. (1970) *The Structure of Scientific Revolutions*, University of Chicago Press, Chicago.

Lahren, S. (1979) 'Rejection of Innovation: The Innovation Decision Process of Family Operated Range Livestock Operations in the Rocky Mountain Region of Montana', unpublished PhD thesis, University of Colorado.

Lawrence, G. (1987) *Capitalism and the Countryside*, Pluto Press, Sydney.

Lawrence, G. (1989) 'The Rural Crisis Downunder: Australia's Declining Fortunes in the Global Farm Economy', in D. Goodman and M. Redclift (eds), *The International Farm Crisis*, Macmillan, London.

Long, N. (1977) *An Introduction to the Sociology of Rural Development*, Tavistock, London.

Marten, G. (1988) 'Productivity, Stability, Sustainability, Equitability and Autonomy as Properties for Agroecosystem Assessment', *Agricultural Systems* 26: 291–316.

Nachmias, C. and Sadan, E. (1976) 'Individual Modernity, Schooling and Economic Performance of Family Farm Operators in Israel', *International Journal of Comparative Sociology* 18 (3/4): 268–279.

Naisbitt, J. (1982) *Megatrends: Ten New Directions Transforming our Lives,* Warner Books, New York.

Newbery, D. (1975) 'Tenurial Obstacles to Innovation', *Journal of Development Studies* 11 (4): 263–277.

Newby, H. (1980) *Trend Report: Rural Sociology,* Sage Publications, London.

Olshan, M. (1981) 'Modernity, the Folk Society and the Old Order Amish: An Alternative Interpretation', *Rural Sociology* 46 (2): 297–309.

Piccone, M. and Schoorl, D. (1987) 'Appreciation and Judgement — The Key Roles of Agricultural Extension', *Proceedings of the Australian Agricultural Extension Conference*, Brisbane, October: 233–243.

Rogers, E. (1962) *Diffusion of Innovations,* Free Press, New York.

Rogers, E. and Shoemaker, F. (1971) *Communication of Innovations: A Cross-Cultural Approach,* Free Press, New York.

Roling, N. (1988) *Extension Science: Information Systems in Agricultural Development,* Cambridge University Press, Cambridge.

Rolls, M., Jones, G. and Garforth, C. (1986) 'The Dimensions of Rural Extension', in G. Jones (ed.), *Investing in Rural Extension: Strategies and Goals*, Elsevier, London.

Saint, W. and Coward, E. (1977) 'Agriculture and Behavioural Science: Emerging Orientations', *Science* 197: 733–737.

Thenes, C. (1979) 'Appropriate Technology. Another Development Policy', *Agecop-Liaison* 47: 17–32.

Van Dijk, D. (1979) 'Developing a Geographic–Geomorphic Approach to Soil-Land Classification', *Proceedings of Tenth New Zealand Geography Conference:* 264–266.

Van Dijk, D.(1980) *Salt and Soil Reaction Patterns in the Tara Brigalow Lands, South-East Queensland*, Division of Soils Divisional Report No. 47, CSIRO, Australia.

Van Es, J. and Pampel, F. (1976) 'Environmental Practices: New Strategies Needed', *Journal of Extension* 14: 10–15.

Warner, K. (1974) 'The Need for Some Innovative Concepts of Innovation: An Examination of Research on the Diffusion of Innovations', *Policy Sciences* 5 (4): 433–451.

Yapa, L. and Mayfield, R. (1978) 'Non-adoption of Innovations: Evidence from Discriminant Analysis', *Economic Geography* 54: 145–156.

Zuber-Skerritt, O. (1990) *Action Research for Change and Development*, CALT, Griffith University, Brisbane.

Zuber-Skerritt, O. (1991) *Action Research in Higher Education: Examples and Reflections*, CALT, Griffith University, Brisbane.

8 POWER RELATIONS IN RURAL COMMUNITIES: IMPLICATIONS FOR ENVIRONMENTAL MANAGEMENT

IAN GRAY

Governments are looking to local communities to initiate and implement rural environment management strategies. The Murray–Darling Basin Ministerial Council (1990: 13) sees local or subregional organisations as the 'building blocks' for community involvement upon which environmental management depends. Such desire in central government for devolution to the local level has often been expressed, for example, in the contexts of welfare and education policies. Local organisations, including many responsible for urban and regional planning, water and electricity supply and vermin control, already have responsibilities for matters which are directly or indirectly related to the physical environment. Yet, devolution has attracted critical comment (see, for example, Bryson and Mowbray (1981) with regard to welfare policy), and little thought appears to have been given to the implications of placing greater responsibility for natural resources management at the local community level. Critics of devolution often point to the risks of spatially unequal results, that is, those areas which have greater resources and organising capacities will fare better than others. This chapter takes a sociologist's view which explores such arguments. It examines some social features of rural communities and their political structures and processes. These suggest that local organisations may unintentionally fail to deliver an essential coordinated and equitable response to environmental problems within their localities since the ideals of community — which central environmental policy makers assume to exist — may not accord with reality.

LOCAL ORGANISATIONS

Local organisations may be small but they can have considerable influence over local destinies. This is most obvious in the planning and development functions of local government. Many local organisations have revenue-raising and punitive powers. The recently formed New South Wales regional Total Catchment Management Committees, as well as the older Pastures Protection Boards, have such capacities. The 'Communities of Common Concern' advocated by the Murray–Darling Basin Ministerial Council as action groups are less formal. They are more like Agricultural Bureaux or local discussion groups in their spontaneity and voluntary participation and compliance. As intervening institutions between individuals and government, all such organisations may take on an importance, as local government has, which is far greater than the significance of their day to day activity. They can become local advocates, holding the key to the extensive resources of state and federal governments. This is what the Murray–Darling Basin Ministerial Council intends for 'Communities of Common Concern', as it includes in their activities 'enlisting Government support and funds to complement their own resources, for their activities, through the NRMS [Natural Resources Management Strategy] and other sources', and 'communicating to Government, their aspirations and concerns for the management of natural resources at the local, regional and Basin-wide levels' (Murray–Darling Basin Ministerial Council 1990: 13).

Local organisations, from local government to 'Communities of Common Concern', are *managers*. They have responsibilities for certain types of activity, often defined by state or federal government, and are accountable to their funding providers — whether they be government or the public (taxpayers, ratepayers, or members) — for the efficient administration and execution of their tasks. They are, however, more than agencies of management. They are also *political arenas*. Each has resources to allocate and costs to spread and each must determine its priorities with regard to the needs and desires of interest groups. Such groups may include people who undertake a particular type of farming, those who live in part of a locality (a neighbourhood), or those who have a particular problem. Resources allocated to one service area, type of activity or group of people cannot also be spread among others. Not all local concerns are likely to be effectively pursued among regional, state or federal institutions at the same time.

This intersection of management and politics has been well illustrated in the study of rural local government. Some research on rural local government has sought to ascertain whether local politics is elitist or pluralist. Elitism is a situation in which one interest group, or even a small group of individuals, dominates local politics, while pluralism is indicated by a number of interest groups all competing for access to local resources. Pluralism is usually considered closer to the ideals of democracy as it holds promise for a balance of power. For local environmental organisations, pluralism would seem desirable, not only because it appears more democratic, but because in the natural environment, problems cannot be viewed in isolation from each other. No local environmental organisation can carry out its function adequately without consideration of all components of the local ecological system.

Organisations are, however, human constructions and they must rely on people for inputs to decision making. If inputs are received in a climate of elitism, important parts of the ecological system may be overlooked. While some people would be deprived of resources needed to solve their particular problem, the whole community may suffer as mutual systematic problems are allowed to persist.

Studies of local government have revealed both elitism and pluralism but evidence for elitism appears stronger. Unfortunately this research has not been extended far into the political aspects of other types of local organisations, although the evidence offered by Dempsey (1990) on participation in organisations shows that even in small towns, where propinquity is more likely to enforce mixing, local organisations can be exclusivist. Research on local government does, however, offer a window on local political structures and processes. It reveals structures in which the range of groups represented is constrained and the range of concerns expressed is kept narrow.

ELITISM

Studies of Australian rural communities have revealed two, overlapping, types of elitism in local government. 'Representational elitism' is that in which only one group of people is represented in the political arena. The group is usually described in terms of a demographic or socioeconomic characteristic of the local population. The elected representatives in Australian local government are often thought to be predominantly male, middle aged or older, and of small business, farm or professional occupation. Research by Atkins (1979), Bowman (n.d. and 1983), Burdess (1984), Chapman and Wood (1984), Power, Wettenhall and Halligan (1981) and Sinclair (1987) supports this view.

Community studies have been contributing to this depiction for a long time. After studying 180 towns in Victoria, McIntyre and McIntyre (1944) found that many inhabitants considered local government to be unrepresentative of local people. Oeser and Emery (1957) found a town's political structure to be dominated by farmers, a similar result to that of Cowlishaw (1988) who revealed grazier dominance in 'Brindleton'. Wild (1983) compared Heathcote's council members with the socio-economic characteristics of their constituents. He too found elitism.

Representational elitism favouring the upper middle class over the working class can be attributed to the substantial personal resources required of councillors. Employed people find it harder than the self-employed to devote normal working hours to civic service. Representational elitism may also be explained by a popular image of local government as a property management institution. The upper middle class can be more readily cast in management roles, while working class people lack confidence, fear being branded as social climbers or are apathetic (see de Lepervanche 1984; Dempsey 1990; Gray 1991). There is also a more intriguing explanation: elites can consciously act to exclude from office those whom they consider to be less worthy. An unsuccessful attempt at such action was revealed by Dempsey (1990: 205).

For local environmental organisations, the class dimension of representational elitism may remain, but one might suppose that education may become an important associated factor. Knowledge and understanding of environmental processes may at least in part determine who participates in organisations and who does not. The research necessary to reveal either representational elitism or pluralism in local rural environmental organisations, and the factors underlying either, is yet to be carried out.

Local government has also been found to be ruled by one small group of individuals: 'clique elitism'. McIntyre and McIntyre (1944) found that some people believed that their local council was run by just one person. Substantial evidence of clique dominance is not as common as that of representational elitism. Evidence of clique activity is obviously harder to acquire, as it could only come from close observation, something which a clique might resist. Such observation was, however, made by Wild (1974) in 'Bradstow' and again by Wild (1983) in Heathcote. The latter's councillors cemented their relationship in the council and resisted external pressure as they controlled the flow of information. Wild found that 'Bradstow's' municipal council was controlled by a male middle class clique, in which the mayor and the town clerk were prominent. They acted in concert to determine the agenda of local politics. Cowlishaw (1988) has offered a brief description of similar activity in 'Brindleton'. Elite cliques are also likely to be representationally elitist, given the pervasive dominance of upper middle class males in local government and their essentially small membership. Again, the extent of clique elitism in rural local environmental organisations is unknown.

While rural community studies have revealed both kinds of elitism, they have also shown that local government can operate pluralistically. Oxley (1978) found a pluralist process in the shire he studied. The business of local government was open and no individual or group sought self-advancement. Oxley attributed this to political ambition among councillors and their desire not to risk loss of public esteem. However, Oxley also offers some evidence that the council had earlier been dominated by wealthy graziers who neglected the urban part of the shire. Historical pluralism has occurred elsewhere. Gray (1991) discussed a period of rapid population growth in which neighbourhoods formed 'progress associations' in attempts to gain assistance from a municipal council with development of recreation and other facilities.

The effects of representation and clique elitism on political processes may be very similar but they are not necessarily either problematic or inequitable. It is quite feasible that an elite could manage a fair allocation of resources and respond to all known problems. Apparently pluralist local government may act consistently in favour of some group at the expense of others if a balance of power is not achieved. The shire council depicted by Gray (1991) had elements of pluralism, but the outcomes of local political issues tended to be elitist, as local government was made responsive to the concerns of farmers and business people. Exclusion of the 'Brindleton' and 'Smalltown' kind was not practised, although working class and women councillors were relatively few. This was a product of the historically small number of working class and women candidates for election and the slow turnover of councillors. The tendency for political outcomes to favour farming and business interests was associated with the informal and unintentional construction of the political agenda, particularly the ways in which it was shaped by ideology, rather than with the deliberate actions of an elite.

IDEOLOGIES AND POLITICAL AGENDAS

Decisions are made in local government in an ideological climate which produces patterned or even stereotyped responses to the issues which are brought before it. Moreover, ideology affects the range of matters which are considered to be valid points for debate and decision. Three ideologies have been found to affect local political relations. They are the ratepayer principle, localism and belief in political neutrality (Halligan and Paris 1984). Consideration should be given to the ways in which ideologies create a climate in local environmental organisations in which stereotyped decision making emerges. Some discussion will demonstrate the effects of ideologies in the local government context.

The ratepayer principle is expressed as a desire to constrain councils to activities which further the interests of ratepayers and to carry out those activities as efficiently as possible. Ratepayers are property owners and as such the interests which may be served by local government are related to property. Hence one sometimes hears of the 'three Rs' of local government activity, roads, rates and rubbish, which sum up, albeit sarcastically, the traditional property-service work of local government. The ratepayer principle dictates that local government services should not deviate from the 'three Rs', because only property-service work can be reasonably funded by rates. This places cultural, welfare and recreational matters beyond the realm of local government. Of course many councils have extended their activities beyond property services, with a large proportion, sometimes half or more, of councils' income derived from sources other than rates, including a share of federally collected income tax. Nevertheless, many councils stay as close as they can to traditional property-service roles. Rural councils have generally been slower than urban councils to expand the range of their activities. While there are exceptions, notably among the provincial cities, many rural councils remain reluctant to offer human services.

This illustrates how the task of an organisation can become popularly defined, with the definition persisting, regardless of the range of interests which exists in a locality. Local people have interests which go beyond property services; and they may look to local leadership to provide for many of those interests. Local control over service provision is often seen to be attractive. Rhetoric supporting local government boasts about the advantages of local self-determination. However, such sentiments are often accompanied by exhortations to keep local government to its traditional realm, implying that local autonomy is good — so long as it serves the interests of property owners. This is a recipe for elitism, in that a narrow range of interests may be served. Such elitism can occur regardless of the range of interests represented among elected members.

Elitism, or at least constraints on pluralism, can also arise within property-services delivery as an unintended consequence of the desire for efficient management. The priority management system adopted by a shire council was viewed by Gray (1991) as discouraging pluralistic participation in the council's decision making. Property services were provided across the shire at a generally high standard. Yet from time to time problems arose and were perceived by some people to be threatening their interests and deserving of greater attention from the council. Sometimes people found it hard to gain such attention.

Local government is sometimes seen to operate on a 'squeaky wheel' model, in which those interest groups who can 'squeak' the 'wheel' most loudly obtain the oil (Bowman 1983). Councils would want, quite reasonably, to avoid this situation. As critics of public participation in local government and planning have noted, high-status groups are better participators than low-status groups (Buller and Hoggart 1986) and response tends to come from an articulate minority (Newby *et al.* 1978) which ultimately reinforces the power of the middle class (Sandercock 1978). Local environmental organisations may also encounter 'squeaky wheels'.

The means of avoiding 'squeaky wheels' which the council described in Gray (1991) chose to adopt consisted of a document called a 'Policy Register', and included a committee to establish the priorities for local works to be entered in the 'Policy Register'. Priorities were determined on efficiency criteria. For example, of the several laneways in the town which remained unsealed, those which cost most to maintain were to be sealed first. When a group of residents sought to intervene in the priority-setting process by asking that early attention be given to a drainage problem, they were disappointed when the shire engineer advised the council that the work required would have been too expensive. The council's discussion went no further. The council had, unintentionally perhaps, risked equity for efficiency. Local environmental organisations may face a similar trade-off.

Debate about property-services matters could be driven by questions of efficiency, especially when one or more councillors attempted to challenge technical recommendations made by engineering consultants on the grounds that they may not have been the most efficient course of action (see Gray 1991). The efficiency of the priority-setting apparatus was not questioned and it certainly appeared to be directing the council property-service work in a least-cost manner. Yet it also tended to discourage, if not stifle, the pluralistic participation which had occurred when, as mentioned above, the progress associations were active. Ironically perhaps, many people, including members of the local ratepayers association, saw the role of organisations like progress and ratepayers groups as ensuring that the council operated efficiently, gaining maximum value for all ratepayers' money. Pluralistic ideals were seldom voiced in election rhetoric. Local environmental organisations will invariably face demands for efficiency which, when responded to, may smother pluralism.

It is important to note that occurrence of elitism does not depend on conscious attempts by some individual or individuals to obtain more than an equitable slice of the local public cake. One sometimes hears stories about the rural shire president whose access road was mysteriously sealed long before other roads in the vicinity. Such stories may contain elements of truth but they also may be part of a mythology associated with a popular assumption that elitism prevails in rural local government. Seeing elitism this way is misguided (to the extent that elitist allocation of local resources can unconsciously and unintentionally be fostered in the ideological climate of local government and country towns) regardless of any acts of self-interest on the part of local office bearers.

Localism in a general sense is 'a set of beliefs about the significance of place' (Strathern 1984: 44). As Halligan and Paris (1984) pointed out, place is very significant to local politics. Local organisations of all types see themselves as catering for the interests of their locality. Those interests which are seen as local are more likely than others to gain expression in the local political arena. Localism fosters an image of a social entity to which all local residents belong, and which, moreover, is a common interest. In this way, localism distracts attention from local social divisions.

The most extensive analysis of localism and belonging in an Australian country town has been carried out by Dempsey (1990). The people of his 'Smalltown' like to think of their community as a big happy family, a metaphor which Dempsey found to be grossly misleading as many people suffered exclusion and marginalisation. Local people quite readily praised their locality, expressing their belief that other towns, and particularly cities, were inferior to 'Smalltown'. They claimed a strong sense of community — one of many positive features that 'Smalltown' was believed to display when compared with other places, including other small towns. Dempsey found expression of these beliefs during activities of local organisations.

Localism is strongest when the locality is seen to be under threat, as occurs when city-based decision makers are found to be neglecting rural interests. Reaction to threats emanating from the cities can be an expression of country-mindedness, when country people see a threat to a general rural interest as well as a threat to their locality. The ideologies of localism and country-mindedness are related, but are usefully separated for purposes of analysis. One can understand the country-minded motivation for cooperation in defence of a general rural interest, as occurs when rural economic crises prompt protests against state and federal governments, and localistic competition for economic development and sporting awards, or the adverse reaction which is sometimes obtained by a business from one town wishing to promote itself in another (for examples, see Cowlishaw 1988; and Gray 1991).

Localism has peculiarly rural characteristics and foundations. Dempsey explained localism in terms of the size of 'Smalltown' (about 2700 people), its slow rate of immigration, its distance from a large town, its socioeconomic heterogeneity and its institutions in which many people interacted. Hall, Thorns and Willmott (1984) proposed, from research in five New Zealand localities, that localism, as expressed in local collective action, is based on relations of kinship, property and propinquity.

Such structures do not provide a complete model, as the actions of people are also important. Dempsey noted that 'Smalltown's' upper middle class worked to foster and maintain localism. In local politics, the significance of localism lies in its denial of internal division. As Dempsey found, localism discourages class consciousness. People who would voice a group interest may find it very difficult to obtain attentive ears if their views appear to be sectional and their interests not apparently equated to the local interest. Conversely, a group which can define its interest as the local interest can become very powerful. Gray (1991) found that farming and business groups were able to do so, as they were seen to represent the ideals of self-reliance, perseverance and entrepreneurship upon which rural ideology places great value. Other groups found that they were popularly ignored, despite having good access to local media and politicians. Local environmental organisations should be cautious of those groups which would, quite sincerely, express their interests as the common local interest, and thereby claim priority.

Again, it is important to note that these processes are not deliberate acts of elitist self-interest. They amount to no more than one or two groups seeking to further their legitimate interests in a legitimate manner. The point is that local people see some interests as more legitimate than others. In an economy which is dependent on small business and farms, the prominence of such interests is understandable. What can be harder to accept, however, are circumstances where other interests which are also shared are not accepted as common.

The third of the Halligan and Paris (1984) ideologies consists of belief that local government should be kept free of politics. This is often seen as a desire to keep political parties out of local government, but it also denies legitimacy to any group which would seek to further its interests, except, of course the powerful group whose interests are seen to be legitimate to pursue. This ideology has been reported in rural local politics by Aitkin (1972), Pandey (1972) and Wild (1974). Opposition to the status quo would threaten the view of a town as a big happy family, a belief which Dempsey's (1990) analysis of 'Smalltown' showed to be held dearly. Local political censorship was revealed long ago in a community study, when Oeser and Emery (1957) mentioned farmers' actions to keep political discussion out of local currency. Gray (1991) has also discovered a general distaste for conflict, especially that which might be associated with the activities of political parties and which might threaten what was thought to be the smooth and efficient operation of local organisations. People who would seek to gain election on the basis of a political platform were thought to be unsuitable for the responsibility that accompanied local office.

The value placed on political neutrality, in terms of a consensus approach to local politics, was illustrated by Wild (1974) in his discussion of attempts to control the affairs of local government by a mayor who was skilful at accelerating decision making. The mayor would remind councillors that they did not want to be seen to be in conflict. Such action, however, is not necessary for the ideal of political neutrality and consensus to maintain the power of an interest group. There may be no desire among potentially competing groups to upset the local consensus. As Gray (1991) showed, there is little prospect for class- or gender-based politics when working class people — and especially women — feel that the status quo should not be upset. They either avoid seeking local office or after obtaining office do not attempt to oppose a male upper middle class regime. The ideologies of local government can entrench elitism without requiring elitism in representation.

One could expect that the value placed on consensus will also guide the operation of rural local environmental organisations. There is no reason to expect that pluralism will be seen as an ideal. This suggests that the ground is ready for environmental groups to be elitist, with some people deriving considerable power from them and doing so quite legitimately and unintentionally.

ISSUES AND POWER

Local political structures and ideologies determine the content of local politics, as some concerns are made into issues while others remain non-issues, and some interest groups are empowered while others are not. In this climate it is seldom necessary for powerful groups to enter into and win conflict to maintain their dominance. As Lukes (1974) argues, it is possible to identify power relations when no conflict is apparent. To do so, however, involves accepting two points: the possibility that non-issues can be visible, and a definition of powerful people as those who are able to extract benefit from a political apparatus.

The latter point is related to the former, but is less problematic. It is not difficult to consider as powerful those people who can make a political system work to their advantage and those who cannot as weak. It is harder to accept a place in political analysis for non-issues, simply because their non-existence logically makes them impossible to detect. Assertion of their reality must be based on assumptions about the nature of interests, particularly assumptions that some groups have interests that are neither being catered for nor being expressed effectively. This raises an important theoretical dilemma, one which disappears in empirical research. Several studies, including those of Bell *et al.* (1976), Crenson (1971), Gaventa (1980) and Gray (1991), have taken the pragmatic step of asserting that it is contrary to anybody's interests to suffer from some identifiable problem, such as poverty or pollution. One can assert that interests are not being defended, or potential issues are not being raised, where it can be shown that no action is being taken to ameliorate an inequitable distribution of some commonly valued good, such as a clean and productive environment. This approach has obvious merit in the context of environmental action, because a great deal of effort is being made to place environmental degradation onto political agendas. People are suffering — and will continue to suffer — unless environmental problems are raised as issues. People who raise them successfully have a chance to acquire resources necessary to solve problems, while those whose concerns remain non-issues have little prospect for obtaining help.

Some potential concerns may remain non-issues because they are not perceived at all. Soil salinity offers an example. Rising levels of salinity in irrigation land have been known for some time. However, the prospect of serious salting problems arising in dryland farming areas has loomed relatively recently. While recognised at a regional level, the raising of salting as an issue at the local level suffers two impediments. One is lack of awareness among those affected or potentially affected by it; the other is the complication which arises when those detrimentally affected by salting are not those on whose land the cause of the problem is situated.

If farmers do not accurately perceive salting they will not be able to raise it effectively as an issue. If they do not understand it they may not be able to communicate their perceptions, consequently failing to raise an issue, for an issue is a matter of public rather than private concern. It must be public if it is to attract collective action. Evidence gathered in several areas suggests that although a large number of farmers *are* aware of the salting or the potential salinity problems which have arisen on their land, fewer understand its causes (Barr and Cary 1984; Vanclay and Cary 1989; Blake and Cock 1990; Dunn and Gray 1990). Less happily, evidence of farmers feeling that, although salting is a serious threat, it will probably only affect other people has also been found (Barr and Cary 1984; Vanclay and Cary 1989; Francis and Gray 1990). This suggests that many may not obtain a perception which motivates action and the raising of an issue. Dunn and Gray (1990) found that although many farmers were aware of a threat from salting, few felt that they knew the extent of the problem in their district. This suggests that little communication had occurred among those who might raise an issue.

While environmental problems like soil salinity may appear to be objective matters, one should consider how ideologies might inhibit their perception. Evidence from observation of local government suggests that the ways in which farmers and others interpret their own environmental interests, and interpret the issues which they already perceive, could be very important. If issues are seen to be the creation of non-local, non-rural people seeking their own ends, they are less likely to be accepted and so would fail to motivate action among local people. In a climate of localism and denial of conflict, the issues that are raised may be constrained to those matters which are accepted as common rather than sectional, and a narrow range of interests accepted for pursuit on organisation agendas. If efficiency is seen to be the matter of common interest above all others, those who are most affected may see remedial action taking place only in situations in which it can be done most efficiently. Benefit would accrue to those who can convince others that their interests are everybody's. In this way some people could place themselves, perhaps unconsciously, in positions of power as they attract resources at the expense of other people.

CONCLUSION

The evidence of elitism in local government, through the monopolisation of participation by some interest groups (including the processes in which ideologies are used to constrain the range of issues before local government), suggests that questions about elitism should be asked with respect to local organisations which have environmental responsibilities. Rural community studies have shown that, even where local people strongly believe their community to be united and cohesive, social marginalisation and exclusion occur. Moreover, social marginalisation and exclusion can be reflected in resource allocation decision making, without conscious intention among local decision makers, when people call on shared beliefs and values to interpret their interests as those of a dominant group, or simply remain unaware of their own environmental interests. This is not to say that all rural local organisations will necessarily be elitist. Many set out to be — and become — responsive to a plurality of needs. Gray (1991) mentions a neighbourhood centre that had a membership which crossed class boundaries, one which tackled a wide range of local social problems. This organisation implicitly assumed the existence of, and responded to, a range of interests in its locality. The view expressed here should not be taken as damnation of locally based organisations; rather, it should be seen as a warning of the unintended consequences of seemingly enlightened localist policies. It is dangerous to assume that all local organisations will be pluralistic.

'Communities of Common Concern', and other rural local organisations which pursue environmental programs, should not be assumed to have a 'common concern', as they are often thought to have under the 'big happy family' image of rural communities identified by Dempsey (1990). There are, and will be, many potential local environmental issues whose implications are plural, as they cross social as well as spatial boundaries within communities. Even small country towns can be socially heterogeneous with some groups within them unaware of the circumstances of others, especially those who are marginalised. Town dwellers vary in socioeconomic status; some people are well educated while others are not; and, while most farms are operated by families or individuals, others are operated by corporations. Populations make a plurality of demands on their environments. Some people have a more immediate interest in, for example, water quality, while others may be more directly affected by soil erosion. Some people place greater immediate demands on their environment than others.

Environmental interests are plural but they are also shared; ecological problems do not stop at property boundaries. Political structures would ideally accommodate both commonality and diversity. As community studies have shown, however, local organisations are often comfortably elitist in membership and leadership. Pluralism may suffer and elitism prevail at the level of participation, with or without any constraining effects of ideology — a point which must remain speculative until supported or denied by research on local environmental organisations.

Allocation of costs and benefits associated with environmental action is a political process which contradicts the view of a singular community of interests. There is a common interest in protection of the environment, but appreciation of the common interest may, as has occurred in the local government arena, cloud the pluralistic nature of individual interests. The best course of action from the community perspective could become very expensive for some of its members, especially those on whose land the cause of what is assumed to be a common problem could be identified, as well as those on whose land an effect of the problem is not perceived.

Australian rural local polities have displayed various degrees of elitism as they operate under assumptions of a unity of interests. They are products both of an elitist tradition in leadership and participation in rural communities and of a persistent set of values and beliefs which rural people use to make sense of their local social environments. In a speech at a field day held on a New South Wales farm late in 1990, the president of an environmental action committee assured the audience that all the members of the committee had agreed to act together rather than represent interests. This was a statement of the political neutrality ideal often expressed for local government. The members of the committee — like many local government representatives — apparently preferred to see themselves as an elite charged with the responsibility to look after their people, a view which attracts the dangers of elitism to an organisation which must be pluralistic, that is, one catering to all interests if it is to achieve its ecological aims. This does not imply that there is no room for leadership and teamwork. Rather, it warns that local environmental organisations should appreciate their political natures and ensure responsiveness to all needs whether expressed or not, while recognising their task of balancing competing interests. The extent of environmental elitism and its effects on the agendas and action plans of local environment organisations has not yet been revealed. However, the potential contradiction between rural community ideals and the plurality of environmental interests at the local level should be borne in mind by those who seek to devolve responsibility for environmental programs to local organisations.

REFERENCES

Aitkin, D. (1972) *The Country Party in New South Wales: A Study of Organization and Survival*, The Australian National University Press, Canberra.

Atkins, R. (1979) *A New Look at Local Governments*, The Law Book Company, Sydney.

Barr, N. and Cary, J. (1984) *Farmer Perceptions of Soil Salting: Appraisal of an Insidious Hazard,* School of Agriculture and Forestry, University of Melbourne, Melbourne.

Bell, C., Newby, H., Rose, D. and Saunders, P. (1976) 'Community Power in Rural Areas', Paper delivered at the 47th ANZAAS Congress, Hobart, May.

Blake, B. and Cock, P. (1990) *Salinity and Community,* Environmental Report No. 30, Graduate School of Environmental Science, Monash University, Melbourne.

Bowman, M. (n.d.) *The Suburban Political Process in Box Hill Melbourne,* Melbourne Politics Monograph, Melbourne.

Bowman, M. (1983) 'Local Government in Australia', in M. Bowman and W. Hampton (eds), *Local Democracies: A Study in Comparative Local Government*, Longman Cheshire, Melbourne.

Bryson, L. and Mowbray, M. (1981) ' "Community": The Spray On Solution', *Australian Journal of Social Issues* 16 (4): 255–267.

Buller, H. and Hoggart, K. (1986) 'Nondecision-Making and Community Power — Residential Control in Rural Areas', *Progress in Planning* 25: 135–203.

Burdess, N. (1984) 'Public Involvement in New South Wales Local Government', *Australian Journal of Public Administration* 43 (3): 296–300.

Chapman, R. and Wood, M. (1984) *Australian Local Government: The Federal Dimension*, Allen and Unwin, Sydney.

Cowlishaw, G. (1988) *Black, White or Brindle: Race in Rural Australia,* Cambridge University Press, Cambridge.

Crenson, M. (1971) *The Un-Politics of Air Pollution: A Study of Non-Decisionmaking in the Cities,* The Johns Hopkins Press, Baltimore.

de Lepervanche, M. (1984) *Indians in a White Australia: An Account of Race, Class and Indian Immigration to Eastern Australia,* Allen and Unwin, Sydney.

Dempsey, K. (1990) *Smalltown: A Study of Social Inequality, Cohesion and Belonging*, Oxford University Press, Melbourne.

Dunn, A. and Gray, I. (1990) *Farmer Perceptions of Soil Salinity in Young Shire, NSW*, Centre for Rural Social Research, Charles Sturt University, Wagga Wagga.

Francis, R. and Gray, I. (1990) *Management Issues in the Upper Billabong Creek Catchment*, Centre for Rural Social Research, Charles Sturt University, Wagga Wagga.

Gaventa, J. (1980) *Power and Powerlessness*, University of Illinois Press, Urbana.

Gray, I. (1991) *Politics in Place: Social Power Relations in an Australian Country Town*, Cambridge University Press, Cambridge.

Hall, R., Thorns, D. and Willmott, E. (1984) 'Community, Class and Kinship — Bases for Collective Action Within Localities', *Environment and Planning D: Society and Space* 2 (2): 201–215.

Halligan, J. and Paris, C. (1984) 'The Politics of Local Government', in J. Halligan and C. Paris (eds), *Australian Urban Politics*, Longman Cheshire, Melbourne.

Lukes, S. (1974) *Power: A Radical View*, Macmillan, London.

McIntyre, A. and McIntyre, J. (1944) *Country Towns of Victoria*, Melbourne University Press, Melbourne.

Murray–Darling Basin Ministerial Council (1990) *Murray–Darling Basin Natural Resources Management Strategy*, Murray–Darling Basin Ministerial Council, Canberra.

Newby, H., Bell, C., Rose, D. and Saunders, P. (1978) *Property, Paternalism and Power*, Hutchinson, London.

Oeser, O. and Emery, F. (1957) *Social Structure and Personality in a Rural Community*, Routledge and Kegan Paul, London.

Oxley, H. (1978) *Mateship in Local Organisation* (2nd edn), University of Queensland Press, St Lucia.

Pandey, U. (1972) 'Power in Barretta: Influence and Decision-Making in an Australian Country Town', unpublished PhD thesis, University of New England, Armidale, New South Wales.

Power, J., Wettenhall, R. and Halligan, J. (1981) *Local Government Systems of Australia*, Advisory Council for Inter-Governmental Relations, Information Paper No. 7, Australian Government Publishing Service, Canberra.

Sandercock, L. (1978) 'Citizen Participation: The New Conservatism', in P. Troy (ed.), *Federal Power in Australian Cities*, Hale and Iremonger, Sydney.

Sinclair, A. (1987) *Women in Local Government*, Hargreen, Melbourne.

Strathern, M. (1984) 'Localism Displaced: A "Vanishing Village" in Rural England', *Ethnos* 49 (1–2): 43–61.

Vanclay, F. and Cary, J. (1989) *Farmers' Perceptions of Dryland Soil Salinity*, School of Agriculture and Forestry, University of Melbourne, Melbourne.

Wild, R. (1974) *Bradstow*, Angus and Robertson, Sydney.

Wild, R. (1983) *Heathcote*, Allen and Unwin, Sydney.

9 SOCIAL ASPECTS OF THE FARM FINANCIAL CRISIS

LIA BRYANT

In order to understand what is happening to the rural environment we need to understand what is happening to the people on the land. The social, political and economic forces that shape society also affect the way members of society use and treat their land. This fundamental concept of individuals interacting with and shaping — and being shaped by — both society and environment has been an oversight in contemporary landcare decision-making processes. Approaches to landcare have been pragmatic with emphasis on altering the direct antecedents to land degradation, that is, the behaviour of farmers, and altering the direct consequences of that behaviour on the environment.

Solutions to land degradation must be aligned to the reality that farming is a political, economic and social activity. It is political because markets are structured by local and international political decisions and the actions of state agents who in turn are influenced by external groups (Newby 1982; Cox, Lowe and Winter 1986). The power of those external groups is related to their structural position in the economy and the political strength of their organisation. It is economic, because farming in Australia is a capitalist activity, it is dependent upon profit and is an integral feature of the Australian economy. Finally, central to the social context of farming is the farm family and the relationship between family and business. These relationships are interactive with structural changes in agriculture affecting the way the farm is run — including the need for work off the farm, the need to cut back spending on the farm and family living and the need to use family labour. Changes in the family structure like ages of children, health, separation and divorce equally affect the family and the business (Gasson *et al.* 1988).

The fact that farming is subject to the 'booms and busts' of the capitalist cycle is not news to agricultural producers. Historically, Australian agriculture has experienced periods of economically low times (for examples, see Yates 1974; Bell and Nalson 1974) characterised by farmer protests, farm poverty and an exodus of farmers from the industry. During the mid-1980s, there was much debate and discussion by farmer organisations, community and political leaders, academics and not least of all the media as to whether Australia was in the grips of a rural crisis. In general, the question was answered in an affirmative or negative manner depending on the political and economic position of those entering the debate.

The early 1990s have also been characterised by farmer protests, further reductions in farm incomes and rural community decline. Any debate as to whether Australia is currently experiencing a farm financial crisis must examine two critical factors. First, since the term 'crisis' denotes an acute and temporary phase, can the conditions of agricultural production which give rise to 'bust cycles' be considered acute and temporary, or have they become an inherent feature of the agricultural economy? Given the General Agreements on Tariffs and Trade (GATT), the Uruguay Round of Talks and subsequent trade discussions, the assumption can be made that protectionist policies are not likely to alter in the near future. Second, farmers are not an homogeneous group (see Lawrence 1987), yet the term 'farm financial crisis' indicates widespread doom for agricultural producers in all industries, all regions and states and for enterprises of all sizes.

The commodity overview provided by the Australian Bureau of Agricultural and Resource Economics (ABARE) indicates that Australia's major rural commodities, wool, wheat, sugar and dairy products, are at risk due to near record levels of world production (Fisher 1991). The increased subsidies of the European Economic Community (EC) and the USA have exacerbated the high levels of produce especially in the wheat and dairy industries.

These market conditions will ultimately result in a greater proportion of farmers receiving a negative income (Harris, Kirby and Fisher 1991). Agricultural economists predict that the aggregate net farm income for the 1990–91 financial year will decline by 33 percent and by a further 21 percent in 1991–92 (Kirby 1991). Farmers experiencing the greatest income difficulties will be those in the wheat and wool industries. The average farm cash operating surplus for broadacre farms is expected to fall in the 1990–91 period from $A53 036 to $A21 582, a drop of 59 percent, and in 1991–92 a further decline to $A15 185 is expected (Fisher 1991). (As discussed in Chapter 3, average net farm income is much lower.)

The size and scope of the 1990s farm financial crisis are more intense than that experienced in the past decade. Unlike the 1980s, interest rates will continue to drop and input prices will rise only marginally. However, savings to farmers will be largely offset by higher debt levels resulting from inordinately low incomes. The Director of the ABARE (Fisher 1991) suggests that in real terms the average cash operating surplus for all broadacre farms is expected to be half that experienced in the 1982–83 drought years. Sheep and wheat farmers in all Australian states are experiencing financial pressure but South Australia and Western Australia are the hardest hit. How the effects of the downturn will be experienced by individual farm enterprises will obviously depend upon their financial position. Farms with high debt, high costs and limited liquid assets will be severely placed. In comparison to the mid-1980s, farmers in trouble in the 1990s will be unable to meet overheads and production costs let alone service their debt (Chambers *et al.* 1991). The farms most likely to fall into this category are small to medium sized family properties.

The term 'farm financial crisis' is not generic. It is an appropriate term when located in a specific historic period and related to specific locations, farm types and industries. For farm families experiencing severe financial difficulty and personal stress, the term 'farm financial crisis' is also appropriate, as for them 'crisis' has become an experience they are forced to live with.

Structural changes in agriculture have heavily impacted on the way farm families are able to live and work their land. The trend in Australia is toward bigger farms with capital becoming concentrated and centralised in the hands of larger but fewer farmers (Lawrence 1987, 1989a). Small family farmers have fewer resources to sustain economic downturn and the impact of the farm financial crisis on them is substantial.

This chapter examines the relationship between the economic and social features of Australian agriculture and land degradation by exploring proposed solutions to land degradation, the economic organisation of society, the social costs of the farm financial crisis and the implications of government policy for the objective of landcare.

POPULAR SOLUTIONS TO LAND DEGRADATION

The most common solutions posed by environmentalists and government agencies to achieve sound land practices may be placed into four broad categories: incentives, technocentric solutions, regulation and ecocentric solutions (Pepper 1984; Roberts 1989). The most uniform incentive provided to farmers to encourage sound landcare practices is a direct tax deduction for landcare carried out. However, this incentive is, of course, only available to those farmers with a taxable income. Technocentric solutions to land degradation are diagnosed by scientists and technologists who determine methods to overcome or 'fix' environmental problems. Technocentric responses also involve environmental managers and planners who educate farmers and others about the environment and who use techniques, like cost–benefit analysis, in an attempt to illustrate to farmers that sound land practices in the long term are remunerative.

Until very recently, environmental laws administered by some Australian states have rewarded what is now considered to be harmful land practices, like the clearing of scrub (Fisher 1980). Government regulation is now exercised in most states via a range of laws which attempt to control soil and land degradation and to conserve vegetation and water. In South Australia a 'soil conservation order' may be taken out against a land owner by a local community conservation board to compel the owner to refrain from certain practices or to take action to carry out specific rehabilitative works (*Soil Conservation and Land Care Act of South Australia, 1989*). Failure to comply with the order may result in the board carrying out the work and then recovering the debt from the land owner in default, or placing a charge on the title of the land.

Finally, the most pervading solutions to environmental problems are ecocentric. These focus on raising the consciousness and knowledge of the public of what is happening to the environment and provide ways for the community to participate in caring for the environment. Ecocentric thought emphasises consensus in the face of a common threat. This supposition is repeatedly reposited in government statements and literature and is typified in *Our Country Our Future* (1989) — the environmental statement of former Prime Minister Bob Hawke.

Technocentric and ecocentric solutions assert the idea that 'education is the greatest resource' (Schumacher 1973: 64, cited in Pepper 1984). These solutions assume that land degradation occurs as a result of ignorance on behalf of farmers or an attitude to farming that is not considerate of the environment. Translated into practice, this means that if we educate our farmers to understand and appreciate environmental issues and teach them environmentally safe farming techniques, problems of land degradation will be solved. However, education would appear to be only a partial solution to land degradation (see Chapter 6).

LAND DEGRADATION AND THE ECONOMIC ORGANISATION OF SOCIETY

The way society is materially organised and the way in which farmers maintain themselves economically determines their relationship to nature. To maximise profits, families in financial difficulty attempt to decrease production costs and increase output to meet debt and avoid displacement. This results in an inability or perceived inability by the farmer to undertake sound landcare practices. There is currently little empirical information outlining the specific experiences of Australian farmers in financial difficulty and the farming practices they adopt. The evidence that is available indicates that farmers tend to increase cultivation by shorter rotations, overuse fertilisers and chemicals and overstock. The environmental cost is soil erosion, salinisation, water pollution, the destruction of animal and plant life and a range of other destructive measures (Lawrence 1987; and see Chapter 3).

A trend becoming evident in South Australia is that farmers are now moving from grazing to cropping as prices for sheep have plummeted and have been predicted to remain low. This means that a greater proportion of farm land is used for cropping, resulting, in turn, in shorter rotations and the reduction of soil nitrogen levels. This process becomes harmful if continued over time and if appropriate methods to replace soil nutrients are not carried out. A method increasingly adopted to replace nutrients is the use of nitrogen fertiliser which can lead, in turn, to soil acidification.

Excessive cultivation also accelerates water erosion and results in the soil losing its protective cover of plant growth. Methods to overcome this problem include minimal tillage, stubble retention, strip cropping and controlled grazing (South Australian Environmental Protection Council 1988). The question must be asked: is minimal tillage possible for crop farmers in severe financial difficulty, particularly when they are unable to rely on sheep production as their alternate enterprise?

Farmers in financial difficulty are also unable to engage in rehabilitative land practices, such as property planning including the redesigning of fences and boundaries to subdivide property according to land capability. Rehabilitative practices can be capital intensive and the payoffs slow.

There is no immediate solution to the economic problems experienced within Australian agriculture (Lawrence 1987) and it cannot be overlooked that the agricultural outlook for the early to mid 1990s indicates that an increasing number of Australian farmers will receive a negative income. The consequence of dramatically reduced farm income coupled with escalating debt means that a higher proportion of income is used toward repayment of debt. Given that the economic pressures on farmers will continue it can be assumed that so will unsound land use practices. Consequently, the labour process — that is, the way farmers transform nature to make their living — 'succeeds in transforming the environment in ways that ultimately make it less productive' (Redclift 1990: 64).

The economic dimension is an important factor which guides and restricts farmers' choices for landcare; however, to define the relationship of the farmer to nature in economic terms alone would be to ignore farmers as actors with values and ideologies that also shape their relationship to the land. For many farmers, farming is considered to be more than a job or a business, it is said to be 'a way of life'. Rural ideology is heavily based upon notions of independence and self-reliance. The importance of family and work contribute very strongly to this ideology and these characteristics induce farm families to remain in farming at all costs.

When placed under financial pressure farmers struggle to keep their business, home and future ambitions and believe that if they work for 'one more year' the drought might break or prices might improve and they would have a chance of recovery (Bryant 1990). To keep the debtors at bay and maintain their land, farm families work harder to increase output and at the same time decrease business and family living costs. An Eyre Peninsula wheat farmer explained: 'You keep going to try to save the land for your son. Your first priority is the farm bills — these have to be paid and you just have to go without.'

SOCIAL COSTS OF THE FARM FINANCIAL CRISIS

The costs of the farm financial crisis are not simply economic or environmental. The social consequences are severe for rural communities and individuals within farm families.

Current levels of farm poverty are related to structural conditions in agriculture. As well as having limited financial resources for family living many farm families are ineligible for government-based financial assistance including study allowances, unemployment benefits, carer's pensions and family allowance supplements due to asset testing. The formula for asset testing is based on an urban notion of a reasonable asset limit and debt structure and thereby can exclude members of farm families in financial need.

Levels of farm poverty rose from an average of 12 percent

in 1975 to approximately 20 percent in 1984 (Lawrence 1987: 45) and, while data are not available, it is likely that farm poverty levels have continued to increase. Since the mid-1980s, at least a third of primary producers in Australia have experienced a relatively sustained negative income and the expansion of the rural crisis to include the wool industry will involve a greater number of farm families and rural people experiencing income deprivation.

Poverty, however, extends further than financial resources to include social power and, in particular, 'equality of access to the bases of power' (Samya-Coorey 1989: 14). For farm families the basis of social power includes access to information and assistance, the ability to enjoy good health and education and political and social representation. The incidence of poverty is greater in non-metropolitan than in urban areas, and social and living costs increase from non-metropolitan to provincial to rural to remote areas (McKenzie 1985, cited in Samya-Coorey 1989). Coinciding with the disadvantage associated with remoteness in terms of access to services for rural people, state and federal governments have rationalised their existing health, welfare and agricultural-based services in country areas. Essential services are being withdrawn from some country hospitals, hospitals and schools have been closed, and there have been across-the-board staff cuts. Some agricultural extension services have also developed a user-pays system thereby limiting the ability of poorer farmers to obtain the latest farming and management information and technologies.

In an effort to increase family income and repay debtors an increasing number of family members are obtaining off-farm work. The youth are leaving their communities to take work off the farm as the farm can no longer support them; and more farm women are obtaining off-farm work to keep the family afloat. In a study of Eyre Peninsula farmers at least half the women interviewed had regular off-farm work, while only one-sixth of the men interviewed were engaged in temporary — and in some instances seasonal — off-farm work (Bryant 1990). These results are comparable to national trends which indicate more women are engaging in work off the farm (Masson 1986; Gibson, Baxter and Kingston 1990). However, in conjunction with off-farm work, women on farms are also often involved in various types of labouring including agricultural production and management tasks, domestic production and community work (Sachs 1983; Donnelly and Smith 1981; Masson 1986; Reimer 1986; Department of Prime Minister and Cabinet 1988). Alston (1990) has raised the concern that in the current economic climate an increasing number of women are being exploited by the demands of the multiple roles they are expected to perform.

Rural families are experiencing high levels of stress which can be linked to increased incidence of stress-related illnesses like hypertension, heart attacks, asthma, ulcers, back and chest pain and an increase in substance abuse, especially alcoholism and the overuse of prescribed drugs (Lawrence 1987; Bryant 1989). Lawrence (1987: 43) reports that 'surveys have revealed that rural people experience 10 percent more illness, and have 28 percent more hypertension and psychiatric disorder, than have Australian urban populations'. There has been a rise in the level of recorded suicides among rural people (see Lawrence 1987; and Bush 1990). A South Australian rural welfare worker reported that there is 'vulnerability toward suicide within the rural sector; . . . some families . . . close off all their interactions within their community and . . . men [in particular] . . . are unable to picture any lifestyle but farming' (Boylan 1990: 22).

The stress of living in extreme financial difficulty also places pressure on relationships, and farmers from the Eyre Peninsula in South Australia reported high levels of anger or withdrawal and the subsequent inability to communicate positively with their spouse, children, neighbours and extended family (Bryant 1989). Increases in domestic violence in rural areas has also been reported (Lawrence and Williams 1990).

The children of farmers also experience emotional and physical reactions to living with economic hardship, stress and uncertainty for their future. Eyre Peninsula farmers stated that rural youth had little formal emotional, financial or practical support. The special needs of farm youth in areas of financial assistance, employment, housing and general life opportunities have not been adequately considered by social scientists or government agencies in the context of the farm financial crisis.

Farm families in financial difficulty have limited adjustment choices and an increasing number are being faced with displacement. Over the past 15 years about 19 000 families have left farming (Cribb 1985, cited in Lawrence 1989a). Displacement from agriculture impacts heavily upon the social structure of the farm family and results in the break-up of the family as a productive unit. Eyre Peninsula families displaced from their farms had to adjust to no longer living and working together (Bryant 1989, 1990). For some, displacement also resulted in the break-up of the family unit as the male former farmer or children moved elsewhere for work.

The Eyre Peninsula study of farm family displacement indicated that most male and female former farmers who sought employment were able to find work, usually unskilled wage-labouring jobs after or just prior to leaving the farm. However, there was a distinct element of underemployment with some farmers engaged in part-time or casual work — something which was not always constant. Interestingly, engagement in part-time and casual work was experienced equally among women and men.

In conjunction with low levels of education, the skills developed by farmers in the process of farming are not highly marketable in an urban labour market. Consequently, the limited life chances available to male farmers reinforces the downward spiral in their class position after displacement. The nature of inheritance of property and farming as an occupation raises questions as to the affect of displacement on employment opportunities for the children of farmers, as entry into farming is gender biased with succession and inter-generational land transfer is most usually that of father to son. Age becomes the determinant for when changes take place, usually as the father draws near retirement (about 65 years of age) and the son matures (about 35 years of age) (see Phillips 1987). Women either move off the farm to find work or marry into another farming operation. All sons in the 12–17 and 18 plus age groupings of Eyre Peninsula former farmers had either planned to work on the farm, had worked on the farm until lack of finances prevented this or had worked on the farm up until its sale. Farm youths/adults attempting to resettle have similar concerns to their parents and also must look for employment and housing. However, if these children are not bona fide partners in the farm business then they are not entitled to re-establishment grant monies under the Rural Adjustment Scheme (an asset-tested grant to help families resettle, currently in the vicinity of $33 000). Eyre Peninsula youths departed from farming with absolutely no financial resources and with limited formal education (usually year 11). The life chances and social mobility of the children of farmers, particularly their sons, is restricted with displacement.

Women on farms have better prospects of employment as a result of off-farm work and higher levels of education. As such, they have at least some prospects of upward mobility after leaving farming.

Farm families displaced from agriculture have also had to come to terms with the fact of their 'lived reality' — something which challenges some of their most deeply held ideologies. They have had to cope with a loss of perceived independence and this has manifested itself in the workplace. As an Eyre Peninsula male farmer said, 'I find it difficult to converse with management and feel frustrated at no longer being able to make my own decisions' (Bryant 1990: 76). Perhaps the hardest aspect of displacement for families has been the reality that hard work is not necessarily financially and socially rewarded. One farmer said 'I've worked for 40 years and there is nothing to show for it. I have to find something else to live for now' — a sentiment often repeated by most male and female farmers (Bryant 1990).

A common response for male farmers who have been displaced is for self-blame — that is, I failed and that is why we lost the farm. Farmers stated that feelings of self-blame derived from the expectations of past generations and expectations for future generations.

The expectation of the small business operator that hard work is socially rewarded is the embodiment of the ideology of Utopian capitalism (Mills 1962). Its value of individualism leads to respect for the 'self-made person' and tend to devalue those doomed to economic failure. The notion of individualism discerningly masks economic and political realities and is tied to the concept of the free market. It translates into a discourse that neatly overrides the reality that markets are shaped by political forces.

In terms of land degradation these values play a part in inducing farmers to carry out practices that will provide short-term economic rewards. These values also reinforce the notion of land degradation occurring primarily as a result of the behaviour of *individual* farmers. This, in turn, gives rise to environmental solutions couched in technocentric and ecocentric terms.

The fact that the labour of small–medium family farmers can be exploited and dominated by others — with banks often holding control and eventually realising assets of farms in financial difficulty (Davis 1980; Mooney 1983) and with transnationals having the ability to influence the setting of prices and manipulate markets (Lawrence 1989b) — is comfortably ignored.

The effects of the rural crisis move beyond the farm family to the rural community. Economic downturn and the decrease in the population base of rural towns impact upon the provision of government and private services and the ability of small business to continue to trade. The Australian Small Business Association estimated that 15 000 small rural businesses (non-farming) failed during the 1989–90 financial year (Austin 1990). Lawrence and Williams (1990) point out that when public and private expenditure decreases in rural towns, off-farm work opportunities for farm families are also reduced.

Aborigines and other farm workers also suffer economic hardship in economically difficult times as they are often the first to loose their jobs. Lawrence (1987: 51) summarises the effects of unemployment in rural areas as those which act:

> to prevent marginal farmers from obtaining off-farm work, to increase the extent of poverty amongst low income rural workers, to disadvantage particular groups such as youth and Aborigines, to exacerbate the problem of hidden unemployment amongst women and to lead to the contraction of the economic and social base of rural towns.

The flow-on effect of these circumstances results in the economic decline of small rural towns. Communities with populations of less than 2000 are likely to experience the closure of schools, hospitals, police stations, post offices and banks in the next decade (Austin 1990). When the economic infrastructure of the town dismantles so does the social fabric; that is, community identity, interaction and social activities deteriorate (see Chapter 10).

BALANCING ANTHROPOCENTRIC AND ECOCENTRIC POLICIES

In the statement on the environment delivered by the then Prime Minister Bob Hawke (1989: 4), the guiding principle underlying the national conservation strategy was seen to be as follows: 'The Australian government recognises the fundamental link between economic growth and environment. It recognises that environmental aspects are an integral part of economic decisions. It is committed to the principle of ecologically sustainable development.' This principle conflicts with the Commonwealth Rural Adjustment Scheme (RAS) and Department of Social Security (DSS) income support policies in relation to farmers. Financial assistance under RAS is available to 'non-viable' farmers in two forms: Household Support and the Re-establishment Grant. The latter is a means-tested lump sum paid to families who leave the industry and its purpose is to aid relocation. Household Support is available to farm families who are ineligible for the unemployment benefit. However, unlike the unemployment benefit it is advanced as a loan with a mortgage over land and other assets being required. Household Support has a very clear purpose: it is an interim measure for non-viable farmers to attempt sale of land. If, however, land is not sold and the family has not left farming all monies are converted to a loan repayable over seven years. As discussed earlier, most social security benefits are asset tested and can exclude farmers in financial difficulty.

The government has an expectation that families in financial difficulty will leave farming. However, farm families are not as economically rational as the government may wish. They have an ideological commitment to farming as a way of life and in practice have restricted options for an alternative occupation and lifestyle — as a consequence of age, limited educational qualifications and labour market skills. In any event, during a rural downturn farmers experience difficulty in finding buyers for their land (see Chapters 3 and 10).

The dichotomy existing in current government policy undermines any attempt to encourage sound landcare practices by farmers. On the one hand sustainable agriculture and sound environmental policies are promoted, and on the other an adequate income support system or direct economic incentives to farmers in financial difficulty are denied. Current incentives for sound land use practices are provided to those farmers who financially do not need it, like the taxation incentive which is of use only to farmers with a taxable income.

In *Our Common Future* (1988: 17) — the report of the World Commission on Environment and Development — the problem of international economic conditions impacting on agricultural practices is highlighted:

> Two conditions must be satisfied before international economic exchanges can become beneficial for all involved. The sustainability of ecosystems on which the global economy depends must be guaranteed. And the economic partners must be satisfied that the basis of exchange is equitable. For many developing countries, neither condition is met. Growth in many developing countries is being stifled by depressed commodity prices, protectionism, intolerable debt burdens, and declining flows of development finance.

Although the connection between economy and ecology is emphasised in respect of developing countries, it is clear that there are parallels between the situation of farmers in developing countries and that of farmers in Australia. Lawrence (1989a: 5) points out that Australia:

> like many third world countries is dependent on the sale of largely unprocessed agricultural and mineral products yet maintains a politico-cultural structure (including a wages system) and has a social history in keeping with that of more 'centre' nations (such as the US, UK and West Germany).

Australian farmers are reliant on exporting their produce and are subject to the dictates of the international and local economy. Although they may be aware of the problems of land degradation they are constantly pressured by short-term economic imperatives to engage in practices that are harmful to the environment (Richens 1984; Allison 1984). A Kangaroo Island farmer explains:

> I can understand [that] the government [does not] . . . care if . . . [a farmer] goes broke . . . but they are forcing farmers to mine their land rather than farm it. The lack of money to fence trees means we can't preserve the factory. Even if they don't care about the social hurt, they must care about what happens to the land (Austin 1990: 40).

A policy for landcare becomes redundant without consideration of the human relationship to the land, especially given that the conditions of the agricultural labour process which create exploitation of the land appear to have become a feature of Australian agriculture.

CONCLUSION

Land degradation does not merely occur as a result of an individual farmer's action but is tied to the economic and social organisation of society. The farm financial crisis is real for those farmers struggling to pay production costs, feed their families and service their debts. They are pressured by economic imperatives, in particular for apparent short-term gains, to undertake practices that are harmful to the environment. They also lack the funds to carry out rehabilitative works. Additionally, the strength of agrarian ideology induces farm families to do all they can to maintain their farm. These elements impinge on land use options and demonstrate that the popular solutions proposed provide an incomplete answer to the problem of land degradation. Placing land use within a social, political and economic framework does not render land degradation as an excusable phenomenon. It does, however, attempt to bridge the conceptual gap between individual, environment and society and holds promise for the evolution of public policies which will treat land degradation in a more sociologically sophisticated manner.

REFERENCES

Allison, K. (1984) 'Conditions of the Pastoral Zone in the Western Division and its Management', Paper presented to Conservation and the Economy Conference, September, Sydney.

Alston, M. (1990) 'Feminism and Farm Women', *Australian Social Work* 43 (1): 23–27.

Austin, N. (1990) 'The Stifled Heartland', *The Bulletin* 31 July: 38–44.

Bell, J. and Nalson, J. (1974) *Occupational and Residential Mobility of*

Ex-Dairy Farmers on the Northern Coast of New South Wales: A Study of Alternative Occupations, Department of Sociology, University of New England, Armidale, New South Wales.

Boylan, G. (1990) 'Social and Emotional Concerns of Farm Families in Financial Difficulty', in L. Bryant (ed.), *Service Delivery to Farm Families in Financial Difficulty*, Technical Report No. 165, Department of Agriculture, South Australia.

Bryant, L. (1989) *The Resettlement Process of Displaced Farm Families: A Study of 12 Families from Eyre Peninsula, South Australia*, Technical Paper No. 27, Department of Agriculture, South Australia.

Bryant, L. (1990) 'Farm Family Displacement', MA (Qualifying) thesis, Discipline of Sociology, The Flinders University of South Australia, Adelaide.

Bush, R. (1990) 'Rural Youth Suicide', *Rural Welfare Research Bulletin* No. 6: 25–27.

Chambers, R., Hall, N., Harris, J. and Beare, S. (1991) 'Outlook for Australian Farm Incomes', Paper presented to the National Agricultural and Resources Outlook Conference, Canberra.

Cox, G., Lowe, P. and Winter, M. (1986) 'The State and the Farmer: Perspectives on Agricultural Policy', in G. Cox, P. Lowe and M. Winter (eds), *Agriculture: People and Politics*, Allen and Unwin, London.

Cribb, J. (1985) 'Small Farmers Fight for the Right to Survive', *National Farmer* November, 47: 54–55.

Davis, J. (1980) 'Capitalist Agricultural Development and the Exploitation of the Propertied Labourer', in F. Buttel and H. Newby (eds), *The Rural Sociology of Advanced Societies: Critical Perspectives*, Croom Helm, London.

Department of Prime Minister and Cabinet (1988) *Life Has Never Been Easy*, Report of the Survey of Rural Women in Australia, Australian Government Publishing Service, Canberra.

Donnelly, L. and Smith, L. (1981) 'The Changing Role and Status of Rural Women', in R. Coward and W. Smith, Jr (eds), *The Family in Rural Society*, Westview, Boulder.

Fisher, B. (1991) 'Australian Commodities — Short and Medium Term Prospects', *Agriculture and Resources Quarterly* 3 (1).

Fisher, D. (1980) *Environmental Law in Australia*, University of Queensland Press, St Lucia.

Gasson, R., Crow, G., Errington, A., Hutson, J., Marsden, T. and Winter, M. (1988) 'The Farm as a Family Business: A Review', *Journal of Agricultural Economics* 39 (1): 1–43.

Gibson, D., Baxter, J. and Kingston, C. (1990) 'Beyond the Dichotomy: The Paid and Unpaid Work of Rural Women', in M. Alston (ed.), *Rural Women*, Key Papers Number 1, Centre for Rural Social Research, Charles Sturt University, Wagga Wagga.

Harris, J., Kirby, M. and Fisher, B. (1991) 'Markets for Australian Commodities — Overview', Paper presented to the National Agricultural and Resources Outlook Conference, Canberra.

Hawke, R. (1989) *Our Country Our Future*, Australian Government Publishing Service, Canberra.

Kirby, M. (1991) 'Commodity Overview', *Agricultural and Resources Quarterly* 3 (1).

Lawrence, G. (1987) *Capitalism and the Countryside*, Pluto Press, Sydney.

Lawrence, G. (1989a) 'The Rural Crisis Downunder: Australia's Declining Fortunes in the Global Farm Economy', in D. Goodman and M. Redclift (eds), *The International Farm Crisis*, Macmillan, London.

Lawrence, G. (1989b) 'The Impact of Change on Rural Society, Regional Communities and the Family Farm', Paper presented to the 18th Riverina Outlook Conference, Charles Sturt University, Wagga Wagga.

Lawrence, G. and Williams, C. (1990) 'The "Dynamics of Decline" ', in T. Cullen, P. Dunn and G. Lawrence (eds), *Rural Health and Welfare in Australia*, Arena, Melbourne.

McKenzie, B. (1985) 'Who Speaks for the Rural Poor?', *Inside Australia* Winter, 1 (3): 26–27.

Masson, S. (1986) 'Women's Work', *Inside Australia* Autumn: 12–19.

Mills, C. (1962) *White Collar*, Oxford University Press, London.

Mooney, P. (1983) 'Towards a Class Analysis of Midwestern Agriculture', *Rural Sociology* 48 (4): 563–584.

Newby, H. (1982) 'Rural Sociology and its Relevance to the Agricultural Economist', *Journal of Agricultural Economics* 33: 125–165.

Pepper, D. (1984) *The Roots of Modern Environmentalism*, Croom Helm, London.

Phillips, K. (1987) 'Family Farming and Property Ownership: A Research Study on Eyre Peninsula', Honours thesis, Discipline of Sociology, The Flinders University of South Australia, Adelaide.

Redclift, M. (1990) 'Economic Models and Environmental Values: A Discourse on Theory', in R. Turner (ed.), *Sustainable Environmental Management*, Westview, Boulder.

Reimer, B. (1986) 'Women as Farm Labour', *Rural Sociology* 18 (2): 143–155.

Richens, D. (1984) 'Trees on Farms, Agricultural Change and Conservation', Paper presented to the Conservation and Economy Conference, September, Sydney.

Roberts, B. (1989) 'The Implementation of Land Care Through Local Group Action', in R. Hampson (ed.), *Proceedings of the 16th National Conference of the Australian Farm Management Society — Management for Sustainable Development*, 29–31 March.

Sachs, C. (1983) *The Invisible Farmers: Women in Agricultural Production*, Rowman and Allanheld, Ottawa.

Samya-Coorey, L. (1989) 'Effects of Rural Poverty on Children', *Welfare in Australia* 14–19.

Schumacher, E. (1973) *Small is Beautiful: Economics as Though People Mattered*, Harper and Row, New York.

South Australian Environmental Protection Council (1988) *The State of the Environment*, Department of Environment and Planning, Government Printer, South Australia.

World Commission on Environment and Development (1988) *Our Common Future*, Oxford University Press, Oxford.

Yates, D. (1974) *The Sociology of Wool Production in 1971*, Department of Agricultural Economics and Business Management, University of New England, Armidale, New South Wales.

10 LAND DEGRADATION AND RURAL COMMUNITIES IN VICTORIA: EXPERIENCE AND RESPONSE

SHARMAN STONE

A number of factors affect the way in which a population interacts with, relates to, and modifies its surrounding physical environment. These factors include the population size and density, society's technology and economy, social values and patterns of behaviour. The European settlers in Australia used land and water management systems that had been developed over centuries in Europe. They also carried with them attitudes towards farming as an occupation and way of life. The imported technology and value systems have been evolving in Australia ever since.

Early farmers needed to adapt to the Australian cycles of drought, flood, fire, low prices and pestilence. To survive emotionally and financially, they developed long-term planning horizons and were prepared to work and invest for the greater benefit of their descendants. Handing on the farm to the next generation became a major aim of farm management. Farmers also came to attribute substantial non-monetary benefits to farming as a lifestyle, valuing highly their ability to lead an independent life, with maximum self-determination in running the farm. They came to believe they were making an important and essential contribution to society and were providing the very best life for their families.

The strength of these convictions, combined with the often poor employment prospects outside agriculture, account for the trauma experienced by some farming families who are forced to contemplate a life away from the farm. These factors also explain their willingness to continue to farm even when returns keep them in poverty (Blanks 1988; Bryant 1989; Ginnivan and Lees 1991).

Populations in rural Australia have not been stable over time. They have been declining since earlier this century in response to the 'weeding out' of those undercapitalised or not sufficiently skilled to survive, and in response to improving technology, communications and transport. While land and water resources provided opportunities for high productivity and while interest rates remained in single figures, most farmers overcame the short-lived disasters of flood, drought, plagues and low prices. Many were both fatalistic and conservative, and accepted and prepared for these inevitabilities.

A small amount of change did take place in rural communities with farmers increasing the size of their holdings, or with the replacement families being slowly integrated into the community. However, long-term environmental problems like soil erosion or salinisation have acted to change the traditional patterns of rural adjustment by reducing the viability of small farmers. They are forced out of farming, thus accelerating the rate of change and the nature of the farm and rural community. Furthermore, as population and prosperity continue to fall, government and other industry services are 'rationalised' or 'centralised', taking with them job opportunities and the non-local people who once worked in rural areas.

The population loss continues as the area's declining productivity and economic future becomes widely acknowledged, up to the point where there are few if any new buyers for farms or businesses. Consequently all but the most essential of government services are usually relocated. The populations of regions in decline become more homogeneous, consisting only of those who are too old to shift, those who are prepared to accept a diminishing income, and those who are locked in by an inability to sell. The community is deprived of the enervating influence of the population mix of earlier times. Gone too are opportunities for interaction, communication and psychological support. These opportunities diminish in line with the loss of social clubs and associations and service industries. The impact is compounded by the emotional and physical strain of longer hours of work with fewer helpers in enterprises where costs continue to out-pace returns by ever-increasing margins.

This chapter considers the processes which have facilitated change in rural communities and how these are affected by irreversible or significant damage to land and water resources. It concludes with an example of how government intervention, which assists in re-establishing the bases for communication and planning, can be a catalyst for community-led recovery.

RURAL SUBCULTURE AND ADAPTION TO THE ENVIRONMENT

Over the last century, Australian farming subculture has developed codes of conduct to help farmers survive, both emotionally and financially, short to medium term environmental disasters. With fire or flood, for example, the neighbours and volunteer fire brigades rush to the rescue, followed by emergency fodder, fencing materials and offers of free labour to reconstruct any damage. The farmers' response to fire and flood — so-called crisis management behaviour — has helped to establish Australian cultural myths which attribute to rural people determined self-reliance, independence, egalitarianism, mateship, conservatism and resourcefulness. By comparison, cities are considered to be polluted and noisy with urban people characterised as selfish and competitive (Kapferer 1990).

In reality, rural settlements have always been dependent on both markets and government services. The agents of wider society have played a critical role in small town development. The myth of rural community and farmer independence persists, however, and has special ideological significance for the nation (White 1981).

Nevertheless, small towns and their social organisation are still very different from what is found in urban communities, reflecting, among other things, the limited economic base, smaller population size, and less diverse ethnic and family backgrounds. The culture of country towns which grew to service farming families evolved as the pioneers banded together to build local institutions and infrastructure. Such community-based services and facilities were largely built by communal effort, organised through a hierarchy of voluntary clubs, committees and associations which, in turn, were a vehicle for people aspiring to local, state, or national political leadership.

Voluntary associations — as many as 300 in a town of 1000 people — organised the social life of the community, raised the funds and organised the communal 'working bees'. Communal work, like building, planting or harvesting, provided important opportunities for social interaction and for inspection of neighbours' machinery and farming techniques. It therefore served an educative as well as a practical function. Once amenities were built, sporting teams, churches and schools allowed opportunities for meeting newcomers and provided regular informal opportunities to discuss farm practices. Observing and talking with successful operators who faced the same environmental difficulties has long been a favourite means by which Australian farmers transferred information.

Small rural communities were proudly independent and parochial, competing with neighbouring communities for scarce government funding, access to natural resources, infrastructure and sporting supremacy. The following editorial comment was made in 1975 by one small-town newspaper editor as he whipped up enthusiasm for yet another community project, which if successful, would divert local government resources away from the neighbouring town.

> Just look around at the pool, trotting track, the amenities block and park, the High School oval, the magnificent sporting facilities etc., etc., — nearly all the result of voluntary labour. To be sure Boort has had a lot of government money to help in these projects, but it has been the government policy to help those who help themselves, and Boort has taken advantage of this (*The Boort and Quambatook Standard Times* 30 September 1975: 1).

A year later the same editor, once again, referred to the communal spirit he saw prevailing:

> Boort has a wonderful record of self-help. Nearly all the town's facilities have been provided by voluntary workers banding together in the wonderful feeling of camaraderie that exists when people are working together to create something (*The Boort and Quambatook Standard Times* 6 July 1976: 2).

Despite the egalitarian myth reflected in these comments, there has always been marked status differentiation within rural communities (see, for example, Chapter 8). Although stratification is rarely acknowledged, the quality and extent of the interaction between different social groups over the years is reflected in the level of innovation in the community and in the provision of new community facilities and services. The successful development of the rural towns depended on the quality of the interaction and integration of the two main groups of people: the vast majority of locals who were 'born and bred' in the area, and those who were not. The first category contained the farmers, some local business people and salaried workers. The other category included the often professionally trained, transitory local, state or federal government employees: teachers, police, postal, water or transport authority personnel and doctors. Only after 30 or more years of local residence would the distinction between the two groups blur.

As participants in local organisations, non-locals advanced new ideas and had different perspectives acquired from their wider life experience and often higher levels of formal education. These non-locals were the agents of wider society not only in their work but in their desire to help change the towns so they or their families could enjoy standards of living, services or facilities similar to those of the city.

Ideas introduced by the non-locals — for example preschooling or the building of a squash court — required substantial community support. The 'local' leadership would need to adopt the idea, then mobilise the community in the time-consuming rounds of fundraising, communal work, media liaison and representation to government. Such activity served a dual purpose in both the life of the community and of the individual. Opportunities were created for travel, interaction and the development of negotiating skills. The long drives to the city and countless dusk-to-midnight committee meetings also provided the opportunity for talk about local farming conditions and the state of the environment.

The interaction between locals and non-locals was critical for the community. While stimulated largely by self-interest, a better quality of life was often achieved and shared throughout the community. The interaction also served to ease the social and geographic isolation of newcomers and their families. A more fulfilling community life encouraged newcomers to stay beyond their originally allotted time. Longer stays meant more stable attendance and membership of local institutions, and more locally influential, experienced, empathetic and committed service providers.

LAND DEGRADATION AND SELECTIVE DEPOPULATION

While the productivity of the land remained substantially the same — and while debt servicing was not the problem it is today — the short-lived disasters of drought, flood, pestilence and low prices were met with the conservative philosophy of the traditional farm family which treasured a 'way of life' more than the prospect of greater financial reward.

There has always been some rural adjustment as the least viable, least skilled or undercapitalised farmers left agriculture. Improvements in transport and farm technology have also meant a steadily declining population with large regional centres growing at the expense of small country towns. These patterns of rural adjustment change, however, when land degradation occurs.

Deteriorating land and water resources require inter-farm action across catchments. This could include reafforestation, the building of community drains, vermin control or stream rehabilitation. Cooperative and coordinated interaction involving local and state governments, as well as farm and business interests in the rural communities, is necessary. Unfortunately, the need for cooperation is greatest at a time when the social, demographic and economic consequences of land degradation have lead to a breakdown in traditional communication networks. This has meant that the project-initiation, planning and implementation processes which once helped communities adapt and change in line with developments in the wider society are no longer being conducted in an adequate manner.

As land degradation begins to affect a rural community, it promotes a different view of the long-term prospects of the area and produces a changed pattern of rural settlement. Farmers who regard their enterprises more as a business than as a 'family trust' will be among the first to respond to land degradation by 'making good their escape' before land values fall. The older, more conservative farmers who are unwilling to go into debt to establish a new enterprise, and have little other prospect of employment, are more likely to continue in farming despite diminishing returns. They survive by reducing personal and farm-related expenditure in an effort to meet the costs of overdraft interest and agricultural inputs. They consider survival to be more important than expansion or innovation.

Compared with flood or fire — events which created a concerted community response — the signs of the land degradation are slow to manifest and communities are slow to respond. A comprehensive strategy on the control of salinity in northern Victoria, an area that has been affected by a high saline watertable for over 10 years, referred to the regional and historic nature of the problem (State Rivers and Water Supply Commission 1975). However, farmers considered salting to be an individual farm problem, related to poor irrigation management or farm layout (Stone 1977). Stone (1977) found that many locals feared that public discussion of the problem would result in declining local land values and would undermine their personal standing as 'good farmers'. The fear of the social stigma associated with salinity meant that the issue was not discussed, and therefore farmers were not aware of the extent to which the salinity was affecting other farmers in the district. Exchange of information may have encouraged farmers to debate the issue and develop response mechanisms. 'Problem denial' is therefore the first response of a community facing severe land degradation. Sawtell and Bottomley (1989: 54) also discuss this refusal to acknowledge the existence of a problem: 'a pump contractor stated that farmers would approach him to put in a groundwater pump, but not in the context of salinity . . . Fear of publicity destroying real estate prices is quite real. As an estate agent said, "All they're trying to do is hide the salt and sell the farm".'

In the days of production expansion, if ownership of a farm did change hands, the replacements tended to be similar to existing farmers. However, people buying into an environmentally degraded area are frequently less experienced, undercapitalised, and lured by the cheaper land prices, or, as in the case of northern Victoria, the prospect of irrigation. They may have no previous farming or irrigation experience. Some degraded properties too small to be viable but close to the town are bought as cheap housing or for their 'investment potential', and are not actively farmed. Other properties, particularly those close to urban areas, are bought as weekenders, or for people wishing to commute — those whose main income is not farming.

A lack of knowledge of, and sometimes an ideological opposition to, active control of noxious weeds and vermin and careful management of water resources result in even greater environmental degradation, which further detracts from the resale value of adjacent farms and creates disharmony in the community.

A similar adjustment process takes place in the towns servicing the declining farmland. The most astute traders and self-employed professionals note the change in community prospects and sell their business to a person with less acumen or with lower expectations of profit or expansion. The town newspaper sells out to a syndicate with regional rather than local interests, leaving the community searching its pages in vain for something more than the supermarket specials in a regional centre an hour's drive away.

As the population and community prospects spiral downwards, government authorities and banks downgrade local branches, reducing the level of management experience and staff numbers. They may only operate part time, or close altogether. With local services being reduced, the employment prospects for locals diminish. Families journey to regional centres more frequently, either for employment or for services no longer available in their home town. While in the regional centres, families shop and purchase items that could be bought in the home town, thus hastening the demise of the remaining businesses in their home town. As the population dwindles, the local institutions close and the clubs and societies run out of members. As fewer of the non-locals remain and the opportunities for them to interact with the locals diminish, the potential for new ideas, new initiatives and new projects also diminishes.

In the study of the social impact of salinity, community service providers identified the social effects of salinity as being 'likely to impact on personal health, psychological well-being, and marital and family relationships' (Sawtell and Bottomley 1989: 48). Consequently, the mental health needs of the community are greatest at a time when the social institutions that service those needs are disintegrating, and government and health services have also been reduced. With fewer non-locals to provide the social skills and political expertise to get recognition of these needs, there is a diminished government response.

Younger local farming men and women, once the backbone of the organisational life of the community — running the sporting teams, churches, service clubs, school canteens and pre-schools — seek salaried employment off the farm, often in distant country towns. With high debt to asset ratios associated with their more recent establishment and their greater need for personal expenditure, younger farmers are particularly exposed to the greater economic stress of diminishing productivity. This alone limits their ability to implement salinity controls. However, there is also a labour crisis as well as a financial crisis on their farms. Farmers must work harder to maintain output and to replace the labour they can no longer employ. The time they are away — working off the farm — is time that they cannot spend working on their own farm, or being involved in community or recreational activities. Consequently, their own farm gets neglected. The lack of social interaction, particularly with other farmers, further isolates these farmers from opportunities to discuss environmental issues and coping mechanisms (Sawtell and Bottomley 1989).

As the environment deteriorates, the traditional role of the older generation of men and women changes. Once spokespersons and stalwarts in various advisory and 'auxiliary' positions, such as local shire councillors, or people engaged in running hospitals, halls and agricultural societies, they are now less likely to have a son and daughter-in-law at home on the land or in a local business enabling them to be released to pursue these important and personally fulfilling roles.

While one rural value — the belief in farmer independence and self-determination — facilitated innovation in times of a healthy environment, when land and water resources deteriorate, the emphasis on privacy and non-interference serves further to isolate farming families. They no longer have access to the social functions and activities that provided them with opportunities for interaction away from the farm. Their unwillingness to discuss personal problems or to admit defeat means that many farmers attempt to hang onto their farms, but have insufficient capital. This furthers environmental degradation and reduces the asset value of their farms and their equity in it.

While family farmers will strenuously support one another in a short-lived and immediate crisis by lending equipment and looking after stock, each is in fact in open and direct competition with neighbours in the supply of goods to a strictly limited market. This, combined with the situation where the costs of controlling land degradation on one property might give the greatest benefit to distant farmers, means that it may be very difficult to convince farmers to alter agricultural practices. Furthermore, the belief in the neighbour's right to independent decision making and the desire to maintain harmonious relations in a small community also limit the action which can be taken to prevent degradation.

If a farmer has little contact with the community because of pressure of work on the farm, absentee ownership or off-farm work, or because community social institutions have declined, then the lack of social support and lack of wider understanding will limit changes in farm management practice.

FACILITATING FARMER INTERACTION

In the 1980s the Victorian government's 'Salinity Program' was implemented at the cost of some $20 million per annum. It was based on a belief that landholders, government departments, interest groups and local government had to be specially facilitated, organised and given incentives to work together to define an area's problems and to implement a solution or set of solutions appropriate to that area.

Some years after the Salinity Program began, areas like northern Victoria developed subregional salinity plans. These were initiated by community-led working groups resourced with full-time staff who came to live in the area, replacing the non-locals of traditional times. The community working groups were locally elected and comprised farmers, and local government and interest-group representatives. All members of these groups who were not government employees had their expenses reimbursed and received sitting fees to free them from their normal work. Specialist technical support was supplied by government departments specially funded to provide information or advice as required and requested by these working groups.

Conditions were set to counter parochialism and encourage cooperation. A salinity strategy for each designated area had to be submitted to the government within a negotiated timeframe (usually two years), and in accordance with the guidelines which identified broader community interests, as well as government-approved methods of costing project proposals.

Obtaining rural community participation, understanding and acceptance of the multifaceted and long-term salinity strategies (which typically included on-farm works and changes to irrigation management) meant that farmers had to accept the rights of others to question their individual farm decisions. They had to replace their long-held values of independence with a notion of the common good of the region, and a full recognition of the impacts of their farm practices on others.

The northern Victorian subregional working groups succeeded in changing these traditional values in part by using specially designed Local Action and Advisory Groups (LAAGs). These automatically included every landholder in an area designated to be in need of salinity mitigation. Called subregions, these areas ranged from 400 to several thousand landholdings.

LAAGs utilise the farmers' traditionally preferred face-to-face interaction in small, informal close-to-home settings. The leader is a respected local chosen by the working group. The LAAG is delineated by dividing the subregion into a patchwork of groups where every farm is included in a cluster of no more than 12 farms. By comprising immediate neighbours only, the LAAG members are conscious of their area's reputation as being environmentally damaged. They have a vested interest in the improvement of their cluster of neighbouring properties. They come to see themselves as collectively threatened by environmental degradation and assess all agricultural practices implemented in their area.

The LAAG leader's role is to act as a conduit for the flow of information. When the working group wants assistance with the identification of issues, prioritising of options or approval of a strategy, each LAAG leader personally contacts every one of his or her LAAG members. Personal communication between neighbours is therefore re-established to echo patterns more typical of the days when community networks and organisations were intact. Furthermore, this process means that every farming unit in a particular area becomes informed about the environmental investigation and planning process. It is empowered and comes to own part of the action, since its approval of the strategy is essential before the strategy can go to government for funding. Typically the LAAG process leads to farmer participation rates of over 90 percent (Stone 1987; Rural Water Commission of Victoria 1989) and has proved equally effective in flood management planning.

With widespread exposure and public recognition of an area's environmental problems, the community's solidarity and some restoration of the local communication, the opportunities for learning from the more successful farmers and the potential for cooperative action substantially increase. New opportunities and venues are established for neighbours to discuss farm practices and joint action in a non-threatening, creative and supportive environment.

The northern Victorian experience has shown that it is possible to re-create and support productive and integrative communication networks. Once they are better informed and working cooperatively, people in regional communities can develop practical, achievable and comprehensive strategies which not only provide prospects for environmental sustainability, but also enhance their quality of life.

REFERENCES

Blanks, R. (1988) 'Adjustment is About People', Paper presented to the National Agricultural Outlook Conference, Canberra, January.

Bryant, L. (1989) *The Resettlement Process of Displaced Farm Families: A Study of 12 Families from Eyre Peninsula, South Australia,* Technical Paper No. 27, Department of Agriculture, South Australia.

Ginnivan, D. and Lees, J. (1991) *'Moving On': Farm Families in Transition from Agriculture*, Rural Development Centre, University of New England, Armidale, New South Wales.

Kapferer, J. (1990) 'Rural Myths and Urban Ideologies', *Australian and New Zealand Journal of Sociology* 26 (1): 87–106.

Rural Water Commission of Victoria (1989) *The Barr Creek Project: A Review*, Rural Water Commission of Victoria, Melbourne.

State Rivers and Water Supply Commission (1975) *Salinity Control and Drainage, A Strategy for Northern Victorian Irrigation and River Murray Quality*, State Rivers and Water Supply Commission, Melbourne.

Sawtell, J. and Bottomley, J. (1989) *The Social Impact of Salinity in the Shire of Deakin and Waranga,* A Report to the Goulburn Regional Advisory Council, Urban Ministry, Melbourne.

Stone, S. (1977) 'Attitudes, Facilities, and Self Determination in the Gordon Shire', unpublished MA thesis, La Trobe University, Melbourne.

Stone, S. (1987) *New Processes for Public Participation in Water Use Planning,* Rural Water Commission, Victoria.

White, R. (1981) *Inventing Australia,* Allen and Unwin, Sydney.

11 PARTICIPATORY ENVIRONMENTAL MANAGEMENT IN NEW SOUTH WALES: POLICY AND PRACTICE

PETER MARTIN, SHANE TARR and STEWART LOCKIE

The heightened awareness and concern for environmental issues in Australia has been primarily centred in urban communities, stimulated by both local issues, such as the logging of old-growth forests, and global concerns, such as the greenhouse effect and ozone deterioration. The degradation of rural lands has occupied a relatively marginal position on the environmental agenda. More recently, however, it has gained more political and community attention. The massive extent of land degradation has been recognised for a number of years but the magnitude of problems continues to increase. Major disasters, such as extensive dryland salinity in Western Australia and irrigation salinity in Victoria, have stimulated urban and rural attention. These community concerns have prompted a political response that has seen rural environmental issues become less marginalised and more politically focused.

Recently, the NSW government has begun implementation of Total Catchment Management (TCM) and has supported the federal Landcare program. These programs attempt, using a coordinated, participatory approach, to address rural environmental issues. This chapter discusses the NSW natural resource management policy (in terms of TCM and Landcare) and attempts to identify those areas of practice that are problematic and contradictory in light of policy that espouses a participatory approach to rural environmental management. Implementation of TCM and Landcare are at early stages and the fluidity of events along with the current heterogeneity of government practice allows only preliminary generalisations to be drawn. We argue that examples of contradictory practice emerge from an ideologically distorted view of what participation means within a framework of economic rationalisation.

BACKGROUND

The importance of land degradation as a national issue has been recognised at the highest levels for over 50 years (Bradsen 1988). Throughout the 1930s and 1940s there was a high level of both political and community awareness and concern for the problem of soil conservation. It was not until 1943, however, that action was taken federally through the establishment of the Rural Reconstruction Commission. The Premiers Conference of 1946 decided that soil erosion should be tackled through state bodies and that the role of the Commonwealth should be to assist and coordinate the activities of those bodies. A Standing Committee on Soil Conservation was established for this function in 1946. Despite considerable pressure for the federal government to take a leading role in soil conservation and provide significant funding, it did not do so. By 1950, land degradation all but disappeared as a political issue, as wool prices boomed and fertilisers increased yields, only to re-emerge in the 1970s as an even more severe problem (Bradsen 1988). Activity increased during the 1970s and in 1983 the National Soil Conservation Program (NSCP) was established and the tax deductible status of land clearance costs removed.

The first state soil conservation legislation was passed in NSW in 1938 with the other states following — South Australia in 1939, Victoria in 1940, Western Australia in 1945, the ACT in 1947, Queensland in 1951 and the Northern Territory in 1969 (Bradsen 1988). Whilst providing for the existence of soil conservation agencies of one form or another, the overriding focus of these organisations was on research and the provision of extension and education services. Soil conservation in Australia has always been an essentially voluntary activity. Other forms of land degradation on the other hand, such as noxious weeds and animals, have been treated with very specific regulatory measures, although it can be argued that they are significantly less complex issues which are consequently easier to legislate.

The extension practice of the NSW state-based soil conservation agency has been dominated by an engineering approach to land degradation problems (Bradsen 1987: 114). Regulatory measures, where they existed, suffered from either a lack of will in application or extreme difficulty in policing. The seeming ineffectiveness of traditional extension and regulation opened the stage for a new institutional approach which espouses the need for rural people to take ownership, responsibility and joint action in dealing with rural environmental degradation. Landcare emerged nationally as this new approach and in NSW it is being nurtured, somewhat uneasily, under the structural wing of TCM.

LANDCARE

The genesis of Landcare occurred in Western Australia and Victoria — states with arguably the most obvious and critical areas of degraded land. In Western Australia in the early 1980s, rural people started forming groups and attempted to work collaboratively on local land degradation issues. These farmer-initiated measures were supported by the state government through legislation (an amendment of the *Soil and Land Conservation Act, 1982*) for the formation of Land Conservation Districts. By 1985, Commonwealth funds from the National Soil Conservation Program were being directed towards these groups (L. Nothrup pers. comm.). In contrast to Western Australia, Victoria's Landcare program was initiated by the state. Victoria's 'LandCare' program was officially launched in November 1986 which enabled the formation of LandCare groups and legitimised their operation. NSCP Commonwealth funding followed to support the program (L. Nothrup pers. comm.).

In the meantime, some farmer groups were appearing in NSW, and TCM was evolving as the umbrella administrative and structural framework for natural resource management. The somewhat slower development of Landcare in NSW could be due to the relative lack of obvious environmental damage in rural areas compared to other states. It seems that the situation in NSW did not provide enough stimulus for the NSW Soil Conservation Service — which has recently been subsumed within the Department for Conservation and Land Management — to overcome its inertia and traditional engineering approach to promote the Landcare initiative.

Accelerated development of Landcare in NSW occurred only after Commonwealth funding became available for the support of Landcare throughout Australia. The impetus for this funding originated through a most unlikely alliance between traditional political adversaries in the Australian Conservation Foundation (ACF) and the National Farmers Federation (NFF). The outcome of this relationship was the production of a 'National Land Management Program' (Toyne and Farley 1989) which was submitted to the federal government emphasising the need for a Landcare approach. The political power of this alliance resulted in the creation of a Community Landcare Sub-Program of the NSCP. Funding was consequently provided to the states to implement their own Landcare programs. As part of the conditions of funding, the states were required to submit plans for their Landcare activities over the next decade. The NSW state government recently released the *Decade of Landcare: Draft Plan for NSW* (NSW Landcare Working Party 1991: 7). The document articulates Landcare as: 'a program and a philosophy concerned with sustainable landuse in all environments by all sectors of community and government. It involves natural resource management in a Total Catchment Management (TCM) framework at the local level of decision making and action.'

TOTAL CATCHMENT MANAGEMENT

The location of Landcare within the coordinating framework of TCM requires some further explanation of the development of TCM as the current and predominant natural resource management policy in NSW.

In contrast to Landcare the development of TCM policy has been mostly confined to NSW government initiatives. Its basic premises and practices have had a long history overseas, as watershed management in the USA since the 1930s, and more recently in Canada and New Zealand (Mitchell and Pigram 1988). It has also existed in NSW, under various guises. Catchment management principles were institutionalised in the Hunter Valley Conservation Trust in 1950, and similar principles were evident in the charter of other organisations such as the Ministry of Conservation, which was established in 1944 with responsibility for coordinating the management of the state's soil, water and forest resources (Burton 1986; Bradsen 1988). The Hunter Valley Conservation Trust was charged with wide-ranging responsibilities for the coordination and management of natural resources within the Hunter River catchment. Other assorted organisations have arisen which have played a catchment management role such as the Murray–Darling Basin Committee and the Nepean–Hawkesbury Joint Rivers Council.

More recently in Australia, the principles of TCM have been the subject of debate around the discipline of 'integrated resource management' and further developed in the water policy arena as 'integrated catchment management' (Laut and Taplin 1988). Within the myriad of different names and perspectives, the central thrust has been an appreciation of the need for a more integrated — and coordinated — approach to the management of natural resources on a catchment basis.

In 1983, with the support of the NSW State Minister for

Agriculture, a steering committee was established by the Commissioner of the Soil Conservation Service to investigate the principles of TCM and to make recommendations about its implementation as a new government initiative (Burton 1986: 1). The result was favourable and was announced as a proposed state policy by Premier Neville Wran in a 1984 pre-election rural policy speech: 'The concept of Total Catchment Management will be comprehensively implemented in each of the major river valleys of this State, protecting the land, improving stream flow, and controlling erosion as a integrated policy.'

Following the return to office of the Wran government in 1986, TCM policy was formally adopted with bipartisan support, along with the State Trees Policy and Soils Policy. The initial steering committee was disbanded after an Interdepartmental Committee for TCM was established to oversee the implementation of TCM. The policy was outlined in *Total Catchment Management: A State Policy* (Soil Conservation Service of NSW 1987). In general, it committed the NSW government to: 'ensure the coordinated use and management of land, water, vegetation and other natural resources on a catchment basis'. TCM was to be achieved by:

- establishing better coordinating mechanisms;
- facilitating the development of TCM strategies;
- requiring consideration of TCM and its strategies in the environmental planning processes of the state;
- fostering the involvement and participation of the community.

After the adoption of TCM policy, a plethora of TCM groups appeared throughout the state (Lake Illawarra TCM Committee 1989). The composition of these groups reflected the integrating, interdepartmental focus of TCM. Most groups were made up of locally or regionally based state government departmental staff and those from local government. Only a few took the notion of community participation seriously and included community members on committees.

In 1989, following a change of government, the Catchment Management Bill was enacted. The Bill established the State Catchment Management Coordinating Committee and provided for a network of regional TCM committees and trusts to coordinate TCM at the catchment level. Regional committees consist of a majority representation of landholders or land users, plus representatives of local government, environmental interests and appropriate government departments and authorities. The major difference between committees and trusts is that trusts have the ability to raise revenue through levying and catchment contributions.

At this stage there were only two trusts in the state, the Hunter Valley Catchment Management Trust and the Upper Parramatta River Catchment Trust, both of which existed as trusts before the passing of the Catchment Management Bill. It is not envisaged at the moment that any new trusts will be formed although there is provision to do this in the legislation. The conditions set down in the legislation for the formation of new trusts state that the catchment must be suffering degradation that is adversely affecting the community, that there is a joint responsibility to deal with the degradation which is most equitably achieved through a trust, and that there must be clear support within the catchment (Catchment Management Bill 1989).

A more detailed discussion of the NSW government position on TCM (State Catchment Management Coordinating Committee 1991) was released in April along with the *Decade of Landcare: Draft Plan*. This document clearly supports the past emphasis on integration and coordination, but ascribes an increasingly important role to community participation:

> We are moving away from the era where Governments exclusively played the role of planners, experts and decision makers. Government agencies are refocusing their roles away from centralised planning and legislative controls towards coordination, facilitation and support of community participation in natural resource management. (State Catchment Management Coordinating Committee 1991: 1).

The catchcry of TCM is 'Community and Government Working Together'. Along with the implementation of Landcare and its focus on local participatory action, community participation has clearly become an espoused cornerstone of natural resource management policy in NSW.

The somewhat different histories and origins of Landcare and TCM have resulted in difficulty in defining their individual identities and the relationship between them. This is problematic for both government agencies and community members. Landcare at the federal level is seen primarily as an institutional approach to rural environmental issues such as soil conservation, revegetation, wetland and habitat conservation and vermin and weed control (Campbell 1989). Interpretation of Landcare by NSW in the draft Landcare plan is more embracing both in the range of environmental concerns (such as inclusion of urban and coastal issues, ground and surface water quality) and in the institutional focus. Although community Landcare groups remain important, considerable attention is given to government agency interests — such as research and development for Landcare — and to agency support, as well as to broader activities such as community awareness and education.

A tinge of TCM pervades the draft Landcare plan and it seems there is an uneasy courting between the two in both this document and in the Natural Resources Discussion Paper. There seems to be confusion about how to facilitate a constructive relationship between the two without losing the identity of Landcare as a farmer-based and farmer-controlled movement. TCM, as a government-imposed resource management structure (and a relatively abstract encompassing policy) is viewed with suspicion by many rural people. This confusion and lack of integration at the centralised policy level is representative of government agencies attempting to find their place within government rhetoric of participation in environmental management. At the interface of government agencies and the community within TCM and Landcare there emerges the contradictory nature of bureaucratic practice within a policy of participatory environmental management.

BUREAUCRATIC PRACTICE AND COMMUNITY PARTICIPATION POLICY

The espoused rationale for community participation in resource management relates to the ethical argument for community ownership of, and responsibility for, sustainable resource use. The instrumental argument is that the magnitude of land degradation problems is beyond the resources of government to address. However, as we will demonstrate, the policy of community participation is tightly encapsulated within a strategic instrumental view of the usefulness of rural support and involvement, within constraints determined by, and in the interests of, government.

The historical technocentric and scientistic orientation of the government agencies involved in resource management impose significant barriers to the recognition of rural people as competent partners in resource management. Whatever reasons ultimately lie behind the adoption of TCM and Landcare, unless government agencies are sensitive to the logic of the participatory process, the results of their overall actions — albeit many individually positive ones — may not yield the desired outcomes.

PLANNING FOR THE DECADE OF LANDCARE

The federal government requires a plan for the 'Decade of

Landcare' from each state. The development of the NSW draft plan was primarily carried out by the NSW Landcare Working Party, comprising staff from relevant resource agencies and representatives of other interest groups, including the NSW Farmers Association, the Nature Conservation Council and the Local Government and Shires Association. The draft plan was released for a public consultation phase of two months. The working party initially had two weeks to incorporate public concerns. During the consultation period public forums were held throughout the state to gather feedback.

The most common message fed back to the working party at these public forums was that very few people had read the document. This does not seem surprising as the 130 page document is a daunting prospect for almost any reader. The translation of participatory policy into consultative practice in the context of the decade plan is indicative of either a lack of agency commitment to the policy, or an ideologically distorted interpretation of participation. Following the initial consultation phase which ended in June 1991, there was a ministerially imposed extension until the end of July in response to public comment (D. Marsden pers. comm.). Consultants were then engaged to design a more participatory process for further development of the plan.

The distortion of the participative process did not just arise through government control of the initial planning. The consultation process following this made little attempt to 'connect' with participants, resulting in a significant distortion in the feedback received (B. Coates pers. comm.). Ironically, particular minority elements in the rural community (centred upon Dorrigo, NSW) seized on the documents as an opportunity to arouse anti-government feeling, by arguing that government natural resource policy represented a threat to their individual freedoms. This seemingly paradoxical interpretation of participatory policy could be interpreted positively, as rural people expressing their views. However, their protest is heard well above the silence of others as result of a consultative practice that alienated a great many potential participants.

The significance of the plan in future decision making at federal and NSW government level is clearly stated by the working party in the public brochure publicising the document's release (State Catchment Management Coordinating Committee 1991: 4):

> Local, State and Federal Governments will . . . refer to the plan
>
> - when making policy or legislation decisions concerning sustainable landuse.

- when assessing applications for funding for specific works or programs.
- when evaluating the success of the Landcare program.

Genuine participation in the future of natural resource management in NSW surely requires an involved process of input into a document that provides the framework for the allocation of scarce resources.

It is difficult to interpret the understanding of participation from the Draft Decade of Landcare document. Terms such as 'community and government cooperation', 'ownership', 'community participation' and 'integrated resource management' are used frequently throughout the document. All of these suggest a new way of thinking about and approaching resource management. There is, however, insufficient clarity in regard to these terms, the assumptions that underlie them, and what they mean in practice. This is understandable, to a point, since government agencies have had little experience with these concepts in theory or practice. However, it would be expected that difficulty with these ideas in practice might stimulate further inquiry or research.

What is clear from the document is the rather shallow interpretation of community participation in the context of traditional government roles such as research and development and natural resource assessment.

The section on research and development did not demonstrate carefully determined priorities but reflected the existing research culture of participating government agencies on the working party. The predominance of research into biophysical environmental issues was not congruent with the supposed direction of Landcare, given the many other problematic issues drawn out in previous chapters of the document. These problematic issues related to extension and information systems and the efficacy of current approaches to soil conservation — in terms of the predominance of structural engineering approaches over management, education and other ameliorating strategies.

Considerable emphasis was put on the development of knowledge in certain areas of the environment. The supposed direction of Landcare and TCM and the issues identified regarding their implementation suggest that research and development is also necessary in the areas of:

- how information can be integrated and used for effective decision making by community Landcare groups and TCM committees; and
- the TCM and Landcare process itself. There is considerable

uncertainty currently in the areas of coordination, decision-making structures, and appropriate relationships between institutions.

Overall, the research agenda very much reflected a technocentric bias and approach to the environment. The linking of research results to the processes of TCM and Landcare are not mentioned. There seems to be an attitude that the social processes and understandings needed for the implementation of TCM and Landcare are either obvious or will just appear with time and do not deserve the intellectual and research effort of the sort devoted to the natural sciences.

Similarly, there was no connection made between research methodology and other facets of TCM and Landcare, such as participation and coordination. Since TCM and Landcare philosophies emphasise participation and coordination, it seems that research methodology should also reflect this position. There are two ways in which this can occur; the first is through increased involvement of groups in the setting of research priorities; the second is through involvement in the research process itself. Certainly, in some areas of research, this is difficult. Research projects that include Landcare and/or TCM groups would be more effective in encouraging Landcare attitudes, behaviours and adoption of results. The involvement of rural people also has the added advantage of utilising the considerable working knowledge of local people. While integration is a key element of these programs, it seems that there is marked resistance to actually facilitating this.

The articulation of resource assessment and monitoring plans also reflects a poor understanding of the role of participation. Great reliance for the monitoring of resources and the development of resource inventories was placed on Geographic Information Systems (GISs) which — at this time in the development of sophisticated data sets — necessarily restricts the input of information from rural people. The predominance of GIS highlights a concern that little thought has been given to linking the information systems developed for TCM and Landcare with the information needs of participant decision makers. Community involvement necessitates information systems that do two things: first, provide information to community members in forms that they can understand and use; and second, take account of the store of knowledge that exists in the community. To provide these means moving away from the notion that information systems are static, or structural things, like computers, to a broader notion of information systems encompassing communication networks between people.

These centralised planning activities for Landcare and TCM call into question the extent of commitment to participation with rural people. The interpretation of participation by government agencies seems to be directed towards strategic management for their own interests, rather than to a communicative approach in which the interests of the whole community are seen as equally important and legitimate (see Martin 1991). Strategic management in this sense can be described as an attempt to 'steer complex systems' and to address the 'management of complexity and uncertainty' (Ulrich 1988). The approach accepts the notion of variation and change in social intentions, but the mediation of interests is not determined by those actually or potentially affected, for example, rural people. Rather, they are mediated by reference to criteria 'owned' by the strategic managers, such as government agencies.

TCM PRACTICE — REGIONAL NATURAL RESOURCE MANAGEMENT

As mentioned earlier, the fundamental thrust for TCM is participatory, integrated resource management. Legislation has been enacted that legitimates a hierarchical, structural framework of committees from the state level to the regional (catchment) level. Charged with the task of ensuring integration, coordination and community participation, committees are composed of a range of representatives from relevant government departments and other interest groups. A majority representation of landholders act as community representatives. It is important to note, however, that these community representatives are appointed by the Minister for Agriculture. Often they are people of high profile with their own political interests. The wider community has no real voice in the selection of any members of the various committees. This notion of participants being chosen by the government is something of a contradiction in light of the logic of the participatory process.

TCM committees and trusts are still in the initial stages of development. It is difficult to make generalisations about the implementation of TCM policy from their present activities. The committees are at present being given considerable freedom to develop their own *modus operandi* within the context of each region (S. Booth pers. comm.). This pluralistic approach by the State Catchment Management Coordinating Committee is congruent with TCM policy and would seem to be an encouraging step towards TCM obtaining a local identity in each catchment.

Much of our research experience has been with a long-established TCM trust, the Hunter Valley Catchment Management Trust (HVCT). The Trust was first established as the Hunter Valley Conservation Trust in 1950 in response to the considerable land degradation and flood damage in the Hunter Valley. With the enactment of the Catchment Management Bill in 1989 the HVCT became a regional TCM trust. The Trust has a long history in TCM-type activities, with analysis of these (in light of the new TCM thrust) providing some insights into potential threats to the participatory process in TCM.

Soil erosion is considered a significant environmental threat in the Hunter Valley (Emery 1989). Recognition of this is reflected in high current and historical levels of soil conservation activity by both the Soil Conservation Service (SCS) and the HVCT (Kaleski 1962; Stewart 1968). These activities, both now and in the past, have been oriented towards the installation of structural works (McGrath 1969).

Most of the activity of the SCS is devoted to providing advice and resources to land users concerning soil conservation. The primary solution then offered to those with erosion problems has traditionally been structural soil conservation works such as dams and contour banks. Structural works on individual farms are planned by SCS staff and carried out using SCS earthmoving equipment under their plant hire scheme at commercial rates. The SCS acts, essentially, as a decentralised service to landholders. Use of the service is completely voluntary and measures to encourage its use are arguably limited. It has from time to time targeted problem areas, generally subcatchments, for integrated works programs and heavily subsidised land users to participate. These activities have met with mixed success (Atkins and Rolfe 1989) and are being phased out (J. Butcher pers. comm.).

The HVCT contributes to the soil conservation effort through three programs, all of which provide a subsidy to land users for the construction of structural soil conservation works. Considerable moneys gathered by the trust from rate levies on the Hunter Valley community are, therefore, indirectly used to support the activities of the SCS. While there is nothing wrong with this (if it is indeed the most effective way of using the money) the effects of this strategy have not been sufficient to turn the tide against soil erosion. A common criticism is that the effects of direct subsidies for works are too diffuse on a catchment scale to offer a significant incentive to landholders not already favourably disposed to carrying out works programs. They may also encourage works solution to problems that might be better solved by other means. More erosion might be prevented in the long term if some of these resources are redirected towards more strategic programs (Lockie and Martin 1991).

We do not argue that there should be no programs of structural soil conservation works, or support for the activities of the SCS or other agencies. However, as pointed out by Mitchell and Pigram (1988), these organisations often have more to gain by identifying and defending their own areas of responsibility than they do by supporting a coordinated approach. The role of the Trust, as a catchment group, is to transcend these sectarian interests, not simply to support their activities. There is a strong possibility that this is an issue of concern for other catchment committees and trusts. A more detailed analysis of allocation of resources to soil conservation works by the HVCT can be found in Lockie and Martin (1991).

The lack of decision-making autonomy of the Trust regarding soil conservation has developed in a number of ways, including:

- the establishment of works programs that require either no prioritisation or prioritisation that is conducted outside the Trust;
- the maintenance of a structural works orientation through the above programs to the virtual exclusion of other strategies such as education and improved land management;
- the dependence on others for information and expertise by supporting spending priorities on conservation works, rather than towards more professional staff for the Trust, or other measures to improve autonomy.

Despite criticism from the outside, it should be noted that the Trust has endorsed this relatively conflict free approach, while the agencies involved have, likewise, used the Trust as a source of additional resources. The pressure exerted by these agencies when this relationship has been threatened has been quite strong. A decision by the Trust in 1988 not to contribute funds to a SCS project in the Hunter Valley because of budget restrictions attracted the attention of the Commissioner of the SCS who urged that the project be funded and the decision was subsequently reversed.

This example highlights the potential problems that can emerge from the government agency–community interface at the regional TCM level. If a policy of 'Community and Government Working Together' is to be achieved in practice, considerable effort needs to be directed towards ensuring the involvement of community members in a way that allows some level of communicative rationality to emerge (Martin 1991).

The power of government agencies through both perceptions of expertise and their control of information and knowledge resources cannot be eliminated simply through the physical inclusion of community representatives on decision-making committees.

The perception of TCM emerging from the Landcare plan consultation process indicates confusion and suspicion of the concept. There is a small and very conservative group of land users who see TCM as a government plot and as a threat to their autonomy, while many others perceive it to be a political stunt to gain community support. These perceptions reflect the difficulties in selling an approach which seems to many to be more concerned with the interests of regional or state-based agencies than with those of individuals or local groups.

The apparent state government concern for the wider spatial and temporal dimensions of environmental problems, reflected in the hierarchical framework of TCM, raises obvious difficulties when also attempting to encourage local action, responsibility and ownership for environmental care. This potentially contentious relationship is described in the Decade of Landcare Draft Plan (NSW Landcare Working Party 1991: 53): 'In order for Landcare groups to get special service or funding consideration, their problem and strategy will need to be compatible with regional and sub-catchment strategies. However, Catchment Management Committees will not control the actions of Landcare groups'.

LANDCARE: BUREAUCRACY AND RURAL PEOPLE AT THE LOCAL LEVEL

The quality of interaction between rural people and bureaucracy at the local level varies considerably throughout the state. Our experience suggests that this is highly dependent on the attitude of local extension officers, the commitment of regional bureaucracy to Landcare, and the motivations of people for forming groups. The complexity of factors contributing to the participatory process at the local level is considerable. To illustrate this complexity, the following example relates the experience of one problematic interaction between the SCS and a Landcare group in the far west of NSW.

This Landcare group was formed to acquire government funding in order to help combat their woody weed problem with blade ploughing. Most members thought that the funding was provided primarily to begin tackling the weed problem, which they also thought the government had a responsibility to help them with. The SCS, on the other hand, saw the funding as being for blade ploughing trials, the benefits being in the information generated, rather than in the actual work completed.

Most members of the Landcare group were highly cynical of government and felt that they were entitled to government assistance because of their contribution to Australia's export earnings. They thought that the amount allocated ($300 000 including SCS contribution) was quite small and that the government should provide more funding to continue the project. Their cynicism was based on resentment about the imposition of higher government charges, monetary policies that led to higher interest rates, and the winding down of government services in their locality.

Most in the group were extremely critical of the SCS, arguing that its district officers could not talk to local people, resented local people having control over government resources, and felt local people did not have the technical skills to comprehend the problem. Other agencies such as the NSW Department of Agriculture were seen to be more production oriented and in tune with farmers. This Landcare group sought to maintain production so they could stay on their farms. The SCS, by focusing on conservation measures, was seen to be acting against their material (economic) interests.

Few members of the Landcare group believed that government agencies were interested in their participation in the decision-making processes that affect their lives. There was some consensus amongst the group that state agencies find farmer involvement a nuisance, except to legitimate their own activities. There was also concern that the agencies saw themselves in the role of experts who had little to learn from farmers. Comments such as 'they want us to listen and agree but never listen to what we say before they disagree with us' were not uncommon. From the other perspective, SCS officers were concerned that farmers had narrow, short-term, interests, and that they often held views which were scientifically and technically unsound.

In the midst of these differences between the group and government agencies, the group itself did not function smoothly as a Landcare group. The group looked upon its formation as a Landcare group as a way to acquire funding for blade ploughing to combat woody weeds. Broader issues associated with bringing disparate land users together in an attempt to take ownership of their collective land management problems have not been addressed by the group. Most members are individualists who are wary of what they see as potential group interference in their private activities. There are also a number of power relationships within the group which constrain discussion and generate conflict. (This is clearly an area for future research.)

The example above is not presented as the most representative form of interaction between government agencies and rural people at the local level. This is a rather extreme case that was flawed in its foundation stage due to the markedly different expectations of the SCS and the local land users. However, the example highlights the complexity of the participatory process in practice at the local level. The basic issues of the legitimacy of rural people's knowledge and experience in the light of participation with government experts and the antagonistic and somewhat contradictory nature of rural people's perception of government intervention create a highly problematic situation for a government espousing a participatory natural resources policy.

The general issues emerging from this example are similar to, though more problematic than, those expressed by other rural people involved in Landcare groups throughout NSW. A review of Landcare in NSW was recently carried out by a colleague in the Landcare and Environment Program at the University of Western Sydney — Hawkesbury (Woodhill 1991). This study included interviews with 80 farmers and 35 extension officers involved with 25 Landcare groups throughout NSW. Woodhill (1991) found considerable variations in perspectives on Landcare from the respondents but identified areas of concern which relate to the interface between government and rural people.

Woodhill (1991: 13) states that a 'significant number of landholders felt their experience and knowledge was not valued and too often they were just told what to do'. Various quotes from land users clearly articulate the problem that exists for them:

> The SCS won't listen to farmers' ideas about earthworks and water flow; we have experience of soils for generations.
>
> I hate this term 'educate farmers'.
>
> They [extension officers] have expertise we haven't got but we've got expertise they can't get from a textbook. They come out here and just don't take in what we have to say.

At the nexus of the current relationship between government agencies and local community members is the long history of expert–layperson interaction that has dominated both the theory and practice of extension in NSW (Russell *et al.* 1989). This approach is based on the assumption that knowledge, produced by researchers, is transferred to farmers via extension officers. Rural land users were seen as recipients of this knowledge and, in fact, barriers to its adoption (Rogers 1983). Research into this process itself, on the other hand, shows that landholders see extension services as only one of many sources of information available to them (Anderson 1981), which explains in part why the traditional research/extension/adoption equation has been ineffective in a practical sense. Extension services have been used by a small proportion of 'progressive' farmers but have had little contact or direct affect on the wider community of rural land users (Russell *et al.* 1989).

With the introduction of Landcare, the expectations of rural land users are changing, and demands to be listened to are being expressed. The role of the government expert is being challenged both by rural land users and by voices within government agencies. Government extension officers are now required to change their extension practice from individual servicing to group facilitation and to collaborate with land users in improving their land degradation problems. The dynamics of this shift are manifested in the diversity of perspectives and understandings expressed by extension officers (Woodhill 1991: 14):

> There are two philosophies, the old and the new, they are like chalk and cheese, change doesn't happen over night (District Soil Conservationist).
>
> Landcare is not a challenge, we've always done it, the same elements apply, group extension should not take priority over individual servicing and other extension techniques (SCS Regional Manager).
>
> We have a very *ad hoc* and reactive approach to extension. There needs to be more focus on developing the relationship between extension staff and landholders. There is a whole new grab-bag of skills needed, at the moment many staff are not keen to develop them (SCS Property Planner).
>
> The idea of old extension and new extension is garbage, good extension officers have always operated in the so-called new way (District Soil Conservationist).

To suggest that the problematic interaction in local Landcare groups is only a function of the perspectives of government officers on Landcare and extension would be to oversimplify the situation. As the example of the far west NSW Landcare group discussed above demonstrates, the diversity of perspectives and the somewhat contradictory perception of the state by rural people can create and fuel conflicts at the local level.

Action to address the problems occurring in agency–community interactions has taken the forms of training programs for government extension officers (Woodhill *et al.* 1990) and a review of Landcare throughout the state (Woodhill 1991). The interests of the state government and its agencies in gaining support from rural people for their programs will not be served by alienating practices in extension. Furthermore, federal funding for the state Landcare program is dependent on effective implementation at the local level.

ECONOMIC RATIONALISATION, TCM AND LANDCARE

Since 1988, the Liberal–National government in NSW has sought to rationalise its activities in rural communities. The state government has closed schools, courthouses, railway stations and hospitals. It has also significantly reduced welfare, police and agricultural services in most rural regions (Lawrence and Williams 1990). Yet while the state government is seeking to rationalise its activities, it also appears committed to Landcare in NSW.

When we examine more carefully the logic of rationalisation, it can be argued that these state government policies may not be as contradictory as they first appear. The rationalisation of government activities has been accompanied by the idea that the user-pays principle should be applied to the provision of government services. By encouraging the development of Landcare, the state government is able to get those involved in such groups to take responsibility for their own measures to alleviate land degradation. The implementation of Landcare can be seen to reinforce the ideology of economic rationalisation and contribute to the legitimation of the government's rationalisation measures.

The situation in NSW, indeed throughout Australia, is quite different from that which prevails in both Western Europe and North America, where, as Blaikie and Brookfield (1987: 231) point out, state intervention has been directed to propping up farm incomes to retain viable rural economies and populations. The mass of subsidies and supports within the Common Agricultural Policy of the European Community (EC), to take but the most obvious example, represents an attempt to preserve rural communities. It can also be stated that in most EC member nations, rural lobby groups are politically more powerful than in Australia. The forms of intervention practised abroad, nevertheless, open up opportunities for conservation measures to be imposed upon farmers to improve their land management. Intervention undertaken abroad as a means of supporting farmers and improving their resource use has a direct negative effect on the incomes of Australian producers. In effect, it reduces Australian producers' ability to pay for conservation measures (Lawrence 1987; and see Chapter 3).

The NFF has sought to prevent EC-style regulatory powers being introduced to the Australian agricultural environment. Rick Farley, Executive Director of the NFF at the time of its 1989 alliance with the ACF, made it quite clear that, based on his overseas experiences, unless farmer organisations involved themselves with mainstream environmental groups the latter would be able to out-manoeuvre them politically on environmental matters (R. Farley pers. comm.). This could, of course, lead to the sort of regulation of farming practices the NFF sees as undesirable.

In its alliance with the ACF, the NFF has sought to emphasise the voluntary nature of Landcare groups. This is consistent with its opposition to greater levels of government intervention and with its support for continued industry restructuring, the lowering of government charges, relaxation of tight monetary policy, taxation incentives and the liberalisation of farm trade (Eliason 1991). The NFF's views on economic rationalisation, in the context of environmental matters, are not wholly user-pays oriented. The NFF advocates the use of tax incentives to promote Landcare, direct grants to assist with farm planning and collective purchase of specialised conservation equipment, and increased funding for Landcare groups (Eliason 1991).

Since its return to office in May 1991, the NSW Liberal–National government has begun an even greater rationalisation of services. Already it has undertaken measures to reduce the size of the public service, reduce expenditure so as to prevent a budgetary blow-out, and proceed with industrial reform (*The Sydney Morning Herald* 3 July 1991). Support for the continued rationalisation of state services is still provided by the NFF. The current President of the NFF, Graham Blight, has reiterated his support for such initiatives. Writing in *The Land* (30 May 1991), Blight argued that although he opposes most forms of state intervention, he recognised that federal government monetary policy and state government charges have impacted negatively on rural producers' incomes. The NFF supports rationalisation of government services but it does not want rural producers to be seen as wholly responsible for environmental problems associated with land management. Landcare has the support of the NFF on the condition that support is forthcoming from the non-rural sector. Increased support in the form of direct expenditure is unlikely in the foreseeable future as the user-pays principle becomes increasingly entrenched in NSW state politics.

Although the interests of rationalisation and local participatory environmental care might partly coincide, the effect of rationalisation on the implementation of regional environmental management (in the form of TCM committees) is more problematic. TCM committees are charged with coordinating, in a participatory manner, natural resource management throughout the catchment. This is expected despite a meagre allocation of resources to these groups and results in their dependence on state agencies for information and advice at a time when agencies are themselves under substantial budgetary pressure. It seems contradictory that the coordinating agency is required to obtain highly interpreted information from others, and is still expected to maintain the autonomy needed to make decisions regarding the allocation of resources. This major policy initiative has been provided with a shoestring budget. The state government wishes to shift responsibility for environmental management towards rural people at the regional level but seems unwilling to supply the resources needed.

While the NSW government is intent on rationalisation measures, its agencies are attempting to maintain or improve their positions by securing a substantial proportion of the federal funds for Landcare. The draft decade plan expresses a desire for resources to support Landcare, government research, resource inventory development and improved extension services. Integration of Landcare into the TCM framework appears to have been seized on by NSW government agencies as a way of improving federal funding for what is a distinct NSW government initiative. The hope is to maintain their activities despite state government rationalisation of their resources.

CONCLUSION

The NSW state government has clearly committed itself to a policy position of participatory natural resource management in rural environments. This position needs to be supported by a redistribution of power, an adequate allocation of resources and a redefinition of the roles of government and rural people within a framework of participatory management.

To assess government commitment to its own policy is difficult at this stage. This chapter has identified numerous areas where practice and implementation seem contradictory to espoused policy. Nevertheless, within the relevant agencies, there are many individuals who are committed to the ideals of participation and effective environmental care. Further, their commitment is paying off, with substantial progress in community involvement in many areas (such as successful Landcare and TCM groups). In his review of Landcare in NSW Woodhill (1991) received considerable positive comment from rural people regarding the benefits of Landcare, especially in regard to learning and awareness of their environment. Rural people are also utilising the collective institution in Landcare to express more forcefully their views to government officers.

Whatever the varied progress at the local level, development is impeded by significant aspects of existing government practice. Although many in the government agencies subscribe to the participatory view, this conception seems to be limited to ideological constraints imposed by a scientistic, technocentric view of environmental management. All the policy and planning documents endorse participation but have little detail about how this might be implemented except through structural changes that involve rural people in decision-making committees. Government research priorities and methodologies have not changed and environmental information is still seen as biophysical descriptors that can be easily digitised. The results of this practice are threefold: participation is interpreted in practice as consultation; only a minor reallocation of government resources has been made for participatory activities; and the dominance of expert knowledge over the experience of rural land users has been maintained.

The common catchcry in government circles is community education — where community is seen as those others outside government agencies. This one-way street reflects past agricultural extension methods and is an exemplar of a hierarchical view of environmental management.

This distortion of the notion of participation is, we argue, also a consequence of its location within the ideological framework of economic rationalism. The user-pays principle is incorporated into the participatory policy of local responsibility and action with little consideration of the considerable economic bind being faced by rural land users. A strong message from rural people by the recent Landcare review (Woodhill 1991) and Landcare Plan consultation process is for increased government funding for conservation measures during the current rural crisis. Although it is acknowledged that rural people should be encouraged to participate and contribute to conservation measures, the government has, in the past, encouraged a productionist, exploitative approach to agriculture: it now has a responsibility to contribute significantly to rehabilitation and the prevention of further degradation of the rural environment.

Whatever the motives and interests of the government in policy implementation and practice, community members are taking the opportunity to express their views in a more collective manner. With the increased opportunity to be involved and to be heard, TCM and Landcare are initiating and providing a platform for the local voice. In some instances, government and its agencies are responding. From our experience, Landcare groups provide the opportunity for people to build confidence and to be more assertive in their demands and criticisms of government. It is possible that the distorted participatory process initiated by the state government might result in a more demanding — and politically cohesive — rural lobby in NSW.

REFERENCES

Anderson, A. (1981) *Farmers' Expectation and Use of Agricultural Extension Services*, Department of Communication, School of Management and Human Development, Hawkesbury Agricultural College, Richmond

Atkins, D. and Rolfe, J. (1989) *Soil Conservation in the Merriwa District,* Final Year Project Report, Faculty of Agriculture and Rural Development, University of Western Sydney — Hawkesbury, Richmond.

Blaikie, P. and Brookfield, H. (1987) *Land Degradation and Society*, Methuen, London.

Bradsen, J. (1987) 'Land Degradation: Current and Proposed Controls', *Environment and Planning Law Journal* June: 113–133.

Bradsen, J. (1988) *Soil Conservation Legislation*, University of Adelaide, Adelaide.

Burton, J. (1986) 'The Total Catchment Management Concept and its Application in NSW', in *Proceedings of Hydrology and Water Resources Symposium*, Griffith University, Brisbane.

Campbell, A. (1989) 'Landcare in Australia — An Overview', *Australian Journal of Soil and Water Conservation* 2 (4): 18–20.

Eliason, P. (1991) *National Focus: Agriculture, the Environment and Economics*, Volume 3, National Farmers Federation, Canberra.

Emery, K. (1989) *The Hunter Valley Erosion Survey — 1983–85: Program Proposals for the Integrated Management of Soil Erosion and Related Land Degradation Issues within the Hunter River Catchment,* Summary Document for the Hunter Valley Erosion Survey, Soil Conservation Service of NSW.

Kaleski, L. (1962) 'Erosion and Soil Conservation in the Hunter Valley', *Journal of the Soil Conservation Service of NSW* 18: 2–9.

Lake Illawarra TCM Committee. (1989) *Proceedings of the Workshop TCM in NSW — Is it Working?*, The University of Wollongong, July.

Laut, P. and Taplin, B. (1988) 'Catchment Management in Australia in the 1980's', in *Proceedings of the National Workshop on Integrated Catchment Management*, Australian Water Resources Council, Melbourne.

Lawrence, G. (1987) *Capitalism and the Countryside*, Pluto Press, Sydney.

Lawrence, G. and Williams, C. (1990) 'The "Dynamics of Decline" ', in T. Cullen, P. Dunn and G. Lawrence (eds), *Rural Health and Welfare in Australia*, Arena, Melbourne.

Lockie, S. and Martin, P. (1991) *Soil Erosion in the Hunter Valley*, Landcare and Environment Program, Centre for Extension and Rural Development, University of Western Sydney — Hawkesbury, Richmond.

McGrath, J. (1969) 'Soil Conservation Activities in the Hunter Soil Conservation District', *Journal of the Soil Conservation Service of NSW* 9: 85–91.

Martin, P. (1991) 'Environmental Care in Agricultural Catchments: Towards the Communicative Catchment', *Environmental Management* 15 (6).

Martin, P. and Lockie, S. (1991a) *Dryland Salinity in the Hunter Valley*, Landcare and Environment Program, Centre for Extension and Rural Development, University of Western Sydney — Hawkesbury, Richmond.

Martin, P. and Lockie, S. (1991b) *A Review of Environmental Information for Total Catchment Management in the Hunter Valley*, Landcare and Environment Program, Centre for Extension and Rural Development, University of Western Sydney — Hawkesbury, Richmond.

Mitchell, B. and Pigram, J. (1988) *Integrated Catchment Management and the Hunter Valley, NSW*, Occasional Paper No. 2, University of New England, Armidale, New South Wales.

NSW Landcare Working Party (1991) *Decade of Landcare: Draft Plan for NSW*, State Catchment Management Coordinating Committee, Sydney.

Rogers, E. (1983) *Diffusion of Innovations*, Free Press, Glencoe.

Russell, D., Ison, R., Gamble, D. and Williams, R. (1989) *A Critical Review of Extension Theory and Practice*, Faculty of Agriculture and Rural Development, University of Western Sydney — Hawkesbury, Richmond, and Faculty of Agriculture, University of Sydney.

Soil Conservation Service of NSW (1987) *Total Catchment Management: A State Policy*, Soil Conservation Service of NSW, Sydney.

State Catchment Management Coordinating Committee (1991) *TCM Discussion Paper on Natural Resources Management*, State Catchment Management Coordinating Committee, Sydney.

Stewart, J. (1968) 'Erosion Survey of New South Wales Eastern and Central Divisions Re-Assessment — 1987', *Soil Conservation Journal* 24: 139–154.

Toyne, P. and Farley, R. (1989) *A National Land Management Program*, A Joint Submission by the Australian Conservation Foundation and the National Farmers Federation to the Federal Government of Australia, Canberra.

Ulrich, W. (1988) 'Systems Thinking, Systems Practice, and Practical Philosophy', *Systems Practice* 1: 137–163.

Woodhill, J. (1991) *Landcare — Who Cares? Current Issues and Future Directions for Landcare in NSW*, Landcare and Environment Program, Centre for Extension and Rural Development, University of Western Sydney — Hawkesbury, Richmond.

Woodhill, J., Wilson, A., Wright, D. and McKenzie, J. (1990) *TCM and Landcare Co-ordinators Workshop: Program Overview*, Landcare and Environment Program, University of Western Sydney — Hawkesbury, Richmond.

12 SUSTAINABLE AGRICULTURE: PROBLEMS, PROSPECTS AND POLICIES

IAN REEVE

In the 200 years since European settlement, agriculture in Australia has see-sawed between prosperity and adversity. While periods of adversity have always precipitated both social hardship and environmental degradation, the latter has become the focus of increased policy effort in recent decades. The notion of a sustainable agriculture — one that maintains or enhances the quality of the land and water resources upon which it depends — is gaining widespread acceptance. However, there remains disagreement as to whether a sustainable agriculture can be accommodated in industrialised countries without fundamental changes to the social and economic fabric of those countries. The proponents of the alternative agricultures (for example, organic, biodynamic, ecological varieties) have to varying degrees long taken the stance that such changes are necessary. While the alternative agricultures have become less marginalised in the agricultural policy process in recent times, the mainstream view still holds that the farm sector, rather than the input sector (agrichemical companies) or the output sector (food processors and marketers) should be the locus of policies seeking improved sustainability in agriculture.

This chapter examines those interactions between the farm, input and output sectors which have tended to drive modern agriculture towards unsustainability.

AGRICULTURE AND THE INPUT SECTOR

In the last 100 years there has been a marked increase in the proportion of agricultural inputs derived off-farm. Advances in the understanding of plant nutrition in the nineteenth century led to the view that organic amendments such as animal manure were not essential to maintain productivity (productivity being defined in the narrow sense of value of production relative to cost of inputs). Instead, soluble manufactured or natural sources of nitrogen, phosphorus and potassium and trace elements were all that were required. The relative ease of mining, transport and application and the crop responses obtained meant that economically viable operations such as the Chilean nitrate trade could be established. This was one of the first steps in the replacement of farm-derived inputs with purchased off-farm inputs. The development of the internal combustion engine and subsequent farm mechanisation, together with high labour costs, resulted in the replacement of farm-fed horses with oil-fed tractors. Further, the extension of the techniques of economics to agriculture generally provided justification for more intensive use in agriculture of off-farm inputs.

The end result is that agriculture has shifted from being a system that used the culturally transmitted wisdom of the agrarian class to transform solar energy into agricultural output to a system dependent on the techno-industrial complex that consumes and dissipates materials and fossil fuel energy and, in the process, transforms a small part into agricultural output. In the USA, for example, agricultural chemical inputs increased thirteenfold and mechanical power and machinery inputs tripled between 1940 and 1980 (Wallace 1987). The trend of increasing input intensiveness is not necessarily a bad thing in itself (except where the rate of dissipation of materials and energy into the environment exceeds its assimilative capacity or where the displacement of farm labour has serious social costs). The serious implication is that many agricultural practices that could ensure sustainability are unlikely to be developed or adopted in the commercial environment of an input-intensive agriculture.

Agricultural chemicals have played an important role in both the increasing input intensiveness and the deleterious environmental impacts of agriculture. For a firm to be able to develop successfully a new agricultural chemical it has to, among other things, ensure that the sales volume over the life of the product will cover the costs of research, development and production. Since the new product will have to compete against other products in the market, the firm will need to ensure that its product does not have any characteristics that would reduce sales. One such undesirable characteristic is complexity and difficulty in use (Vereijken 1989). A product that requires the user to learn new skills, think in different ways or suffer inconvenience will not compete well against one that does the same job without these demands. Furthermore, as Berry (1987) has pointed out, convenience goes hand in hand with simplified representations of the problem. A good example is the control of aphids in faba beans. A number of studies have shown that the luxury consumption of nitrogen by faba beans that occurs with applications of soluble nitrogenous fertilisers results in raised levels of free amino acids and increased nutritional value of phloem fluids to aphids. Consequently, faba beans fertilised with soluble nitrogenous fertilisers are susceptible to heavier infestations of aphids than faba beans that have been prevented in some way from luxury uptake of nitrogen (Kennedy 1958; Pollard 1973; Patriquin *et al.* 1988). Such infestation can be prevented by reducing levels of application of nitrogenous fertilisers, by using appropriate organic sources of nitrogen, by intercropping with cereals, or by allowing limited weed competition. These techniques keep the levels and timing of aphid infestation below the thresholds at which economically significant yield reduction occurs. The usual practice, however, is to use pesticides to control the levels of infestation. There would be absolutely no point in any marketing campaign for an insecticide to draw attention to the complex interactions between luxury uptake of nitrogen, increase in phloem nitrogen levels at time of flowering as leaf nitrogen is remobilised and exported to reproductive organs, attractiveness to aphis, aphid nutrition and reproduction. Rather the market penetration required for commercial viability will depend upon appealing to farmers' fears of economic loss or their preference for reactive solutions ('You got a problem with aphis? Hit 'em with Aphkill before they run you out of business').

Another aspect is whether or not the firm can ensure that the product is not copied by its competitors. The patents system and trade secrets legislation gives the firm protection against this. However, the system is such that it tends to bias innovation towards synthetic or manufactured substances and away from substances that already exist in the environment, including approaches that involve manipulation of the farm system with no additional inputs rather than those fostering high input use. This tendency even extends to the producers of biological inputs such as seeds. The increasing use of hybrids is only partly due to their having some superior agronomic characteristics. More important to the profitability of the producers of hybrid seeds is the fact that purchasers or competitors are unable to produce the same variety without access to the parent varieties which can be maintained as a trade secret (Wallace 1987).

As a result of these factors, the agricultural chemical industry has little option but to produce synthetic substances for which property rights are clearly definable and enforceable, and which are convenient for the user. The study by MacIntyre (1987) showed that despite the principles of biological control being well known since at least the 1930s and the problems of pest resistance being well understood by the late 1950s, the industry in the USA continued to develop and market chemicals with short spans of usefulness that were overused as pest resistance built up.

The agrichemical industry is an example of what Brooks (1986) has termed 'technological monocultures'. Technological monocultures have their beginnings when, for reasons that can be entirely fortuitous, one technology that performs a particular function gains a competitive edge on other technologies that could perform the same function. Once this technology has the greater share of production volume and infrastructure develops that is dependent on this technology, it is difficult for other technologies that may subsequently be seen as preferable to become established. Brooks identifies a number of undesirable consequences of technological monocultures:

- As the dominant technology matures, research effort is increasingly devoted to marginal refinement at the expense of imaginative innovation.
- This narrowing of research effort often occurs just at the time that the higher-order social or environmental costs of the widespread application of the technology become apparent. Solutions to these problems require preventative innovation rather than curative marginal refinements.
- Technological monocultures may become vulnerable to unexpected side-effects or environmental feedback at critical levels of application. The reaction of those with a stake in the technology is often to attempt to modify the environment in which the technology is applied to reduce its vulnerability rather than devote resources to the development of superior technologies.

The approaches to pest control developed by the agrichemical industry also initiate reactions in soil and crop ecosystems that are perceived as further problems in need of agrichemical solutions. For example, Edwards (1987), in a summary of technical research outcomes, has documented examples of how fertiliser applications can increase disease incidence and susceptibility of plants to insects; the removal, with herbicides, of weed species can increase the incidence of pest and disease attack on crops; insecticides can reduce populations of non-target insect species which feed on weeds and so result in weed growth; fungicides can kill beneficial soil fungi that control populations of insects and nematodes; and, finally, fungicides can reduce the populations of beneficial soil organisms.

In summary, it can be seen that the combination of normal commercial practice in the agrichemical industry and the biological reactions of crop and soil ecosystems predisposes modern agriculture to input intensiveness. There are, however, also interactions in the output sector that can threaten agricultural sustainability.

AGRICULTURE AND THE OUTPUT SECTOR

The interactions between agriculture and the output sector occur along the chain from farm gate to processor to distributor to consumer. The nub of the problem is that the nature of the food system has, for some time, made it very difficult for the consumer to make any connection between food products and the agriculture that produced them. The nature of this difficulty has been recently demonstrated with respect to 'environmentally friendly' consumption.

The wave of 'green consumerism' that appeared in 1988 and 1989 was greeted with enthusiasm by manufacturers, marketers and consumers alike. Here, it seemed, was a phenomenon that would enable consumer preferences expressed in the marketplace to make industry less polluting and less environmentally damaging — a phenomenon to warm the heart of the neo-classical economist. The less wary made extravagant claims of the probable impact of green consumerism.

However, by mid-1990, market surveys began to show that the proportions of consumers purchasing, or prepared to pay higher prices for, 'green' products was declining (Pearce 1990; Cooke 1990). Although no detailed studies appear to have been undertaken as yet, a number of reasons for the decline have been advanced. There is evidence in both Australia and the UK that a number of firms responded to consumer demand by merely changing the image conveyed by packaging and labelling, while making no change to the product itself. While this course of action is obviously a cost-effective means of obtaining short-term increases in market share, it has resulted in growing cynicism and customer resistance among more discerning consumers. Consumer confusion has also been advanced as a possible cause of the decline of consumer interest in 'green' products. Consumers do not generally have access to an objective assessment of the totality of the environmental impacts of the resource extraction, energy consumption and waste generation associated with the manufacture, use and discarding of the product. Marketers seeking to increase market share are likely to seize upon only one aspect of 'environmental friendliness'. For example, cotton might be promoted as a 'natural' fibre despite the pesticide intensiveness of its production; or CFC-free refrigerators might be promoted as 'ozone friendly', but because of poor insulation and high electrical consumption be 'greenhouse unfriendly'.

These problems have led critics to suggest that consumer goods should carry some sort of environmental friendliness rating. Desirable as this might be in terms of influencing consumer decision making, such a rating system would be extremely difficult to implement because of the uncertainties inherent in predicting environmental impact. Added to this are the difficulties inherent in complex product chains where a large number of firms contribute to the final product, each firm in the chain being free to change the sources of its inputs according to the dictates of the market. Even if the environmental friendliness rating were technically measurable, it would be constantly changing for a single product from batch to batch and with time.

However, it is not only complex marketing chains and the consumers' inability to assess the environmental impact of a product that has been responsible for the lack of environmentally favourable price signals at the farm gate. Some product attributes which can be assessed by the consumer seem to be responsible for price signals which do reach the farm gate. These signals, however, have caused particular farm sectors to become more environmentally antagonistic.

Take, for example, the issue of visual appearance of unpackaged fresh fruit and vegetables. Pest damaged, undersized and malformed fruit and vegetables originate on the farm and are visually apparent in the unprocessed product at every stage from the farm gate to the supermarket shelf. For most consumers, visual appeal of fresh fruit and vegetables is the only criterion upon which their decision to purchase can be based. Criteria such as presence or absence of pesticide residues or nutritional quality are not available at the supermarket shelf. As it is technologically impossible, at any stage along the marketing chain, to transform damaged or malformed fruit or vegetables into a visually appealing fresh unprocessed product, price signals have penetrated to the farm level and agricultural practices have changed as a result.

The advent of a new generation of pesticides after World War II provided producers with a technology that was easy to use and could guarantee the visual appeal demanded by consumers. Low prices for blemished produce were a powerful incentive to adoption and to overuse when pest resistance developed. The irony of the situation — and the important principle demonstrated — is that the combination of consumers behaving in what they perceive to be their best interests and a marketing chain that readily transmits price signals has resulted in an outcome that may not be in the long-term best interests of either consumers or farmers.

In summary, it is clear that the roots of modern agriculture's environmental crisis do not lie solely within agriculture. Much of the problem stems from the industrial societies within which agriculture functions. Given this, it is not surprising that the concepts of sustainable agriculture that are now emerging are also shaped by social norms and values.

SUSTAINABLE AGRICULTURE: THE EMERGING CONCEPTS

The concept of sustainability is the expression of a deep and understandable desire on the part of human beings for their species to continue to flourish with an acceptable quality of life beyond this generation.

On the best present assessment, future generations will require, as we do, intact and functioning biological and atmospheric systems that provide clean air, water, productive soils, protection from ultraviolet radiation, maintenance of the temperature range essential for life, access to materials and energy and the knowledge to transform these into goods and services to meet basic and perceived needs. It is becoming increasingly obvious (and has been so to some for at least several decades) that no bequest of wealth — be it the optimist's techno-industrial cornucopia or, be it more realistically, a mixture of useful knowledge and environmental bad debts, inescapable obligations and Faustian bargains — will be a fair substitute for functioning planetary life support systems.

If our desire to treat future generations as we would like to have been treated by previous generations is to be achieved, then a start has to be made in moving our treatment of the planet from a mode that is now generally accepted as not being sustainable to one that is. Our state of knowledge is such that it is not possible either to specify a comprehensive and detailed blueprint of how the human species can guarantee the use of the planet in perpetuity, or to specify a feasible path to this desirable end. Indeed, there may be many different social and economic systems that would secure the future of the human species, and many different paths that could be taken from where we are now. No person working from any disciplinary basis, intuition or ideological dogma can prescribe a course of action that is certain to lead to a social, economic or agricultural system that will endure for all time. Sustainability is the goal that the prescriber hopes may be achieved. Only time will tell if it is the outcome.

Because knowledge and disciplinary tools are less than adequate, any prescription for a sustainable agriculture inevitably contains assumptions that reflect the ideologies and values held by the prescriber. These seem to fall into two groupings. A number of prescriptions or concepts of sustainable agriculture such as low-input sustainable agriculture in the USA and conservation farming in Australia are founded on the ideologies of conventional agriculture and the modern industrialised society. Batie and Taylor (1989) suggested the following precepts:

- increased production is good;
- expanding foreign markets are to be pursued;
- materialism and profit orientations are appropriate;
- *homo sapiens* should dominate nature; and
- science and technology are linked to progress.

Other prescriptions or concepts of sustainable agriculture, such as the alternative agricultures, are founded on various precepts, some novel, some that are simply the obverse of those above and some that were more widely held in the past. Batie and Taylor (1989) list these as:

- family and group self-reliance;
- harmony with nature;
- a global village view of the world;
- a respect of nature and ecosystems;
- a voluntary simplicity; and
- a belief in the 'goodness' of the family farm.

It can be seen from these descriptions that, depending upon one's beliefs, sustainable agriculture can be judged to be either possible or not possible within the existing status quo.

Some parts of the agriculture of modern industrialised nations would appear to be able to be sustained for long periods — some conservatively stocked non-arid rangelands, for example. Other parts are obviously not sustainable as they depend upon the availability of abundant and cheap fossil fuels and their derivatives. The agriculture of some developing countries is clearly not sustainable as it irreversibly depletes nutrients and loses topsoil. On the other hand, subsistence agriculture on some of the major river deltas and in volcanically active regions has survived for some thousands of years and appears to have every chance of continuing to do so provided the geological processes supplying nutrients continue (Fyfe *et al.* 1983). Some of the alternative forms of agriculture, such as organic, biological or ecological agriculture, may be able to be sustained over long periods, in so far as they attempt partly to emulate natural ecosystems which have proven their desirability and resilience. However, if these forms of agriculture

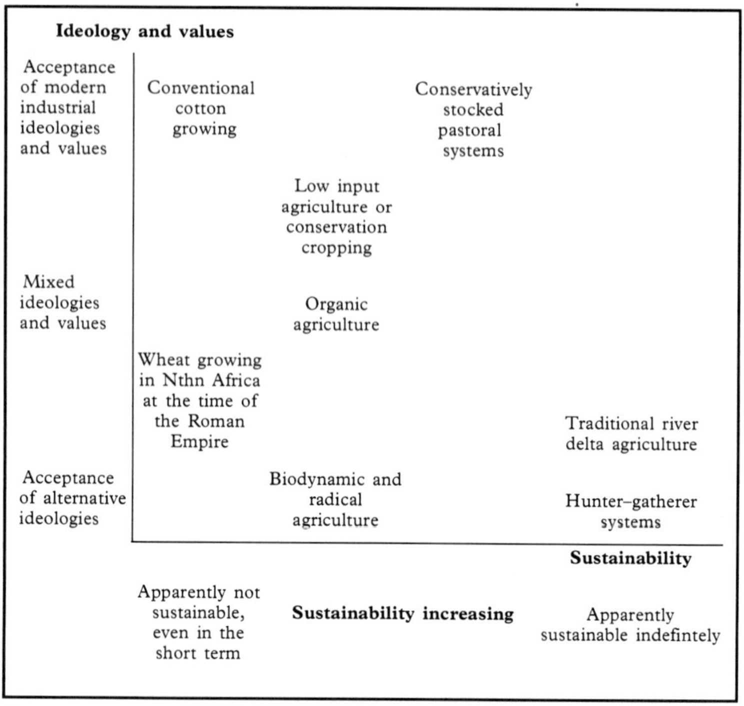

Ideology and values

Acceptance of modern industrial ideologies and values

Conventional cotton growing

Conservatively stocked pastoral systems

Low input agriculture or conservation cropping

Mixed ideologies and values

Organic agriculture

Wheat growing in Nthn Africa at the time of the Roman Empire

Traditional river delta agriculture

Acceptance of alternative ideologies

Biodynamic and radical agriculture

Hunter–gatherer systems

Sustainability

Apparently not sustainable, even in the short term

Sustainability increasing

Apparently sustainable indefintely

Figure 12.1 A mapping of agricultures on the dimensions of sustainability and ideology

are net exporters of nutrients or users of fossil fuel then they are clearly not sustainable indefinitely.

The examples above show that all forms of modern industrialised agriculture are not necessarily non-sustainable. Likewise, the fact that a system of agriculture is different from the mainstream or is biologically or ecologically oriented is insufficient to guarantee sustainability. A particular ideological basis does not guarantee either sustainability or non-sustainability. The ideological basis and apparent sustainability are, however, two useful dimensions that can be used to plot the various forms of agriculture discussed above. This is shown in Figure 12.1.

Different definitions of sustainable agriculture reflect various ideological bases and the multitude of conditions that may be conducive to achieving the intended sustainability. Owing to the multitude of necessary conditions for sustainability and the requisite brevity of definitions, it is a frequent occurrence that one person's or discipline's definition will not capture the essence or priorities considered important by another. Reviewing the various definitions in use in the alternative agricultures and in forms of agriculture being promoted as sustainable led Reeve (1990) to propose the definition shown below. This definition is largely consistent with the goals of sustainable agriculture that have been proposed by the Ecologically Sustainable Development Working Groups — Agriculture (1991).

> Sustainable agriculture is an agriculture that attempts, in the ways suggested by current knowledge and understanding to ensure that present use of agricultural land resources does not detract from their usefulness to future generations. At present it appears that such an agriculture will need to:
>
> - be profitable for the individual farmer,
> - produce adequate quantities of food and fibre for a stable global population,
> - produce food and fibre of a quality preferred by the population,
> - conserve the agricultural resource base,
> - minimise use of non-renewable resources, and
> - minimise deleterious off-farm environmental impacts.
>
> Currently, it appears that such an agriculture will have the agro-ecosystem characteristics of resilience and diversity. Among the forms of agriculture currently being promoted as sustainable are those based on modern industrial ideologies and values (such as low input sustainable agriculture and conservation farming) and those based wholly or partly on alternative ideologies (e.g. the 'alternative agricultures' such as bio-dynamic, biological and organic agriculture). (Reeve 1990: 77–78).

FUTURE POLICY DIRECTIONS

When we look at the conditions that favour sustainability in agriculture it would appear that Australia suffers a number of disadvantages. These are that:

- The agricultural soils of the continent are not well endowed with reserves of available nutrients. Many of our agricultural systems continue to deplete soil organic matter. Land degradation remains a threat to a significant proportion of agricultural and pastoral land.
- Availability and quality of water continue as constraints on production which are likely to worsen as the full implications of the long-term response of hydrologic systems to vegetation changes become apparent.
- Our agriculture is biologically alien to the ecosystems of the continent. We have little in the way of genetic resources in our native biota that can be readily applied to conventional agriculture to soften its impact on the environment.

- We do not have the benefit of a surviving peasant agriculture where some traditional practices may once again be applicable.
- We have a highly urbanised population distant from our agricultural heartlands such that any possibility of sewage (nutrient) recycling is virtually precluded.
- Our agricultural economy is strongly export oriented, further preventing nutrient recycling.
- Our agriculture is highly dependent on fossil fuel for transport of nutrients into and within Australia, for transport of exports and for farming operations.
- Our farmers have a long history of reluctance to engage in education (which is not necessarily the fault of farmers — it may equally be due to a lack of short-term payoffs to education, or reflect the irrelevance of educational offerings to the farmers' situation).
- Our emigrant and pioneering origins gave rise to a rural culture that condoned exploitative agriculture and placed a high value on the right of the individual land owner to determine land use. The persistence of these attitudes is a major obstacle to internalising the external costs of agriculture.

Working in Australia's favour, on the other hand, is the fact that its agriculture is, in comparison with other industrialised countries, less intensive and uses fewer inputs.

Just as the roots of non-sustainability have been shown to be in many areas outside agriculture itself, so the policy effort will need to span the input sector, the output sector, and public sector research, technological assessment and education.

The environment in which firms in the input sector operate is not conducive to the development and marketing of products that would contribute to the sustainability of agriculture. Where agriculture is structured such that the farm sector is a food and fibre transformer for the products of the input sector, individual farmers, no matter how concerned they may be about sustainability, have their choice restricted to whatever the input sector is able to produce profitably. Sustainability will require that the farm sector reduce its dependence on a materials (and energy) intensive input sector.

There are two main directions from which this problem can be approached. First, action can be taken to ensure that farmers have a choice of products, expertise and technology to enable them to choose sustainable agricultural practices. Providing this choice has been one of the main justifications for the Low-Input Sustainable Agriculture Program in the USA. Given that the private sector has difficulty in responding to the need for sustainability in agriculture, it is a matter of some concern that public sector research and extension agencies in Australia are under pressure to adopt private sector modes of management, accountability and cost recovery. In the short term, it would appear that only the public sector and the voluntarily staffed alternative agriculture organisations have the capacity to start widening the choice of agricultural practices available to farmers.

The second, and probably more difficult, approach would be to alter the conditions in which the input sector operates to induce a shift away from synthetic products and oversimplified representations of pest problems. There would appear to be almost insurmountable difficulties in providing conditions under which existing firms could successfully market agricultural practices that were largely knowledge-based manipulations of agroecosystems and involved minimal use of purchased inputs. This type of service would seem to be far more feasibly provided by local consultants (management of agroecosystems being highly situation specific). It is not inconceivable, then, that sustainability in agriculture will require major structural adjustment in the input sector away from the present levels of economic concentration and dependence on value adding to raw materials and fossil fuel towards a more dispersed, knowledge-intensive input sector.

Within the output sector, certification systems are being adopted and are operating successfully in many countries. Australia, as an exporting country, will need to coordinate its activities with overseas systems, a task that has already been started by the Organic Produce Advisory Council. Certification systems and simpler marketing chains have the potential to harness consumer preferences to move some parts of Australian agriculture towards improved sustainability. However, because high-quality foodstuffs do not necessarily mean that the agriculture that produced them was sustainable, and because some mainstream food industries may choose the defensive strategy of blurring the distinction between organic and other foodstuffs, there is a real danger that, with time, certification may suffer the same problems as green consumerism. There seems to be a tendency at present to see a government-sanctioned certification system as a panacea for domestic retailing problems and as an opening to lucrative export markets.

There is an urgent need to look beyond these immediately obvious short-term benefits and examine the wider ramifications. An important area to be assessed is the impact of the strategic response of the mainstream food industries should the certification system result in loss of market share. The second important area to be examined is the extent to which the existence of certified producers amid farms operated conventionally will generate conflicts over 'the right to farm' that will only be able to be resolved via inefficient common law litigation.

One of the major factors that has contributed to public awareness and political will with respect to sustainable agriculture has been the serious environmental consequences of unforeseen side-effects of new technologies in agriculture. The input sector has developed since World War II into an agrichemical technological monoculture. There are signs that this will soon be replaced by another technological monoculture based on biotechnology and genetic engineering (see Chapters 3 and 16).

The reasons for the unforeseen side-effects in the agrichemical monoculture are easily identified with the wisdom of hindsight — technological assessment concentrated mainly on operator and consumer health effects. It is only comparatively recently that environmental impacts have been included in the assessment process. In addition to this, the bulk of the assessment effort went to research prior to release of the product to the public and very little to environmental monitoring after the product was in use — the assessment procedures were assumed to be, or represented as being, totally reliable. In comparative terms, the amount of effort devoted to the detection of early warning signs in the environment is still very small.

Our current technological assessment procedures suffer from scientific chauvinism that cannot admit the small but finite probability of incorrect assessment. Assessment by firms during the research and development phase is becoming increasingly expensive — and the greater the expense, the greater the stake the firm has in seeing that the product reaches the public. Furthermore, the more exhaustive the assessment, the more likely it is that any remaining undetected side-effects will be subtle and so detectable only in the long term.

Since the reality is that synthetic inputs will continue to be required in agriculture for at least some decades, and since the cornucopic promise of biotechnology and genetic engineering is likely to outweigh anticipatory prudence, there is an urgent need to increase the amount of effort devoted to the early detection of unforeseen environmental side-effects. This type of activity has traditionally been regarded as a public sector responsibility, but there may be grounds for regarding it as an external cost generated by the users of technology, and so open to policy instruments to internalise the cost.

Since the environmental and energy crises of the 1970s, much has been written about the need for multidisciplinary and interdisciplinary research and education. Scientific research benefits in many ways by being reductionist, narrow and specialised, so it is not surprising that multidisciplinary research is still in the minority. Environmental problems in agriculture seem to arise when research results are translated into prescriptions for farm practice. One possibility might be that agricultural research should be subject to some form of social and environmental impact assessment in a similar fashion to the animal welfare assessment currently applied when experimental animals are involved.

It is sometimes suggested that the focus on agricultural research on increasing productivity is a root cause of the tendency to non-sustainability in some parts of conventional agriculture. Certainly selection for yield and weight gain in plants and animals respectively has often been at the expense of disease resistance or the ability to perform in adverse environments. Much of the body of knowledge gained in the pursuit of productivity would, however, be equally applicable to the pursuit of sustainability. Some specialisations, such as soil biology, soil ecology and arthropod and nematode ecology and population dynamics, will require more emphasis than has been accorded them in the past.

The call for multidisciplinary approaches has probably had a greater effect in education than in research. There is certainly a wide range of environmentally oriented programs now available in Australian educational institutions. The availability of these programs may in fact have reduced the pressure for change within agricultural programs. With one or two notable exceptions, very few agricultural programs have shifted emphasis away from the production focus. It is, of course, difficult for institutions to lead industry if their performance is to be judged on the immediate employability of their graduates. The challenge for educational institutions is to provide their graduates with the skills that will guarantee them employment in today's non-sustainable agriculture, and the commitment, depth of understanding and adaptive competence to take an active role in the transition to sustainable agriculture.

Finally, the move to a more sustainable agriculture will place considerable demands upon farmers. It is becoming generally accepted that agricultural practice aimed at sustainability requires more exacting management skills than conventional practice. It has also been shown in many studies that farmers face additional problems during the transition period from conventional to alternative agriculture. In a period of globalisation of the markets for agricultural commodities, increased pressure for product specification and declining terms of trade, the addition of managing for sustainability places an extreme burden on many farmers.

While this burden can no doubt be eased by appropriate extension and education, the progress towards sustainability will be mediocre unless supporting policies are implemented in the input and output sectors in the areas suggested in the previous sections. No amount of research, education and extension support to the farm sector will achieve agricultural sustainability unless farmers are operating in an environment where the reduction in dependence on a materials and energy intensive input sector required for sustainability does not run counter to the price and other informational signals they receive from both the input and output sectors.

NOTE

Sections of this chapter are reprinted with the permission of The Rural Development Centre from: Reeve, I. (1990) *Sustainable Agriculture: Ecological Imperative or Economic Impossibility,* TRDC Publication No. 169, The Rural Development Centre, University of New England, Armidale.

REFERENCES

Batie, S. and Taylor, D. (1989) 'Widespread Adoption of Non-conventional Agriculture: Profitability and Impacts', *American Journal of Alternative Agriculture* 4 (3 and 4): 128–134.

Berry, W. (1987) *The Unsettling of America: Culture and Agriculture,* Avon Books, New York.

Brooks, H. (1986) 'The Typology of Surprises in Technology, Institutions and Development', in W. Clark and R. Munn (eds), *Sustainable Development of the Biosphere,* International Institute for Applied Systems Analysis, Loxenburg.

Cooke, A. (1990) 'Survey Shows Buyers Distrust "Green" Labels', *Australian Rural Times* 19, 17 July.

Ecologically Sustainable Development Working Groups — Agriculture (1991) *Draft Report — Agriculture,* Australian Government Publishing Service, Canberra, August.

Edwards, C. (1987) 'The Concept of Integrated Systems in Lower Input/Sustainable Agriculture', *American Journal of Alternative Agriculture* 2 (4): 148–152.

Fyfe, W., Kronberg, B., Leonardos, O. and Olorunfemi, N. (1983) 'Global Tectonics and Agriculture: A Geochemical Perspective', *Agricultural Ecosystems and the Environment* 9: 383–399.

Kennedy, J. (1958) 'Physiological Condition of the Host Plant and Susceptibility to Aphid Attack', *Entomology, Experimental and Applied* 1: 50–56.

MacIntyre, A. (1987) 'Why Pesticides Received Extensive Use in

America: A Political Economy of Agricultural Pest Management to 1970', *Natural Resources Journal* 27: 533–578.

Patriquin, D., Baines, D., Lewis, J. and Macdougall, A. (1988) 'Aphid Infestation of Faba Beans on an Organic Farm in Relation to Weeds, Intercrops and Added Nitrogen', *Agricultural Ecosystems and the Environment* 20: 279–288.

Pearce, J. (1990) 'The Consumers are Not So Green', *New Scientist* 16 June: 13.

Pollard, D. (1973) 'Plant Penetration by Feeding Aphids (*Hemiptera aphidoidea*): A Review', *Bulletin of Entomological Research* 62: 631–714.

Reeve, I. (1990) *Sustainable Agriculture: Ecological Imperative or Economic Impossibility,* TRDC Publication No. 169, Rural Development Centre, University of New England, Armidale, New South Wales.

Vereijken, P. (1989) 'From Integrated Control to Integrated Farming, An Experimental Approach', *Agricultural Ecosystems and the Environment* 26: 37–43.

Wallace, L. (1987) *Agriculture's Futures: America's Food System,* Springer-Verlag, New York.

13 FARM AND CATCHMENT PLANNING: TOOLS FOR SUSTAINABILITY?

ANDREW CAMPBELL

Two threads of Australian agriculture have emerged over the last 30 years, evolving separately at first, but becoming more intricately entwined in the 1990s. Property planning has almost achieved the status of a coherent discipline within agriculture, while sustainability has become a catchword as Australian agriculture has begun to respond to growing environmental awareness and concern throughout the community. This chapter briefly sketches the converging evolutionary paths of the property planning and sustainable agriculture movements within Australia, pointing to potential future developments.

PROPERTY PLANNING

Property planning in Australia emerged in a formal sense during the 1950s. Soil conservation departments in Victoria and New South Wales introduced farm planning services in 1951 and 1957 respectively (Soil Conservation Authority 1961; Junor 1987). These plans were primarily aimed at soil erosion control and were largely prepared by government staff, using land capability assessment as the basis for plan development. Consequently, these early plans focused on physical erosion control works and — to a lesser extent — property layout, water conservation, tillage methods and pasture development.

Also during the 1950s, P. Yeomans, a visionary farmer, surveyor and engineer, developed Keyline (Yeomans 1981) which he described as:

> a set of principles, techniques and systems coordinated into a plan for the development of farm and grazing landscapes — a master plan for the elaboration of a 'replacement' for the natural or existing landscape. A principal aim of Keyline is to increase both the depth and fertility of the soil so that the soil of farming and grazing land is safe and permanent and capable of continuous improvement . . . It includes new cultivation techniques; a method of farm subdivision and layout; planning for timber and scrub clearing and water conservation and irrigation. All are planned to facilitate or assist in the production of fertile soil.

Note the language used to describe the aims of Keyline. Yeomans was ahead of his time in the use of terms such as 'safe', 'permanent', 'capable of continuous improvement', 'facilitate . . . production of fertile soil', in attempting to integrate land and water conservation with improved soil fertility, as well as in recognising the importance of biological activity within soils and the role of remnant vegetation on farms. He also contended that farmers could follow his methods to prepare and implement their own Keyline plans.

Most property planning activity during the 1960s and 1970s was led by state soil conservation agencies, although farm management consultants began to offer production-oriented farm management plans, and plans focusing on surface hydrology became more common in irrigation districts (Campbell 1989a).

The 1980s saw a resurgence in property planning activity in most states of Australia. A private initiative which made a significant contribution to this resurgence was the Potter Farmland Plan. This project established 15 demonstration farms in western Victoria from 1984 to 1988, funded by the participating farmers and the Ian Potter Foundation. It aimed to show how ecological considerations could be incorporated into farm planning and land management so as to improve productivity and redress land degradation. This project, and the 'whole farm planning' process which it spawned, are described in detail in Campbell (1991). The project was based on the following key assumptions:

- That land degradation 'problems' are symptoms of inappropriate land management, and are most likely to be fixed if the required management changes benefit the land user. In other words, conservation and productivity must be complementary. Conservation measures — developed for their own sake — are unlikely to be widely implemented.
- That any plan is ideally best prepared by the people who have to implement it, which means that the best people to be preparing farm plans are farmers. This does not preclude the benefits of consultation with specialist technical advisers, family members, neighbours or consultants.
- That farmers are generalists, who are used to integrating technical, financial and social information from diverse sources in decision making, so the farm planning process must be capable of dealing with more than just the physical layout of the farm.
- A farm plan is not an ideal map of the farm, but simply an

expression of the current state of a planning process, which is dynamic, responsive, ongoing.

These assumptions were supported several years later by a review of the New South Wales Soil Conservation Service farm planning scheme, summarised by Junor (1987), which concluded among other things, that landholders generally did not understand the farm planning concept, that greater farmer involvement and ownership was required and that inputs from disciplines other than soil conservation should be encouraged. Other government agencies also broadened their thinking on farm planning in the late 1980s. The Western Australian Department of Agriculture developed a comprehensive computer-based farm planning package called 'Landman' (*sic*) which integrated land management plans with financial management plans and mathematical programming models to answer 'what if ?' questions to help farmers quickly evaluate the physical and financial impact of any planning decision (Western Australian Department of Agriculture 1988). State government agencies in Western Australia, Victoria and Queensland began to develop self-help farm planning courses and resource materials (Campbell 1989b).

The Potter project moved on from the demonstration farms to develop and run a series of short courses in whole farm planning starting in 1987, at which groups of farmers, usually from the same district, were guided through the farm planning process together. The interaction between course participants was enlightening. It exposed the benefits of looking with fresh eyes at another's problems and the willingness of farmers to be more adventurous in their exploration of possibilities for the management of land other than their own.

THE DEVELOPMENT OF LANDCARE

The late 1980s saw the blossoming of the landcare movement, which had its origins in the Land Conservation District Committees established in Western Australia from 1982, and in the 'LandCare' program initiated in Victoria in 1986. Landcare has spread rapidly to all states, boosted by additional funding under the National Soil Conservation Program and by 1990 there were more than 600 Landcare groups in Australia involving approximately 15 000 farmers. The development of the Landcare movement is discussed in detail in Campbell (1990). Typically, Landcare groups are local groups of people in rural areas with common concerns about land degradation and the improvement of land management practices, who have banded together to try to tackle these issues.

One of the most common activities for Landcare groups is property and catchment planning. Most of the land degradation problems which concern groups cross property boundaries and are thus more suited to catchment-based approaches. Groups are also better able to attract resources from government and private sources to run farm planning short courses and to assist in the preparation of catchment plans. As more groups define their own needs and approach the same task in their own way, the evolution of different approaches to farm and catchment planning has accelerated. Some groups are using computer-based Geographic Information Systems (GISs), others have developed very simple processes based around laser-copied enlarged aerial photographs, and still others have made very effective use of private consultants as hunters and gatherers of information and as 'the voice of the catchment'. Land users are starting to collect and monitor information which was largely the province of specialists five years ago.

The last point is critical. Landcare groups and some individual land users are now familiar with technology such as piezometers, neutron moisture probes, aerial magnetrometric surveys and electromagnetic detection of potentially saline areas. Innovative community-based education programs like 'Saltwatch', 'Streamwatch', 'Frogwatch', 'Kids for Landcare' and 'Plantscan' are involving students (at primary and secondary levels) directly in basic data gathering and analysis. These data can of course be integrated with the practical experience and intuition of land users in preparing farm and catchment plans, ensuring that plans recognise the ecological impact of farming practices. However, the major value of such programs is the speed and effectiveness with which they transmit local environmental knowledge through communities, and teach people to observe and monitor the health of the land around them. 'Land literacy', reading the land, is a term which has been coined to denote this area of environmental education, and which may become a subject in its own right over the next decade.

This discussion has briefly described the recent evolution of farm and catchment planning in Australia. More important than the details of any particular approach, however, are the trends along this evolutionary path which emerge from this historical treatment. They include:

- a move away from 'fixing' land degradation problems towards developing better land management systems;
- greater emphasis (albeit with a long way to go) on integrating the production enterprise and financial management into

the property planning process, rather than confining it to the physical layout of the property — what happens between the fences is just as important as fence location;
- a continual shift in the centre of property planning gravity from public servants and consultants to land users;
- accelerating acceptance of catchment and/or district plans which encompass broader ecological issues (for example, remnant vegetation, river management, groundwater systems), and which are just starting to recognise (if not integrate) social issues;
- an increasing emphasis on process (recognising the importance of involvement and ownership) and flexibility of output — the presentation of the plan is less important than the changes which have occurred inside the planners' heads and those which subsequently occur on the ground;
- institutions, in particular government land conservation agencies and agriculture departments, are learning to respond to requests for planning assistance, rather than designing and running their own planning services according to their own policies and capacities;
- the artificial lines drawn between researcher, extension agent and land user are being blurred and in some cases dissolved.

THE SUSTAINABLE AGRICULTURE MOVEMENT

In parallel with the evolution of farm and catchment planning in Australia, driven by some of the same factors, the sustainable agriculture movement has also developed rapidly over the last three decades. Several agricultural systems have emerged with differing characteristics, all purporting to be more sustainable than conventional agriculture. These include conservation farming, organic farming, biodynamic farming and permaculture. These systems are all still developing and are still practised by a minority of land users. Rather than debate their relative merits, it is more useful to examine the concept of sustainability, and how it relates to property planning.

At a basic level, the sustainability concept can be likened to 'living within one's means' and 'putting back what is taken out'. According to the International Federation of Agriculture Producers (Anon 1991), sustainable agricultural systems are generally:

- Stable. They do not disrupt ecological systems or overexploit natural resources. There is a rational use of renewable resources, the physical condition of the soil is maintained, there is no build-up of weeds, pests, diseases, acidity or toxic elements. Genetic resources in plant and animal species are conserved and options for future generations as to how to use the natural resource base remain open.

- Regenerative. The minerals and nutrients removed by crops and livestock are replenished in the soil.
- Productive and profitable. They are capable of continuous reliable production levels — creating surpluses above family needs for minimum survival.
- Resilient. They can absorb changes, retaining characteristics in the face of disturbances such as climatic extremes, or attacks by pests and diseases.
- Appropriate. They reflect and adapt to the needs, skills, training and finances of land users as well as to the environmental constraints of climate, soils and topography.
- Self-reliant. They are based on the efforts and ideals of the farmers themselves on a regional level — minimising the dependence on non-renewable, often imported resources.
- Non-disruptive. They do not destroy the sociocultural environment by, for example, forcing people to adopt practices against normal behaviours and traditions, or forcing rural people to migrate to the cities.

These characteristics provide a broad checklist against which to evaluate existing farming systems. A superficial assessment of the extent of land degradation in Australia, the economic crises in most agricultural sectors, and the extent of external assistance required after the floods, droughts and fires which are expected if not predictable in this country would suggest that, in aggregate, Australian agriculture could greatly improve its sustainability.

Such a checklist could be refined further at different spatial and temporal scales. Lefroy, Hobbs and Salerian (1992) propose a theoretical framework for assessing the sustainability of farming systems in the Western Australian wheat belt, attempting to integrate agronomic, ecological, social and economic perspectives, at paddock, farm, landscape and national scales. They suggest the identification and consideration of ecological parameters such as water-use potential, water quality, soil loss, soil biological activity, nutrient leaching, energy inputs, solar energy interception, diversity of species and forms; and of economic parameters such as profitability, cash flow, equity and debt, net present value, practicality and acceptability. Social parameters are more difficult to define but are the logical extension of this approach.

As suggested by Lowrance, Hendrix and Odum (1986, quoted in Lefroy *et al.* 1992), some of the confusion of attempting to integrate social, ecological and economic parameters can be removed by creating a hierarchy of dominant constraints and goals. At the paddock level, agronomic considerations predominate over a timescale of several seasons. At the farm level, the survival of the farm business over several generations is dominated by microeconomic constraints. At the catchment or landscape level, ecological concerns of maintaining life support systems over hundreds of generations predominate, while at regional and national levels the dominant constraints of macroeconomic and planning horizons are limited by politics and economics.

Too many property plans have been prepared which essentially rationalise existing farming systems, rather than rigorously analysing these against specific criteria, setting long-term goals and outlining a path to achieve them. Equally, most attempts at developing more sustainable agricultural systems have been piecemeal and *ad hoc*, concentrating at the paddock scale, rather than evolving from a planning process which integrates social, economic and ecological parameters at paddock, farm and catchment levels.

There is clearly a point of convergence between the developing concepts of property and catchment planning, and of sustainable agriculture. Property and catchment planning are processes which can assist individual land users at the paddock and farm levels, and groups of land users at the farm level, to gather, analyse, synthesise and apply information to move towards sustainability. However, an analysis of the characteristics of sustainable agricultural systems is essential to provide the conceptual framework which dictates what sort of information is required, and also how the performance of the system can be assessed. It is difficult to detect a design ethic or tradition in Australian agriculture. Disciplines other than agricultural science, with a stronger planning base, may prove more useful in refining property planning and, in particular, catchment planning processes. McHarg (1969) has produced a classic text which has influenced a generation of landscape planners, and the compendium of permaculture ideas in Mollison (1988) has a great deal to offer even mainstream land users.

The convergence of sustainable agriculture and rigorous property planning is still a long way off for the vast majority of Australian land users, and even for the agricultural education establishment. Most land users still do not have a property plan and would be unaware of any catchment plan in their area — certainly very few have been involved in the development of catchment plans. Most property plans concentrate on physical property layout, with little integration of financial and social constraints and opportunities. The theoretical framework for current planning processes is more often land capability assessment than the elements of sustainability listed above. This is understandable given the lack of hard data in most areas needed to fill in the conceptual matrix suggested by Lefroy *et al.* (1992).

However, the congruence of these two trends in Australian agriculture is very important for policy and decision makers. The context in which land users will be seeking and applying information over the next decades will be critical for research and extension and in particular for the interaction between research and extension.

Sustainability is not absolute or discrete. There are relative degrees of sustainability determined by a range of parameters, which, in turn, define a path to sustainability. Property and catchment planning, at its best, shows land users where they are on this path and provides direction for designing and implementing change.

REFERENCES

Anon (1991) 'Sustainable Farming and the Role of Farmers' Organisations', *World Farmers' Times* 1/91.

Campbell, A. (1989a) 'Bridging the Gap Between Conventional and Sustainable Agriculture — The Role of Whole Farm Planning', *Australian Journal of Soil and Water Conservation* II (2): 43–46.

Campbell, A. (1989b) 'Landcare in Australia — An Overview', *Australian Journal of Soil and Water Conservation* II (4): 18–20.

Campbell, A. (1990) *Landcare — Progress Across the Nation*, National Landcare Facilitator Annual Report, National Soil Conservation Program, Department of Primary Industries and Energy, Canberra.

Campbell, A. (1991) *Planning for Sustainable Farming — The Potter Farmland Plan Story*, Lothian Books, Melbourne.

Junor, R. (1987) 'An Evaluation of 30 Years of Farm Planning in New South Wales', in *Profitable Farms, Productive Land — The Future?*, Ballarat Conference Proceedings, Soil Conservation Association of Victoria, Victoria.

Lefroy, E., Hobbs, R. and Salerian, J. (1992) 'Integrating Economic and Ecological Considerations', in R. Hobbs and D. Saunders (eds), *Re-integrating Fragmented Landscapes*, Springer-Verlag, New York.

Lowrance, R., Hendrix, P. and Odum, E. (1986) 'A Hierarchical Approach to Sustainable Agriculture', *American Journal of Alternative Agriculture* 1: 169–173.

McHarg, I. (1969) *Design with Nature*, Doubleday, New York.

Mollison, B. (1988) *Permaculture. A Designers Manual*, Tagari Publications, Tyalgum.

Soil Conservation Authority (1961) *Annual Report*, Victorian Government Printer, Melbourne.

Western Australian Department of Agriculture (1988) 'LANDMAN, A New Approach to Land Management Planning', *Technote* 12/88.
Yeomans, P. (1981) *Water For Every Farm: Using the Keyline Plan*, Second Back Row Press, Katoomba.

14 THE DILEMMA OF CONSERVATION FARMING: TO USE OR NOT USE CHEMICALS

NEIL BARR and JOHN CARY

In 1987 the United States tightened its procedures for detecting organochlorine residues in imported foodstuffs and soon discovered organochlorine residues in beef imported from Australia. With a major market threatened, Australian authorities tightened controls on the use of organochlorine chemicals and instituted 'traceback' procedures to detect the sources of contamination. Fifteen hundred properties with contaminated soil were quarantined (Australian Parliament 1990: 228). No cattle could be sold from the properties until the contamination was removed or biodegraded. Many of these properties will be quarantined into the next century. How was it possible that such a major embarrassment occurred?

Australian agricultural exports had been threatened once before by organochlorine residues. In the early 1960s the United States reduced the level of organochlorines it would accept in food products. Their tests showed some Australian dairy products contained organochlorine residues above the acceptable limit. Dairy farmers were spraying organochlorines on pastures to control grass grubs, armyworms and cockchafers. Cattle eating the sprayed grass concentrated the organochlorines in their milk. In response, the Australian regulatory authorities banned pasture spraying with organochlorines and temporarily safeguarded exports.

Throughout the following decade the United States progressively banned the agricultural use of organochlorines. Regulatory authorities outside the United States disagreed with the United States' interpretation of toxicological evidence on the safety of organochlorines. In Australia the limited agricultural use of organochlorines continued on farms. Though no longer allowed to spray pastures with organochlorines, Australian graziers could still use organochlorines to protect fence posts from termites. Potato farmers mixed dieldrin with fertilisers to control wireworm, a small native worm which eats potatoes. Because neither potato plants nor grasses which grew following a potato rotation absorbed dieldrin, there was no expectation that this agricultural practice would lead to significant organochlorine residues in either potatoes or meat. However, some contamination was inevitable. Cattle eat soil attached to the roots of grasses they uproot and may, individually, swallow one to two kilograms of soil per day.

Although by the 1980s, organochlorine use was banned in United States agriculture, the United States Customs service tolerated a very small percentage of imported carcasses with low-level organochlorine contamination (Australian Science and Technology Council 1989: 56–64). In 1986, this tolerance to residues disappeared. Protestations that the United States was responding to political pressure rather than scientific evidence were of little practical value. The Australian export beef industry was threatened unless drastic action was taken to eliminate all residues from beef exported to the United States. Australian authorities quarantined farms with organochlorine-contaminated soil.

From a wider perspective, this story of the politics of the regulatory process is not a satisfying explanation of the reasons for the quarantining of 1500 farms. A deeper question remains. How had organochlorine chemicals become accepted as tools of trade on Australian farms? For some understanding of this question we need to explore the development of what the organic agriculture movement calls 'chemical agriculture'. In this chapter we will trace this development in the horticultural and cropping industries and explore some of the dilemmas that farmers face in making decisions about the use or non-use of agricultural chemicals.

THE UNWELCOME IMMIGRANTS

For early Australian orchardists the 'tyranny of distance' was a blessing. The limitations of transport acted as a kind of quarantine. No fruit could be imported from overseas and trees established from imported seed were mostly free of overseas diseases and pests (French 1902). Apple orchards were free of the codling moth (historically referred to as codlin moth) and the black spot fungus. There was no sign of oriental fruit moth, brown rot or leaf curl in stone fruit orchards.

The gold rushes signalled the end of this relative freedom from pests. Immigrants brought new pests and diseases. The codling moth made its first appearance in Tasmanian orchards in 1857. From there it quickly spread to South Australia and then to the remaining eastern states (Olliff 1890). The moth caused extensive damage. Black spot fungus appeared in the east of Australia a few years after the codling moth's arrival.

The fungus attacks the leaves of the apple tree early in the season, later moving to the fruit, leaving large black marks, cracks and deformed, unsaleable fruit. Black spot and codling moth devastated apple orchards, destroying whole regional industries (McAlpine 1904).

Orchardists tried to keep the codling moth under some control by cleaning the loose bark from the trunks of trees and the ground underneath and by picking up damaged or fallen fruit which might harbour grubs. On some orchards, pigs or poultry patrolled the ground, eating dropped apples (Rigg 1924). Growers also wrapped bandages around the trunks of the trees. Instead of overwintering in crevices and loose bark, the moth hid under the folds of the cloth. Growers removed the bandages and boiled them to kill the hiding moths (Anon 1898a; Quinn 1898a).

Other unwelcome immigrants made stone fruit growing a risky business. In 1856 peach leaf curl appeared in southern Australia. Each spring this fungus destroyed the newly emerging leaves of the peach, weakening the tree and crops. Peach production became a matter of luck, dependent on the season. Oriental fruit moth arrived sometime before 1900. The moth has a life cycle similar to the codling moth, but instead of burrowing into apples it burrows into growing shoots of peach trees and into peaches. What is worse than the direct damage it causes is the fact that the oriental fruit moth helps to spread another unwelcome migrant, the fungal disease brown rot. Brown rot attacks the blossoms of peach trees and other stone fruit, infecting the embryo fruit. The infection reactivates in the fruit when it ripens. The spore overwinters on dried rotten fruit which hang on the tree and spread the spore to start a new generation of rot infection the following year. Picking and destroying mummified fruit and infected branches offered the only possibility of limiting the spread of infection.

INORGANIC CHEMICALS AND THE RISE OF CALENDAR SPRAYING

Immigrant insects like codling moth and oriental fruit moth quickly became pests because they were set free in a new land without predators to keep them in check. The quick spread of fungal diseases had another cause: the gradual commercialisation of fruit production. By 1900 social pressures had changed the face of horticulture. Australia believed its future lay in the development of a rural yeomanry who would farm the new Closer Settlement and irrigation developments. Orcharding was no longer a cottage industry. The Closer Settlement policies could only succeed if the large increase in production could be sold overseas. In the artificial world of the commercial orchard where trees of the one type are planted close together, there is sometimes little to stop an infestation spreading from one tree to the next, infesting a whole orchard. Commercialisation helped spread infection between farms. Growers supplied the local market with fruit packed in recycled pine boxes. The boxes recycled fungal spores and insect pupae between farms.

Though commercialisation helped to spread pests and diseases, it also increased the pressure on growers to control those diseases. Apple orchardists needed to produce high-quality, blemish-free, uninfected apples which would travel, without deteriorating, to distant European markets. Orchard cleanliness was a step towards this goal. State governments passed laws requiring growers to inspect, pick and destroy infected fruit once a week (Anon 1898c). Government inspectors enforced cleanliness on those who did not maintain clean orchards. Disinfecting steam baths operated at the central fruit and vegetable market to clean fruit boxes. These laws helped create a clean orchard ethos that persisted in following generations. Orchardists were judged by peers on how their orchard looked. A good orchard was neatly ploughed, without a sign of weeds and with trees clean and neatly pruned in perfectly straight rows.

Despite government advisers' claims that 'cleanliness was next to prosperity' (Anon 1899), cleanliness was not a panacea for the orchardists' problems (Olliff 1891a; Anon 1898d; Krichauff 1898; Quinn 1898b). Keeping a clean orchard was time consuming and only partly effective. One South Australian grower complained that despite picking 6000 caterpillars from trunk bandages and picking 13 000 infected apples, the codling moth numbers were as bad as ever (Anon 1898b; Laffer 1898). Other solutions were needed and, at the turn of the century, there were high hopes for natural pest control. In each state, economic entomologists advocated the use of natural controls for pest insects (Olliff 1891b; Quinn 1898a). The Victorian government entomologist printed 9000 copies of a guide to natural controls (Olliff 1891c). The entomologists were constantly seeking economically useful predators or parasites (McAlpine 1903). They tried to convince growers that the insect-eating birds living around their orchards should be protected from the sporting shooters and tree clearing. Growers were sceptical of this advice, responding to it with tales of extreme losses because of fruit-eating birds. Many considered it safer to shoot all birds (Krichauff 1898; Anon 1898c).

The remaining weapon in the orchardists' armoury was

spraying. Very soon after first settlement farmers had used tobacco-derived nicotine sprays to counter insect pests, but the success of these sprays had diminished as pest resistance increased (Pescott 1910). By the turn of the century new and more effective chemical sprays had been discovered overseas. The simple mixture of lime and copper sulphate, named 'Bordeaux' in honour of its provincial French discoverers, had an impressive effect on black spot, mildew and peach leaf curl, though it did little to control brown rot. Arsenic compounds were found to be extremely effective in controlling leaf-eating insects. Some control of aphids could be achieved by spraying unrefined light petroleum 'red oil' on dormant, deciduous fruit trees to kill aphid eggs. Because 'red oil' damaged green leaf tissue, citrus orchardists resorted to the most dangerous of the inorganic insect controls to control scale: they gassed trees with hydrogen cyanide.

Growers learned to rely increasingly on these inorganic chemical sprays (Lang 1903). Many apple growers abandoned the labour-intensive practices of wrapping cloth around trunks, relying instead on spraying and frequent ploughing to bury infected fruit and leaves. Farmers controlled pests with 'calendar spraying', spraying at a particular time, rather than when indicated by the pest populations. Spraying came to be seen as an insurance which demanded a regular premium payment (McAlpine 1904).

Over the following decades the inorganic sprays became less effective. The export markets were increasingly selective regarding fruit quality; but it was becoming harder to produce quality fruit. The insect pests were becoming less susceptible to most of the inorganic chemicals used for their control. The response of most farmers in Australia, Europe and the United States to increased pest resistance was to increase the strength or frequency of spraying. Consumers were unimpressed by apples marketed with the lead arsenate still clinging to the skin of the fruit. In Europe this concern stimulated the development of the organic and biodynamic agricultural movements whose members advocated a return to agricultural practices which did not use artificial chemicals. In Australia, economic entomologists advised farmers to use less lead arsenate by spraying only when absolutely necessary and relying on natural predators (Pescott 1933). The motivation for this advice was overwhelmingly financial. Britain had threatened to ban Australian apples if the arsenic residue levels did not decrease.

Advice to reduce dependence on chemical sprays was unlikely to have much influence. Where the biological controls advocated by the economic entomologists did work, they did not eradicate the pests, but merely limited their numbers. The economic entomologists had to convince farmers to accept some infestations and consequent damage (Davey 1929). Export markets still demanded unblemished fruit, and other advisers argued against spray moderation because of the risk to fruit quality (Hammond 1932). These advisers advocated, instead, the careful washing of apples to remove lead arsenate residue. For many other diseases including brown rot, peach curl and codling moth there was no biological control and farmers remained dependent on chemical control measures.

THE ORGANIC PESTICIDES

One of the darker aspects of World War II and the cold war which followed was the acceleration of research into toxic chemical weapons. Out of this morbid pursuit came new families of pesticides based upon complex hydrocarbon molecules. The most famous was the organochlorine family which included dichlorodiphenyl trichloroethane (DDT). DDT appeared to be nowhere near as poisonous to humans as lead arsenate; and yet small doses killed insects with great efficiency. Another group of chemicals, the organophosphates, were not as long lasting as the organochlorines, but were far more toxic. By the 1960s, the number of agricultural chemicals had proliferated. There were over 300 agricultural chemicals registered in about 10 000 formulations and trade names across the world.

The new chemicals could replace lead arsenate. Many farmers integrated them into their farm management regimes. The use of various organochlorines was widespread, frequent and at times indiscriminate. DDT, in particular, was widely promoted as the answer for a multitude of pest problems. History was being repeated. Just as with the inorganic chemical revolution, a decade of widespread use of the new organic chemicals revealed some familiar problems.

First, pests developed resistance to the most widely used new chemicals. DDT-resistant codling moths appeared. Second, the new pesticides also encouraged new pests. Two spotted mites flourished as DDT both killed their natural predators and stimulated mite egg production. Any orchardist who used DDT or other insecticides to control leaf-eating pests had to include a miticide in the spray program.

A third problem was farmers' growing sense of familiarity with the new chemicals. Many farmers were apparently unaware of the danger of the chemicals they handled. Some of the new organophosphates were toxic nerve poisons which were easily absorbed through the skin or by breathing spray droplets. Many orchardists did not know the toxicity of the chemicals they used, did not understand the labels, did not have antidotes or were indifferent to safety. Cases were known of farmers scooping up chemicals or stirring vats with their bare hands. Many growers ignored the warnings and sprayed without full protective clothing (Wilson 1960; Nancarrow 1977, 1979). Free blood testing at agricultural field days showed many cases of subclinical poisoning (Commission of Public Health 1976).

INTEGRATED PEST MANAGEMENT

Pest resistance, predator imbalance and farmers' health are important to agriculture but they are not things which normally stir public opinion. It was another matter which brought the use of the new pesticides to public attention: the cumulative effects of DDT and other organochlorines on wildlife and the wider community. In 1962, *Silent Spring*, written by the American author Rachel Carson (1962), gave an account of the environmental impact of organochlorines. The book caught the public's attention in its first pages with an apocryphal scene of destruction.

> . . . a strange blight crept over the area and everything began to change. Some evil spell had settled on the community: mysterious maladies swept the flocks of chickens; the cattle and sheep sickened and died . . . The farmers spoke of much illness among their families. In the town the doctors had become more puzzled by new kinds of sickness appearing among their patients. There had been several sudden and unexplained deaths, not only among adults, but also among children, who would be stricken suddenly while at play and die within a few hours (Carson 1962: 1).

This book's tales of indiscriminate agricultural chemical use and the unintended effects of organochlorines rallied the environmental movement and urban consumer concern in the United States and then elsewhere in the world. *Silent Spring* precipitated the United States' progressive prohibition of agricultural uses of organochlorines over the next 15 years, a process which culminated in the quarantining of the 1500 Australian grazing properties in the late 1980s.

The progressive prohibition of organochlorine chemicals had little effect upon pome and stone fruit orchardists. Other chemicals replaced DDT and Endrin Oil in spray programs to control codling moths, oriental fruit moths and aphids. While this substitution eliminated the environmental risks of organochlorine pesticides, it was only a small step towards Rachel

Carson's vision of an agriculture which used biological rather than chemical means to control pests.

Rachel Carson's book was an apocalyptic revival of the arguments of the economic entomologists. Today the economic entomologists' methods are promoted as 'integrated pest management' (IPM). IPM strategies seek to reduce the use of agricultural chemicals by following four principles: use non-chemical controls wherever possible; spray only when the level of pest infestation reaches a threshold of economic significance, or when the climatic conditions threaten a disease outbreak; avoid using sprays in a way that limits the effectiveness of natural controls; and do not spray regularly with the one chemical, because this will increase pests' rate of natural adaptation.

By the 1980s, effective IPM strategies had been developed to control two spotted mites, oriental fruit moths and brown rot in peach orchards. Oriental fruit moth spreads the brown rot fungus by puncturing peach skins and infecting the fruit under its protective outer layer. Regular insecticidal spraying controls the moth at the expense of massive increases in the two spotted mite populations, necessitating regular miticide sprays. The first step in the IPM alternative is a 'confusion pheromone'. Trees are hung with strips impregnated with an artificially synthesised pheromone. This is the same pheromone released by female oriental fruit moths to attract male moths. The strips release sufficient pheromone scent to confuse male moths and prevent them from finding female partners. This interrupts the breeding cycle. The oriental fruit moth population declines. There is no need to spray to control the moth. Brown rot infestations decrease and the predator mite population increases. Growers can abandon calendar spraying to control two spotted mites, spraying instead only when regular observation shows the two spotted mite population exceeds trigger thresholds. The risk that this judicious spraying may destroy useful predators can be reduced by introducing miticide-resistant predator mites to the orchard.

The rhetoric of IPM promises a profitable agriculture with minimal pesticide use. It can claim to be a fulfilment of at least a portion of Rachel Carson's vision. Few would disagree with the objectives of IPM systems, but it is generally accepted that IPM strategies are not widely adopted on Australian farms (Australian Parliament 1990: 149). Superficially, it would seem, there are obvious incentives to reduce agricultural chemical use. Many farmers are concerned about the effect of chemicals on their health, their soil and especially their markets. In a recent survey of Victorian farmers, 88 percent of

Table 14.1 Agricultural and veterinary chemical sales in Australia: 1987–88

Agricultural chemicals	1987 $ million	1988 $ million
Insecticides	98	120
Fungicides	35	37
Herbicides	284	367
Other	20	23
Total	437	547

Source: Agricultural and Veterinary Chemicals Association, cited in Australian Parliament (1990: 228).

horticulturists described themselves as 'edgy' about the long-term effects of the sprays they used. Only a third believed all the chemicals released were safe to use on farms. Most saw risks to themselves. One in five horticulturists saw significant risk of soil residues (Agrimark Consultants 1988).

The dilemma these horticulturists face is that most believe there is no practical alternative to the present pattern of chemical use. The same survey showed that most farmers saw major difficulties in eliminating chemicals from their farm system. Eighty-five percent of horticulturists had used herbicides in the last 12 months, 85 percent had used insecticides and 90 percent had used fungicides (Agrimark Consultants 1988). The evidence of chemical sales supports this conclusion. The promotion of IPM seems to have had little influence on the scale of chemical purchases made by farmers. The value of agricultural and veterinary chemicals used on Australian farms continues to increase (Australian Parliament 1990: 117).

One of the reasons for the limited use of IPM methods is the small number of demonstrated practical IPM systems available to farmers (Australian Parliament 1990: 150–151). The integrated system for the control of mites, rot and fruit moth available to stone fruit orchardists is atypical. It is a well-developed IPM strategy. Few other systems have been developed to a similar degree of sophistication. Codling moth control still depends upon chemical spraying despite progress with confusion pheromones. Organic apples are difficult to produce and market supplies are thin (Clarke 1988).

Critics claim that research aimed at developing IPM systems has until now been initiated only in response to major crises of pest resistance or consumer confidence. There has been too little strategic development of these systems (Australian Parliament 1990: 150). Unlike chemical management systems, research into non-chemical means of control will not often result in the creation of a marketable product and so offers little attraction to commercial research companies.

Even where IPM systems have been refined and shown to be practical, widespread adoption is not guaranteed. IPM programs are more complex than simply spraying at predetermined times of the year. Farmers need a greater understanding of the farm ecosystem. Systems of biological control are slow and financially risky. Pest populations increase early in the season before the predator numbers can increase sufficiently to control the pest. A temporary population increase is acceptable if the pest does not directly attack fruit, as is the case with mites. However, where a pest directly attacks fruit, as does the codling moth, large financial losses can occur in a short time. Despite the professed interest of the consumer in chemical-free produce, most consumers prefer to purchase blemish-free conventional produce instead of organic produce when faced with the two products on adjacent shelves (see Chapter 12). This suggests that for most consumers, pesticide residues appear to be only marginal in purchasing decisions (Hassell and Associates 1990: 58). Blemished produce sells poorly, and in oversupplied markets it does not sell at all without an 'organic' label. Even an organic label may be insufficient. Organic peach producers can receive less for their product than conventional producers (Hassell and Associates 1990: 44). The use of chemical controls normally achieves an immediate short-term gain, even if the long-term effect is unclear. Many horticulturists are loath to risk crop losses in the hope that biological controls will become available.

CONSERVATION CROPPING: WHAT IS ENVIRONMENTAL SAFETY?

Chemical use poses dilemmas outside the horticultural industries. The widespread use of herbicides in cereal cropping is a recent phenomenon (Ewers 1990). Historically, cultivation has been a fundamental management tool on Australian wheat farms. In the original 'bare fallow' system the cereal grower cultivated a field in autumn and repeated the cultivation after each rainfall to keep it bare of weeds until the following autumn when the crop was planted. Farmers observed that bare fallow gave higher yields and reduced financial risks from low rainfall. The conventional explanation of these observations was that bare fallow conserved subsoil moisture. Bare fallowed paddocks harboured no weeds to transpire soil moisture, so rainfall seeped through the soil to the lower root zone where it could be used by the following season's wheat crop. In grey soil districts bare fallow saved the equivalent of 100 mm of rainfall. This translated into roughly 15 extra bushels of wheat per acre (or an extra one tonne per hectare).

In fact, increased subsoil moisture storage can only explain some of the yield gains associated with bare fallow. Bare fallow increases the amount of nitrogen available to the wheat crop by facilitating the breakdown of organic matter. Bare fallow limits weed infestation of the following crop by encouraging the early germination and destruction of weed seeds. It also provides a break on the proliferation of root diseases of the wheat plant, particularly the root fungus 'take-all'. Wheat, barley and most grasses are hosts for the disease. Growing wheat crops one after the other leads to an explosion in the population of the fungus and declining yields. The best method of control is to keep paddocks free of host grasses and susceptible cereals for at least two years until the 'take all' fungus dies. Croppers using the traditional management system achieved this by growing a crop of oats before the bare fallow. Because oats are a poor host for 'take all', this rotation gave a two year break without a disease host. The wheat crop which followed the break was relatively free of root disease. In some soils this latter advantage probably far outweighed any benefits from increased subsoil water storage.

Over time, continued cultivation has had a deleterious impact upon many of our more fragile soils. In the mallee wheat belt of Western Australia, South Australia and Victoria, paddocks of bare fallow became a high wind erosion risk. Dust storms came to be accepted as an unavoidable feature of wheat production. Bare fallow also increased the rate of recharge and hence the rise of saline watertables. In the wetter wheat growing districts of eastern Australia, bare fallow farming created a serious water erosion hazard and degraded the structure of cropping soils. Constant pulverising by implements hardened layers of soil beneath the plough and crusted the surface layer. Crusted soil surfaces hindered both the infiltration of water into the soil and the emergence of seedlings out of the soil. Hard pans reduced the root depth available to plants and limited soil drainage, creating waterlogging and flooding in winter and premature dry soil conditions in summer.

Modern conservation cropping practice encourages farmers to spare the soil by using the 'plough in a drum', that is, using herbicides to replace cultivation. Where farmers cultivate to save soil moisture, herbicides offer the alternative of a 'chemical fallow'. Weed kill achieved with herbicides helps prevent soil erosion by leaving a covering of dead grass or cereal stubble to hold the soil surface together and maintains the subsoil moisture capacity. Farmers who decide to use chemical fallow can operate at any of three stages of conservation cropping.

The first stage is 'minimum tillage farming'. Any fallowing is achieved by spraying herbicides. The chemical-fallowed paddock is left until the autumn rains herald the sowing season. The ground is then cultivated once or perhaps twice to form a seed bed and to destroy weeds prior to sowing. After harvesting the wheat stubble is burnt. Minimum tillage farming reduces both the impact of ploughing on soil structure and the period the soil is left exposed to wind or water erosion.

The second stage is where all tillage is eliminated and direct drilling is utilised. A farmer burns the wheat stubble before the autumn rains and sows seed directly into the unploughed paddock after the autumn break. Herbicide spraying controls weeds. Direct drilling aims to eliminate both the risk of erosion and the impact of ploughing on soil structure by reducing the number of times heavy tractors and implements cross paddocks.

The most complex form of conservation cropping is trash farming. The farmer does not burn the stubble and drills seed directly into the soil while the stubble is still standing. Sheep may be grazed on the stubble crop before sowing. Trash farming aims to gain all the benefits of direct drilling as well as the added benefit of increasing soil organic matter by retaining the wheat stubble to break down in the soil. The standing stubble further reduces erosion risk.

In wetter cropping districts where the past advantages of bare fallow have been due to root disease control rather than saving soil moisture, selective herbicides now offer the option of the complete elimination of fallow in continuous cropping systems. Continuous cropping is an IPM system which aims to control root disease and weeds and offers the additional benefits of erosion prevention and reduced recharge of the watertable. Selective herbicides allow growers to kill grass weeds growing amongst grain legume and brassica crops. Grain legumes and brassicas can then be used as a disease break crop to replace fallow for the control of root disease.

Herbicides are the key to the conservation cropping alternatives now available to farmers. With them cropping farmers can design crop rotations to balance many competing aims: the income needs of the farm business, the need to control root disease and the need to protect the fertility and structure of the soil. Without herbicides there is no option but continued cultivation. This places cropping farmers in a deeper dilemma than the horticulturists. Like horticulturists, many cropping farmers say they are seriously concerned about the safety of chemical herbicides and the effects of possible chemical residues in the soil (Agrimark Consultants 1988; Harvey and Hurley 1990; Conacher and Conacher 1986). But they also say they are concerned about the health of their soils and understand the need to protect them from erosion and structural decline (Cary, Wilkinson and Ewers 1989; Ewers *et al.* 1989; Harvey and Hurley 1990; Ewers 1990).

The relationships between the sometimes conflicting beliefs of farmers about the use of chemical herbicides and conventional cultivation can be seen in a belief map using Woelfel and Fink's (1980) multidimensional scaling technique (see Box 14.1).

Most farmers view cropping as a trade-off between one environmental danger and another. In a survey of cropping farmers in north-eastern Victoria (Cary *et al.* 1989) farmers believed, on the one hand, that cultivation practice using 'chemical weed control' was effective for killing weeds, although they did not believe that chemical weed control was 'safe for the environment' (see Figure 14.1). On the other hand, these cropping farmers knew that 'weed control by cultivation' was not associated with 'better soil structure'. To achieve a more effective weed kill, there is a trade-off between using chemical herbicides or increasing cultivation. Farmers are caught in the middle. Concerns about environmental safety may be a barrier to the use of chemicals as a means of weed control; concerns about soil structure are a major reason for rejecting mechanical cultivation (Goddard and Nash 1990).

Current research evidence indicates that the practicality of chemicals has won out over the less-tangible environmental dangers of chemicals (Harvey and Hurley 1990). As with other IPM systems, what farmers have rejected is management complexity, uncertain outcomes and increased risk. Very few cereal farmers have rejected herbicides after trying them on their farm. Herbicides are simple to integrate into an existing system of fallow farming. Minimum tillage is now a common feature of many cropping programs.

Other, more complex, components of the conservation cropping package have not been so well accepted. In north-east Victoria it seems that for every two farmers who have successfully adopted direct drilling there is one farmer who has abandoned it. For every farmer who has successfully adopted stubble retention, there is a farmer who has abandoned the practice (Cary *et al.* 1989; Harvey and Hurley 1990). Many of those who have abandoned these practices still hold positive attitudes to the new methods. They believe that, in general, the advantages outweigh the disadvantages, but not always on

Box 14.1 Belief Maps

A belief map is a method of illustrating, on paper, the way people view complex issues. Relationships between the parts of a complex issue have many dimensions, and a belief map displays as much of this complexity as it is possible to display in two dimensions.

We could ask a person to estimate the distances between each of the six state capital cities of Australia. For the six cities we would obtain 15 inner-city distances. Standard mathematical methods can then be used to convert these distances into a map which shows the perceived location of each capital city.

When a group of people is asked to estimate these distances, the estimates can be averaged and the same mathematical techniques can be used to produce an average map to show how the group perceives the location of the capital cities.

The same method is used to build belief maps. In a belief map, a belief is defined as the perceived distance between objects which comprise the belief. We determine which objects are central to people's beliefs about a particular complex issue. Then we ask them to estimate the distances between these various objects. A belief map can be produced from the average perceived distances obtained.

Belief maps show how people understand complex relationships. For example, belief maps can be used to help us understand the often complex decision about which car to purchase. If people believe car X to be luxurious and expensive, in a belief map 'car X', 'luxurious' and 'expensive' will all be close together. If people believe also that car X is fuel inefficient, 'car X' will be some distance away from 'fuel efficient'. Belief maps also help us to show people's understanding of even more complex and abstract issues such as land management.

When people are asked to estimate how far they see themselves from various belief objects, the position of the average person can be mapped in relation to the belief objects. We then have a map of the average person's attitudes towards the belief objects.

their own farms where conventional techniques are often more practical. The crucial problem is not chemical safety but crusting of direct drilled soils and the difficulties of using sowing machinery in paddocks of standing stubble.

COURTING THE CAUTIOUS CROPPER

Cropping farmers who face the dilemma of chemicals and conservation are courted by the competing advocates of chemical-free agriculture and conservation cropping. The Australian subsidiaries of overseas chemical companies have for many years mounted sophisticated long-term marketing strategies to

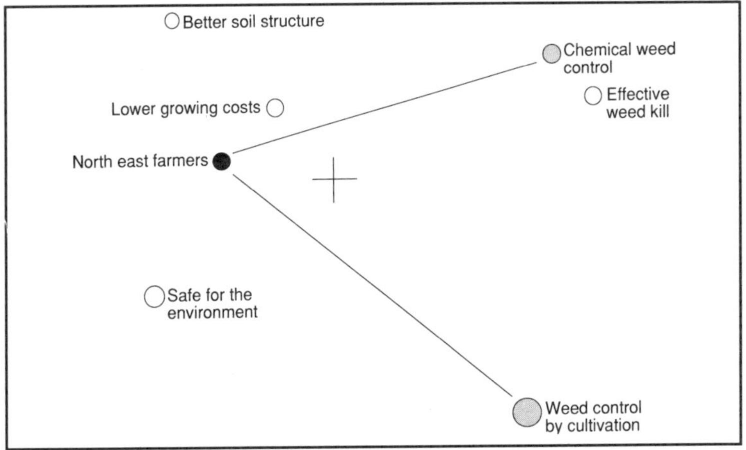

Figure 14.1 Farmers' beliefs about chemical weed control

develop conservation cropping farm management systems and build a market for their products. In southern New South Wales and north-east Victoria, ICI's market development of the herbicide product 'Sprayseed' included the provision of a 24 hour technical advisory service and the fostering of farmer discussion groups (Ewers 1990). The groups had much in common with the more recent Landcare movement. Through group and individual advice, ICI spread messages of product effectiveness, higher yields, greater profits and the satisfaction of looking after the soil. ICI personnel worked with farmers to solve problems as they arose and develop a practical farming system for the region.

Organic farming advocates are critical of the trend towards greater chemical use in Australian cropping. By European standards, Australian wheat farmers have been conservative users of insecticides, herbicides, fungicides and artificial fertilisers. Extensive wheat farms and clover pasture-based rotations have ensured that the use of nitrogen fertilisers is almost unknown in the wheat belt. In North America, this dependence on legume rotation is seen as a major step in the path towards sustainable agriculture (Committee on the Role of Alternative Farming Practices 1989). The trend towards reduced tillage and continuous cropping in Australia has increased cropping farmers' dependence on herbicides and insecticides. This trend is most obvious in continuous cropping programs which incorporate grain legumes. Legumes will often be sown after a herbicide has been sprayed to control grass weeds. The seed will be coated with a fungicide before planting for protection against leaf spot disease. Earthmites and fleas, loopers, heliothis or cutworms may have to be controlled by insecticides. Viral diseases, such as cucumber mosaic virus, are spread by aphids. The most effective control is spraying the crop with an aphicide. Farmers may also spray a selective herbicide to control grass weed competition for the legume and prevent the hosting of root diseases which could infect following wheat crops. Advocates of chemical-free farming question the long-term sustainability of this style of agriculture.

The 'sustainable agriculture' movement has three main philosophical strands in Australia — organic agriculture, biodynamic agriculture and permaculture. The organic agriculture movement eschews the use of artificial fertilisers and sprays, and advocates a farming strategy aimed at indirectly feeding plants by maintaining a healthy soil. The movement promotes a return to traditional cultural practices such as ploughing, composting, crop rotation and companion planting, and the use of natural chemicals and biological control. The movement believes the best way to fight pests is with a healthy plant, and the healthiest plants are produced by feeding plants only on the nutrition created by the action of soil microbes. Soil deficiencies are made good with ground mineral rocks rather than artificial fertilisers.

The biodynamic movement, based upon Rudolph Steiner's philosophy of social and natural relationships, shares the same concern for soil health as organic agriculture but its scope is widened to include a more explicit concern for the relationship between the farmer, the soil and society. Biodynamic farmers view the farm as part of a whole web of relationships, both organic and social. Ideally, farmers make a lifelong commitment to their land with the aim of healing the damaged earth. In contrast, chemical farmers are seen as seeking the maximum yield from their land with little regard for the ecological or social consequences. There are few major differences between biodynamic and organic farming. The biodynamic philosophy has not always been as strongly anti-chemical as that which forms the basis of the organic movement. Some Australian biodynamic advocates have contended that judicious use of artificial fertilisers can be necessary to achieve the appropriate balance on a farm (Podolinski 1985). Another major difference between the two movements is the biodynamic farmers' use of biodynamic formulae, particularly Formula 500 and 501. Formula 500 is manufactured by packing cow manure into cows' horns which are then buried in orientation with the moon. The manure is subsequently retrieved and sprayed over pasture at a time determined by phases of the moon. Biodynamic farmers apply Formula 500 at an extremely low application rate, approximately 100 to 150 grams to the hectare. It is claimed this acts not as a fertiliser, but as a catalyst to unlock the organic nutrients in a biodynamic soil. Formula 501 consists of ground quartz which is sprayed onto growing pasture to 'concentrate light onto plants' to 'stimulate the life forces' (Podolinski 1985; see also Stonehouse 1981; Oelhaf 1978).

The third major strand of alternative agriculture is Bill Mollison's permaculture. Permaculture aims to eliminate the monocultural design of commercial agriculture which has contributed to the development of pest and disease problems. In a permaculture system, trees are planted to recreate something resembling a natural ecosystem, with microenvironments created to benefit particular species groups. The aim is the establishment of a permanent, self-sustaining, regenerating productive system. The initial emphasis of permaculture was towards: 'small groups, living on marginal land available cheaply, where the ethics of farming are aimed at the future, and different, lifestyle, and where regional self sufficiency is more important than cash cropping for export, or monoculture for commercial gain' (Mollison and Holmgren 1978: 1).

Permaculture has had strong appeal to the refugees from urban life who have settled on country properties. Bill Mollison believes most of the land needed to feed Australia can be found within existing towns and that the only long-term future for the large agricultural property will be water supply, forest or restricted meat production. In this vision there is no role for broadscale cropping which is seen as unsustainable (Mollison 1990). It is unlikely this prognosis will have any appeal for the commercial broadacre Australian farmer or for governments.

Organic farmers believe they offer an alternative method of grain farming which seeks to protect both soil and consumers by eliminating artificial chemicals from farming systems. They promote a more traditional style of grain growing using old-style rotations and green manure crops. Crops dependent on insecticides are avoided. Weeds are controlled by ploughing in pastures as green manure crops before they set seed. The paddock is then left fallow for several months over summer until planting. Weeds which germinate after rain are controlled by cultivation. Pastures and wheat are fertilised with ground rock phosphate rather than processed superphosphate. The preference for rock phosphate is based on concern that the sulphuric acid used in the manufacture of superphosphate will damage worms, bacteria and other life in the soil. Stubble is grazed and then lightly cultivated to form a mulch on the surface and protect the seedbed of the next crop (Druce 1988;

Wynen and Edwards 1990; Wynen and Fritz 1987). Conservation cropping supporters criticise organic farming because ploughing is an integral component of its management systems. Organic wheat growers argue that, because they run their farms with longer pasture rotations, the extent of ploughing over a number of years may be little more than on a conventional minimum tillage farm.

Advocates of chemical-free farming run regular seminars, field days and conferences, mostly on members' farms. Their seminars offer a promise of profitability on organic farms, together with an increased sense of satisfaction gained from looking after the soil. Organic and biodynamic wheat farmers claim to have created a profitable and sustainable farm business. They achieve lower yields because of their weed control, phosphorus and pest control choices, but their costs are also lower (Wynen and Edwards 1990). Sometimes organic wheat producers can take advantage of a market premium.

Both the sustainable agriculture and conservation cropping systems have their ardent spokespersons claiming increased soil organic matter and increased earthworm activity as evidence of the success of their methods. Conventional cropping farmers are attracted to practical farm solutions, no matter who offers them. They are sceptical of the claims of the safety of chemicals and herbicides associated with conservation farming (Agrimark Consultants 1988; Goddard and Nash 1990). But as we have seen, this scepticism has not outweighed the perceived practical advantages of herbicides which are now commonly used on cropping farms (Agrimark Consultants 1988; Cary *et al.* 1989; Harvey and Hurley 1990). By contrast, organic cropping is uncommon. Despite claims that organic agriculture is one of the fastest growing forms of agriculture, it is in reality a very small industry. The average Australian consumes 5 cents worth of organic produce each week (Hassell and Associates 1990).

One of the barriers to entry to organic production is the difficulty posed by the period of initial adjustment as chemicals are removed from the farm system. Advocates of organic cropping maintain that it takes time for a farm ecology to adjust to the loss of chemicals. During this weaning period there may be financial difficulties (Hassell and Associates 1990: 32). Conventional farmers are sceptical of even the short-term financial sustainability of such a change. Conventional farmers are also likely to be suspicious of the mystical elements associated with some organic farming systems.

Playing a middle role between the chemical companies and the opponents of agricultural chemicals are governments. It is clear that government Departments of Agriculture believe that any dramatic regulatory reduction in the use of agricultural chemicals would lead to significantly lower agricultural production and a lower standard of living for the country (Australian Parliament 1990). Governments are also concerned about the impact of pesticides upon the environment and public health, and the advantages of being seen by overseas markets as a source of 'clean' produce (Australian Parliament 1990). Recent government policy has been to promote 'clean agriculture'. A 'clean agriculture' is an agriculture which uses enough agricultural chemicals to maintain production yet minimises the risks associated with agricultural chemical usage. For most governments the components of a 'clean agriculture' program are research and promotion of IPM, education programs to encourage responsible chemical use and the regulation of those involved in selling, promoting and using agricultural chemicals.

One of the motivations for 'clean agriculture' programs is to maintain public confidence in current patterns of agricultural chemical use. Market research suggests that many consumers believe that chemical sprays, even when properly used, are a danger to public health, that it is practical to grow our food without pesticides and that chemical residues in food are increasing (Crawford, Worsley and Peters 1984; Thompson *et al.* 1988). Though government policy makers can marshal convincing evidence that each of these statements is untrue and that the purchase of organic food is no guarantee of freedom from residues (Mellab P/L 1990) their fear is that a public loss of confidence in current agricultural practices could lead to increased regulation and decreases in production. The basis for dismissing consumers' concern about agricultural chemicals is a faith in the current system of chemical regulation. Government officials believe the results of residue monitoring justify their faith (Love 1991).

THE POISONS IN OUR FOOD

Regulatory authorities seek to protect the public by establishing a maximum residue level for each pesticide. The level is based upon experimental testing to discover a 'no observable adverse effect' threshold. This threshold is the highest dose level which will produce no observable adverse effect in the most sensitive test animal. Toxicologists use the no observable adverse effect threshold to calculate an 'acceptable daily intake', the average daily dose which will produce no appreciable effect during an entire lifetime. The acceptable daily intake, together with information about dietary habits, is used to set a 'maximum residue level' which will ensure acceptable daily intake levels will not be exceeded. The maximum residue level may be set lower than that indicated by toxicological work if a lower residue level can be achieved with good conventional agricultural practice.

The setting of maximum residue levels depends crucially upon the setting of the 'no observable adverse effect' threshold. There is a great difficulty in defining this threshold when searching for sublethal and carcinogenic effects. This threshold is often determined by experimentation with animals. The ability to detect a carcinogenic effect depends critically upon the size of an experiment. A small experiment with only a few animals will only detect carcinogenic effects caused by large doses. With increasing experimental size it is possible to detect toxic effects from smaller doses. Thus, no matter how large the size of an experiment, scientists who claim to have detected a no observable adverse effect threshold are left with the question of whether a still larger experiment would have detected a toxicological effect. One of the largest recent toxicological studies used 24 000 mice over five years at a cost of US$7 million, only to leave this question of low-dose responses unresolved (Ministry of Environment 1989). To claim a scientific proof of a no observable adverse effect threshold is misleading. It could equally be argued that only a few molecules could be carcinogenic and the failure to find a no observable adverse effect threshold is merely a reflection of an inadequate experiment. The existence of this threshold cannot be proven by experimental techniques. It can only be inferred. That is the nature of biological science. Likewise, it is difficult to establish epidemiological evidence that there are no chronic effects from long-term low dose exposure to chemicals (World Health Organisation 1990).

It is superficially attractive to conclude that because science cannot show the synthetic chemicals to be absolutely safe, they should be eliminated. The same arguments can be applied to natural chemicals. Is a cancer risk from a natural chemical any better than a risk from an artificial chemical?

Until recently natural chemicals were rarely tested for carcinogenicity. However, new tests of carcinogenicity and mutagenicity have revealed research fakery by some companies and weakness in old data. Some artificial chemicals once thought safe have been banned (Schneider 1983). The new tests, used on natural chemicals, show that nature is not benign. Many of the natural flavour constituents of common foodstuffs have been shown to be carcinogenic or mutagenic. Peanuts contain highly carcinogenic aflatoxins; alfalfa sprouts contain the natural and carcinogenic pesticide canavanine; celery, parsnip, parsley and figs contain carcinogenic furocoumarins; pepper, mustard and horseradish contain carcinogenic compounds such as safrole, piperine, isothiocyanates and phorbol esters (Ames 1983). Toxicologist Bruce Ames estimated that humans consume between 1 and 2 grams of these natural carcinogens each day. This is 10 000 times the average intake of artificial pesticides. Many of these chemicals, if manufactured synthetically and presented as new food additives, would not be accepted for registration. The natural chemicals are quite capable of producing their own 'silent spring'. In California, a sudden run of birth deformities in one family involving a child, a litter of goat kids and a litter of puppies was blamed on herbicide spraying. Later investigation showed that the pregnant mother and dog had been drinking milk from the family goats which had been feeding on lupins. Alkaloids in the lupins had been concentrated in the goat's milk, causing the birth defects (Ames 1983).

While wholesome healthy-looking vegetables or fruit may have these hidden risks, damaged or 'off' food can be much worse. Many plants increase their toxin concentrations when damaged or under attack. The attackers, in turn, manufacture their own array of toxins to overcome their prey. Moulds generate aflatoxins and sterigmatocystins. These are highly carcinogenic compounds. The ergot fungus which attacks rye is a very potent psychoactive drug which can cause brain damage. Consumers' historical preference for undamaged food can be explained by damaged food being more likely to be poisonous. In a sense, quality is safety.

It is clear we cannot eliminate natural carcinogens from our diet. Perhaps we should do our best to lower our intake. Anything which increases the level of intake of natural pesticides should be monitored. Breeding for disease resistance, a strategy of non-chemical pest control, is likely to increase our intake of natural pesticides. It creates plants with higher concentrations of these chemicals. Breeders are trying to increase the disease resistance of domestic lettuce by crossing it with wild lettuce to transfer the gene which produces a mutagenic natural pesticide in the sesquiterpene lactone family. In the United States, pest-resistant varieties of potato and celery have been taken off the market in recent years because they exceeded maximum desirable levels for glycoalkaloids and psoralin derivatives. If we applied the conservative low-dose argument which is used against artificial pesticides to these natural pesticides we would come to an interesting conclusion. We would support breeding to reduce the level of natural toxins in food — which, not surprisingly, has been going on for centuries. Potatoes were once much more toxic than they are today. The Inca used to process them to remove the poisonous chaconine and solanine to make them safe to eat. Today's potato is the product of a long period of selection to reduce toxicity. In this sense, the safe potato is an unnatural plant, as are many of the staple foods in our diet. Yet, if we were to reduce the level of natural pesticides in our food, we would need to use more artificial chemicals for pest control.

CONCLUSION

There is some wisdom to be gained from reflecting on Rachel Carson's revealing vision of death and destruction caused by agricultural chemicals. In the developing world this vision has proved at times depressingly accurate. One World Health Organisation (WHO) report has estimated that each year some 220 000 people die of agricultural chemical poisoning (World Health Organisation 1990). Most of these cases occur in the developing world. We, in the developed world, hear of gruesome accidental tragedies such as the chemical leak at Bhopal or the mass poisoning in Iran caused by the milling of pesticide-treated grain. It is suicide, however, which is the major cause of the fatal poisoning by agricultural chemicals, estimated to be over 90 percent of all pesticide fatalities. WHO estimates that occupational exposure accounts for 6 percent of fatalities and food contamination accounts for only 3 percent (World Health Organisation 1990). How do we balance these losses against the life-saving gains made in disease control and the associated dramatic changes to the quality of life in tropical countries? How do we balance them against improved security of food production or food storage? Despite the extent of poisoning detailed, the WHO report we have cited does not recommend the banning of pesticide use. It advocates instead the adoption of stricter legislative control, training programs, strengthened monitoring and greater epidemiological research to improve current estimates of the extent of pesticide poisoning which are based upon limited hospital surveys and animal experimentation. In the developing world decisions about wisely using pesticides will be complex without simple, black and white solutions.

In Australia, we have been spared the tragedies of the Third World. The 'evil spell' predicted by Rachel Carson has not 'settled on the community'. While some would argue Carson's vision was exaggerated and overstated, it nevertheless played an important role in changing the emphasis of conventional pest control towards IPM. We have not been spared the difficult decisions posed by pesticides. The question we face now, 30 years later, is how far we travel down the path of Rachel Carson's alternative. If we are to accept the evidence of most farmers' behaviour, it seems *Silent Spring* overstated the viability of the non-chemical alternative. Unlike consumers, most farmers who are concerned about agricultural chemicals see no clear alternative, only difficult choices and ambivalent decisions.

There will be no end to the difficult choices farmers face. A lesson from any historical assessment of pest control is that reliance on any one control measure is ultimately doomed to failure, whether the control be artificial or natural. Pests develop resistance to natural or artificial chemicals. They adapt to plants bred for insect resistance. Natural selection usually adapts to subvert controls aimed at a pest's life cycle or behaviour. It is hard to see any pest management system being permanently sustainable. Flexibility and adaptation are the key to biological survival. Inflexible commitment to any idea or ideology is the first step to unsustainability.

REFERENCES

Agrimark Consultants (1988) *Attitudes and Behaviour of Users and Retailers towards Rural Chemicals in Victoria*, unpublished report for Department of Agriculture and Rural Affairs, Agrimark Consultants Pty Ltd, Melbourne.

Ames, B. (1983) 'Dietary Carcinogens and Anticarcinogens', *Science* 221 (4617): 1256–1264.

Anon (1898a) 'Notes and Comments', *Journal of Agriculture and Industry of South Australia* 2: 174.

Anon (1898b) 'Norton Summit Agricultural Society Report', *Journal of Agriculture and Industry of South Australia* 2: 131.

Anon (1898c) 'Transcript of the 1898 Agricultural Congress', *Journal of Agriculture and Industry of South Australia* 2: 199.

Anon (1898d) 'Balaklava and Naracoorte Agricultural Society Reports', *Journal of Agriculture and Industry of South Australia* 2: 150–151.

Anon (1899) 'Notes and Comments', *Journal of Agriculture and Industry of South Australia* 2: 470.

Australian Parliament (1990) *Senate Select Enquiry into Agricultural and Veterinary Chemicals*, Australian Government Publishing Service, Canberra.

Australian Science and Technology Council (1989) *Health, Politics, Trade: Controlling Chemical Residues in Agricultural Products*, Australian Government Publishing Service, Canberra.

Carson, R. (1962) *Silent Spring*, Houghton Mifflin, Boston.

Cary, J., Wilkinson, R. and Ewers, C. (1989) *Caring for the Soil on Cropping Lands*, School of Agriculture and Forestry, University of Melbourne, Melbourne.

Clarke, R. (1988) *The Retail Industry for Organic Fruit and Vegetables*, Market Report Series, Department of Agriculture and Rural Affairs, Melbourne.

Commission of Public Health (1976) *Annual Report*, Victorian Government Printer, Melbourne.

Committee on the Role of Alternative Farming Practices in Modern Production Agriculture (1989) *Alternative Agriculture*, National Academic Press, Washington.

Conacher, J. and Conacher, A. (1986) 'Herbicides in Agriculture: Minimum Tillage, Science and Society', Occasional Papers of the Department of Geography, University of Western Australia, *Geowest* No. 22: 100.

Crawford, D., Worsley, A. and Peters, M. (1984) 'Is Food a Health Hazard? Australians' Beliefs About the Quality of Food', *Food Technology in Australia* 36: 9.

Davey, H. (1929) 'Current Orchard Work', *Journal of Agriculture Victoria* 27: 719.

Druce, A. (1988) 'Sustainable Agriculture: A Case Study', Paper presented to the Sustainable Agriculture: Farming for the Future Conference, Benalla.

Ewers, C. (1990) 'Innovation in Response to Environmental Problems', unpublished M.Agr.Sc. thesis, School of Agriculture and Forestry, University of Melbourne, Melbourne.

Ewers, C., Hawkins, H., Kennelly, A. and Cary, J. (1989) *Onion Growers' Perceptions of Soil Management Problems in Northern Tasmania*, School of Agriculture and Forestry, University of Melbourne, Melbourne.

French, C. (1902) 'Economic Entomology and Ornithology', *Journal of Agriculture Victoria* 1: 59.

Goddard, B. and Nash, P. (1990) 'Farmers' Attitudes and Intentions Towards Conservation Cropping Practices', Paper presented to the Fifth Australian Soil Conservation Conference, Stable Cropping Systems Workshop, Perth.

Hammond, A. (1932) 'Common Orchard Diseases: Control Measures Recommended', *Journal of Agriculture Victoria* 30: 261.

Harvey, J. and Hurley, F. (1990) *Cropping and Conservation: Changes in Cultivation Practices in Victorian Grain Growing Areas 1984–89*, Regional Studies Unit, Ballarat University College, Ballarat.

Hassell and Associates (1990) 'The Market for Australian Produced Organic Food', Report prepared for the Rural Industries Development Corporation, Canberra.

Krichauff, F. (1898) 'Chairman's Address to the 1898 Agricultural Congress', *Journal of Agriculture and Industry of South Australia* 2: 186.

Lang, J. (1903) 'The Codlin Moth', *Journal of Agriculture Victoria* 2: 59.

Laffer, G. (1898) 'Experience with Codlin Moth 1898', *Journal of Agriculture and Industry of South Australia* 2: 190–193.

Love, K. (1991) *Victorian Produce Monitoring: Results of Residue Testing 1989–1990*, Research Report Series No. 105, Department of Agriculture, Melbourne.

McAlpine, D. (1903) 'A Fungus Parasite of Codlin Moth', *Journal of Agriculture Victoria* 2: 469.

McAlpine, D. (1904) 'Black Spot Experiments', *Journal of Agriculture Victoria* 2: 761.

Mellab P/L (1990) 'Organic Food Survey', Press Release, Melbourne.

Ministry of Environment (1989) *Pesticides: Issues and Options For New Zealand*, Ministry of Environment, Wellington.

Mollison, B. and Holmgren, D. (1978) *Permaculture One: A Perennial Agriculture for Human Settlements*, Corgi, Melbourne.

Mollison, B. (1990) 'Appropriate Scales for Food Production', *Acres Australia* 3: 28–29.

Nancarrow, R. (1977) 'Practices and Precautions with Pesticides', unpublished Diploma of Agricultural Extension Project, School of Agriculture and Forestry, University of Melbourne, Melbourne.

Nancarrow, R. (1979) 'Applying Adult Education Principles to Agricultural Extension Programs: A Case Study of Pesticide Extension', unpublished M.Agr.Sc. thesis, School of Agriculture and Forestry, University of Melbourne, Melbourne.

Oelhaf, R. (1978) *Organic Agriculture: Economic and Ecological Comparisons with Conventional Methods*, John Wiley and Sons, New York.

Olliff, A. (1890) 'Codlin Moth', *Agricultural Gazette of New South Wales* 1: 4.

Olliff, A. (1891a) 'Entomological Notes', *Agricultural Gazette of New South Wales* 2: 385–386.

Olliff, A. (1891b) 'Insect Friend and Foes', *Agricultural Gazette of New South Wales* 2: 63–66.

Olliff, A.S. (1891c) 'Economic Entomology in Victoria', *Agricultural Gazette of New South Wales* 2: 489–491.

Pescott, E. (1910) 'Orchard and Garden Notes', *Journal of Agriculture Victoria* 8: 478.

Pescott, R. (1933) 'Codling Moths: Experiments at Harcourt', *Journal of Agriculture Victoria* 31: 487.

Podolinski, A. (1985) *Bio-dynamic Agriculture Lectures*, Volume 1, Gavemer Foundation Publishing, Sydney.

Quinn, G. (1898a) 'Orchard Notes for August', *Journal of Agriculture and Industry of South Australia* 2: 31–32.

Quinn, G. (1898b) 'Use of Bordeaux Mixture in South Australia', *Journal of Agriculture and Industry of South Australia* 2: 33.

Rigg, W. (1924) 'Poultry in the Orchard', *Journal of Agriculture Victoria* 22: 186.

Schneider, K. (1983) 'IBT — Guilty: How Many Studies Are No Good?', *Amicus Journal* 5 (2): 4–7.

Stonehouse, B. (1981) *Biological Husbandry: A Scientific Approach to Organic Farming*, Butterworths, London.

Thompson, H., Ashley, J., Stopes, C. and Woodward, L. (1988) 'Consumer Survey', Research Notes No. 6, Elm Farm Research Centre, Kintbury.

Wilson, H. (1960) 'Poisonous Organic Phosphorous Insecticides: Care Needed in Their Use', *Journal of Agriculture Victoria* 58: 453.

Woelfel, J. and Fink, E. (1980) *The Measurement of Communication Processes: Galileo Theory and Method*, Academic Press, New York.

World Health Organisation (1990) *Public Health Impact of Pesticides Used in Agriculture*, WHO, Geneva.

Wynen, E. and Edwards, G. (1990) 'Towards a Comparison of Chemical Free and Chemical Farming in Australia,' *Australian Journal of Agricultural Economics* 34: 39–55.

Wynen, E. and Fritz, S. (1987) *Sustainable Agriculture: A Viable Alternative*, National Association for Sustainable Agriculture, Sydney.

15 THE GROWTH OF AGRIBUSINESS: ENVIRONMENTAL AND SOCIAL IMPLICATIONS OF CONTRACT FARMING

DAVID BURCH, ROY RICKSON and ROSS ANNELS

Western agriculture has been transformed by the rise of agribusiness which largely controls the supply of farm inputs and the processing and marketing of farm outputs. Agribusiness corporations have dramatically altered the social organisation of agriculture and changed the nature and source of agricultural inputs, the production technologies employed, the character and types of crops grown, the level of off-farm processing of agricultural produce, and the marketing of agricultural products. Furthermore, the farm and the farm family have been progressively integrated into the corporate structure of the agribusiness corporation. They have assumed a subordinate and dependent role as a production unit in a movement toward the so-called 'industrialisation of agriculture'.

The extent of this incorporation may be judged by comparing an 'ideal typical' model of traditional agricultural systems with current structures. In the ideal model, the social organisation of agricultural production is seen as an aggregate of the actions of autonomous, individual landholders (family farmers). Crops are grown for (typically) local and regional markets, according to the biophysical properties of their land and the demands of local and regional economies, and mediated through competitive commodity markets. The logic of agribusiness, by contrast, requires that farm production be linked to both the supply of inputs and the demands of processing and marketing facilities, which are geared to the demands (both managed and created) of an international marketplace. Agribusiness companies, operating within a framework of monopoly capitalism, are therefore driven to integrate — vertically and horizontally — and to coordinate this whole system so that they may control the overall process 'from the seedling to the supermarket' (Thiel 1985).

In contrast to the 'conventional' industrial model of vertical integration, the agribusiness corporation does not engage directly in on-farm production through the establishment of large corporate farms. Instead, the central mechanism of control by agribusiness is contract farming, a system in which companies involved wholly or partly in the processing, marketing or retailing of agricultural goods enter into contractual arrangements with farmers for the supply of a particular commodity. These might include vegetables for freezing, fruits for canning or meats for processing.

The contract system now covers a wide range of commodities produced by Western agriculture and, increasingly, by the Third World as well (Minot 1986; de Treville 1987). By 1980, contract production in the United States dominated sugar beet (98 percent of output) fluid milk (95 percent), broiler chickens (89 percent), processed vegetables (85 percent), seed crops (80 percent), citrus fruits (65 percent), turkeys (62 percent) and eggs (52 percent) (Winson 1990: 377). In Australia, 85 percent of chickens and nearly 100 percent of peas are produced under contract (Luckhurst 1984; Burch, Rickson and Annels 1992).

The structural transformation of agriculture which has given an impetus to contract production in recent years has also incorporated new participants in a process of production, processing and distribution which has become highly complex and globally focused. Originally, contracting was undertaken by canning and freezing companies, such as Heinz, McCain or Edgell–Birds Eye, which processed farm outputs and marketed them under their own brand names. Increasingly, however, marketing and distributing organisations — in the form of supermarket chains and fast food outlets — have come to contract directly with sources of supply for 'generic' or 'own brand' products, or for the specialist inputs they need to meet market demand. In the process contracts may be negotiated and issued anywhere in the world as processors and/or retailers seek the cheapest suppliers. It is not unusual to see on the shelves of Australian supermarkets tins of tomatoes from Thailand, beetroot and green peas from Hungary, green beans from New Zealand, mushrooms from China, corn from Taiwan and pineapples from Thailand. Similarly, Singaporeans purchasing a McDonald's hamburger will be consuming beef patties produced and processed in Australia.

This new structure has far-reaching social and environmental implications at a number of levels — for the individual farmer, for the family farm, for the rural community, for the urban consumer and for the wider social sphere. In this chapter, we will focus on a number of these outcomes. In particular, we will analyse the shift in the locus of decision making in operational and management matters implied by contract farming, and consider the consequences of this and other changes for environmental outcomes and policies.

In order to address these issues, we have arranged discussion around a number of analytical categories. These categories are used to demonstrate both the nature and extent of the structural transformation of agriculture and to evaluate the environmental and social outcomes of contract farming. The seven categories are (1) responsibility, (2) profitability, (3) intensity, (4) flexibility, (5) mobility, (6) complexity, and (7) sustainability.

These should not be considered to form an exhaustive typology, or to delineate particularly discrete categories. Rather, the intention is to use these to focus analysis on what we consider to be the major social and environmental impacts.

RESPONSIBILITY

Overall, the system of contract farming has resulted in a transfer of responsibility for many production decisions (and those about soil and water conservation) from the individual farmer to the processing/marketing company which offers the contract (Nelson and Murray 1967; Davis 1980). For example, in the case of frozen vegetables, the contract will typically specify the variety of seed to be planted, the extent of cultivation, the date of planting, the cultivation practices to be followed, the approximate date of harvesting and the standards of grading. In return, the farmer is paid a guaranteed price based on quality and harvested volume (Burch, Rickson and Thiel 1990). Similarly, in the case of chicken production the processing company supplies day-old chicks, feed, medication, field services and some credit. But at no stage does the farmer own the chickens; instead, the grower is paid a service or husbandry fee on a 'per bird grown' basis (Luckhurst 1984).

In situations such as these, farmers experience a considerable loss of autonomy and responsibility in moving from market to contract-based production, and the grower's role in the production process undergoes a significant change: 'At its most extreme, [contract farming] may reduce the farmer to a wage earner on his own land — a piece-worker who provides his own tools and works under supervision to produce commodities which he does not own. He sells his labour power instead of chickens, apples, beans or beets' (Davis 1980: 142).

Farmer compliance with the terms of the contract is ensured

by the appointment of field officers employed by the processing/marketing company who, in the case of vegetable contracts, have rights of access to a farmer's property. These field officers can require farmers to undertake specified procedures, for example crop spraying or other operational and management practices, and a failure to comply may result in the processor refusing to harvest or take delivery of produce. Such action will almost certainly jeopardise the prospect for future contracts.

Decisions about cropping, chemical use (fertilisers, pesticides and herbicides) and soil conservation, in contract farming, are no longer the exclusive prerogative of the farmer, but are the result of an accommodation between growers and agribusiness managers. The relationship is an unequal one with the farmer becoming increasingly dependent upon the corporation; the farmer's decisions are therefore increasingly subject to company control (Burch *et al.* 1990).

Since agribusiness corporations operate in a complex and highly competitive global environment, there is constant pressure on companies to maintain control over sources of supply (the farmer) so as to keep prices low. In this situation, on-farm management practices become an extension of the corporate structure and decisions about resource use will, consequently, reflect corporate interests. In general terms, agribusiness corporations stress the economic goals of the maximisation of growth, efficiency and short-term profits, while the 'ideal typical' farmer who produces for a local market will normally seek to optimise returns over the long run, within the biophysical properties of the land. Of course, in practice, farmers do not always act so rationally; nevertheless, they are more likely to utilise physical resources that can sustain a reasonable, rather than a maximum, return. For these reasons, agribusiness corporations are less likely to invest in land conservation than individual farmers.

While some recognise that contract farming has transferred the responsibility for many production decisions from the individual farmer to the contracting company, it is not yet understood that responsibility for environmental impacts has also shifted. At all levels of government, the legislative framework governing policies on Landcare, soil erosion and soil salinity still assumes that the individual farmer has full control over the production process. The regulatory framework designed to induce adoption of soil and water conservation methods, for example, is focused mainly on the farmer with the explicit assumption that problems associated with soil loss or salinity, depletion of water supplies or excessive uses of chemical pesticides are the farmer's responsibilities. These problems are seen to be best solved by pressuring individual producers to change their values and behaviours. Neither conservation law nor regulatory policy sufficiently recognise the increasing influence of the agribusiness corporations on a farmer's decision making through mechanisms such as production contracts.

Agribusiness corporations need not, in the short term, be concerned about problems on any given farm or area because, unlike farmers, they are not dependent upon maintaining either the quality or quantity of land or water resources in any particular locale. As we shall see shortly, processing/marketing companies have considerable flexibility when it comes to their sources of supply, and corporations may choose to exercise this flexibility rather than the responsibility which attaches to their role as major participants in decisions over resource use (cf. Bates 1987).

Contract farming, for the above reasons, poses a dilemma for official policy on rural land degradation since it implies a separation of land ownership from power to make decisions about either crop production or land conservation. Importantly, this fact has not yet been reflected in the legislation, and only barely acknowledged in the voluntary actions of corporate bodies. The only known exception to this arises out of the soil conservation policy adopted by CIG Pyrethrum, which contracts with some 70 Tasmanian farmers for the production of pyrethrum over 380 hectares.

Pyrethrum is a bush-like perennial produced for its flowers which contain a range of natural insecticides called pyrethrins. When letting the contracts for 1990, the company (a subsidiary of Commonwealth Industrial Gases) made compulsory the adoption of soil conservation measures, such as the establishment of contour drains to harness water and prevent it from washing gullies through crops (*Tasmanian Star* 19 November 1990). This case illustrates the degree of *de facto* control that corporations exert over on-farm practices. Given that this is the only case of corporate environmental protection it also reveals how little responsibility that sector has accepted for the environmental impacts of farming practices they promote through the contract farming system.

PROFITABILITY

The competitive environment of agribusiness requires that farmers, processors and retailers maintain or improve their returns through the pricing mechanism. Each party is strongly motivated to maximise or at least improve the prices received for their outputs and reduce the costs for inputs to production. Under the contract system, the farmers' outputs are the processors' inputs, while the processors' outputs are the retailers' inputs. In this situation, farmers are in a 'double bind' as they are usually price-takers not price-setters, and are unable to reduce the costs of their inputs or guarantee increased returns from their output (Lawrence 1987).

Farmer subordination in the contract system largely derives from its structure and the fact that there are many more farmers than processors. Moreover, in a system of monoculture in which large areas of land are specifically devoted to growing a single crop and consigned to a particular processing plant, grower dependence on the processing company is greatly increased. In these conditions, farmers can be forced to accept reduced prices and falling real incomes to serve the interests of the local processor and the parent company. Farmers can be asked to subsidise processors by accepting lower prices (Tasmanian growers, for example) to enable companies to establish their operations or upgrade processing facilities (*The Australian* 19 August 1991).

In the Lockyer Valley of Queensland, contract farmers have had to accept significant reductions in returns in recent years, rather than the assured income which is held to be one of the benefits of the contract system. In the case of pea production, growers have not only been squeezed by increased costs which have not been reflected in increased prices, but have also had to accept significant price reductions on a number of occasions. In 1985, for example, Edgell–Birds Eye, a subsidiary of Petersville Sleigh, cut the price paid to Lockyer Valley pea producers by 10 percent which, in 1985–86, flowed through to Tasmanian producers (Burch *et al.* 1990). Improved prices were negotiated subsequently, but by 1987 increases of 20–23 percent in the Lockyer Valley only succeeded in restoring prices to their 1984–85 level. In 1991, Edgell–Birds Eye again succeeded in reducing prices, this time by 12.5 percent. On this occasion, growers believed that there was a danger of the company withdrawing from processing in Queensland — something which undoubtedly influenced their acceptance of a reduction in price (*The Courier-Mail* 19 June 1991).

It has been suggested that processing companies use Queensland growers as a bargaining chip to force price reductions on farmers in Tasmania and elsewhere (*The Courier-Mail* 22 May 1991; Burch *et al.* 1992). Producers in Queensland account for only about 13 percent of total Australian pea production and are therefore marginal to the overall operations of the processors. However, negotiations on price usually occur first in Queensland and prices set there significantly affect the price paid to Tasmanian producers, who account for 75 percent of Australian production.

In both 1985 and 1991, Edgell–Birds Eye claimed (with some justification) that the price reductions were forced on them by a flood of 'cheap imports'. In addition, this and other processing companies have found themselves under increasing pressure from supermarket 'own brand' or 'generic' labels sourced from all over the globe. The point is that in this highly competitive situation, there is constant pressure on all participants in the chain of production, processing and marketing to reduce their costs and increase their prices. Inevitably, though, this bears down most heavily on the individual farmer, who can neither reduce the cost of inputs or increase the price of outputs.

Growers cope, in these situations, by trying to reduce other costs, or to increase productivity, so that they may continue participating on the terms laid down by the processor/retailer. However, this not only implies a continuing investment of time and capital on the part of increasingly marginalised farmers, but also increases the heavy demands already placed on agricultural ecosystems. The contract structure therefore encourages growers to push the land beyond reasonable limits at a time when falling incomes already imply fewer resources available for conservation and preservation.

INTENSITY

Intensive cultivation is a basic characteristic of modern contract farming. Production decisions in 'industrial' agriculture are highly centralised and closely associated with corporate goals, that is, the maximisation of profits, growth and efficiency. These goals, supported by the policies and structure of the contract system, promote farm practices associated with maximisation of short-term production from land and water resources rather than sustained yield. It is, therefore, a structure of production that discourages farmers from adopting measures for conserving land and water resources. It fosters as well the use (and perhaps overuse) of chemicals such as pesticides and herbicides.

A consequence of a general commitment to growth, that is, maximisation of short-term yield, is the increased application of intensive techniques such as 'improved' seed varieties, greater reliance upon fertilisers and pesticides, and the widespread establishment of irrigation facilities. All of these factors place great stress on the local environment and expose it to degradation. Hybrid seeds, for example, are designed to generate high yields and/or display characteristics desired by the processor/retailer, such as standard size, robustness in handling and long shelf life. Typically, this process of 'improvement' of seed varieties involves a narrowing of the genetic base of agriculture because of monocultural systems (Mooney 1979; Burch *et al.* 1992).

Since new varieties or hybrids are especially vulnerable when taken from the laboratory and released into the open environment, plant breeders rely heavily upon chemical protection of new plant types. Increases in the level of application of pesticides and other chemicals may result in residues leaching into soil, water sources or food chains. Monoculture and the related narrowing of the genetic base may also reduce soil quality, especially when combined with an intensity of production which minimises crop rotation and fallow periods (Burch *et al.* 1992).

Contract farming and agribusiness control intensifies pressure upon water resources. Modern vegetable varieties are bred to grow optimally under irrigation and there is evidence that at least some vegetable varieties used for processing require higher levels of irrigation than equivalent market varieties (Burch *et al.* 1992). In Australian conditions, continued and intensive use of water resources, in conjunction with other contemporary agricultural practices, are closely linked to increasing salinisation. In Queensland's Lockyer Valley, contract vegetable growers rely on groundwater resources for irrigation, and their actions are implicated in the salinisation problems there, although there is some debate as to the causes of salinity (Gardner 1985). In drought periods, however, the aquifer levels drop dramatically. As a consequence, farmers experience a drop in bore capacity and an increase in salinity. During the 1970s drought, 68 percent of the bores in the Lockyer Valley yielded water with salinity exceeding the tolerance level of beans (Talbot *et al.* 1981).

The choice facing contract farmers is thus stark: continue to rely on irrigation — which will degrade the land; or to seek alternatives to contract farming which allow greater flexibility in their responses to the threat of salinisation. In a survey of contract farming in the Lockyer Valley, 16 percent of ex-contract farmers cited inadequate or saline water as the reason for opting out of contracting (Burch *et al.* 1992).

The question of intensive farming and resource use also takes on particular significance in the contract system because of the need of the processing/retailing companies to ensure a constant throughput of raw materials so that their plant and equipment are fully utilised. For example, in the case of pea or bean production for freezing, processors will arrange for phased planting, so that harvesting can also be planned in a way that ensures optimal utilisation of the processor's capital equipment.

However, a system of phased planting cannot guarantee sufficient throughput to the processing plant if, for example, crops yields are reduced because of the weather. A way of overcoming this potential problem for the processor is to offer contracts which specify planting by area but harvesting by volume. For example, a processor may contract for 20 hectares of peas, but undertakes to harvest, say, 2000 kg per hectare, this figure being calculated on the basis of the volume of raw material needed to maintain productive capacity at the processing plant. However, if a farmer has a good year and, through the application of intensive techniques, produces a yield of 3000 kg per hectare, the optimal needs of the processor will be met, but without incurring any obligation to harvest and process the extra volume. Instead, the extra 1000 kg may be 'bypassed' and ploughed into the ground, with the farmer denied the extra income that this yield might have produced. Some soil benefits may, unwittingly, be obtained in this way — but the cost of the green manure crop will be significant.

As long as the processor agrees to harvest the agreed volume, it can be argued that all contractual obligations have been fulfilled and the grower is no worse off as a result. However, the grower will be worse off, not simply as a result of losing a 'windfall' income associated with a bumper year, but also because the costs of all the inputs — water, seed, fertiliser, pesticides and labour — associated with the unharvested portion of the crop that will not yield any return. It is a particular source of antagonism between farmer and processor if the processor's field officers had ordered the spraying of crops for reasons of 'quality control', with all the expense involved, only to have the processing company refuse to harvest the increased output which, in part at least, would have resulted from its ordering the use of such intensive practices.

Farmers interviewed in the Lockyer Valley have voiced the suspicion that the companies engaged in the processing of peas and beans routinely 'over-contract', either by area or the number of contracts, in order to ensure an optimal throughput of raw materials (Burch *et al.* 1992). Since they contract by area but harvest by volume, processing companies can do this without guaranteeing to harvest more than they need. If yields are low then the fact of 'over-contracting' will compensate and the processor may take all the produce grown; but if yields are high, the processor will 'bypass' much of the crop, taking only what is needed. This was a critical issue in the Lockyer Valley in 1990, when Edgell–Birds Eye proposed to harvest only 3 tonnes per hectare, a decision which would have left about 50 percent of the crop to be ploughed in. Following representations to the company, the figure was increased to 3.7 tonnes per hectare, but this still left much of the crop unharvested (*Gatton, Lockyer and Brisbane Valley Star* 2 October 1990).

The issue of 'bypassing' causes a good deal of conflict between grower and processor. But it involves more than a loss of farm income if intensive agricultural practices are encouraged for the sole purpose of ensuring that the processor's capital equipment is kept occupied. What is also involved is an unnecessary use of resources and further environmental stress in intensively cultivating crops which may well be ploughed back into the ground.

FLEXIBILITY

Contract farming implies a dominant role for the processor and retailer (the two are often combined) that narrows farmers' decision-making options. The structure facilitates corporate flexibility at the price of reducing farmers' flexibility and adaptability which, together, increase farmers' dependence upon the company. The more farmers' decision making is tied to the corporate structure, the less will be their ability to adapt to general problems and opportunities in agriculture. Some decisions farmers may wish to make may be incongruent with the short-term economic goals of the corporation, for example those relating to emergence and desirability of new technologies, development of new markets, and changed government policies.

For example, recent expansion of potato cultivation in Tasmania has been spurred on by the growth in demand for frozen chips for sale in fast food outlets such as McDonalds and Kentucky Fried Chicken. The area devoted to such increased output will be organised around a particular processing plant and farmers will have little alternative but to continue to deal with the owners of that plant. Added to this is the fact that increased production of a new crop will often involve farmers in considerable investment in land and specialist machinery, which may lock them into the contract system. Under these circumstances, growers have little bargaining power which, as noted earlier, makes them vulnerable to pressure from processors to keep prices down.

Contract production, then, institutionalises and legitimises a reduction in operational flexibility, which is likely to make it difficult to reorientate production along alternative lines of development when economic circumstances or environmental considerations dictate this. Thus, contract production of vegetables requires farmers to use particular technologies, such as chemicals for pest control. While this technology is widely used in the non-contract sectors of agriculture, such use is not mandatory and farmers in these sectors are free to adopt alternatives. Indeed, it is arguable that the renewed interest in organic agriculture will lead to a search for new techniques of pest control which avoid the environmental degradation resulting from chemical use (see Reganold, Papendick and Parr 1990). However, in the case of contract farming, chemical techniques of pest control are both determined and enforced by the terms of the contract, and growers have no choice but to apply chemicals as required.

Contract farming also implies a pattern of monoculture and, as argued above, a narrowing of agriculture's genetic base through more intensive cultivation. Two consequences are declining flexibility or adaptability by farmers and the decreasing diversity of agricultural ecosystems. As agricultural ecosystems are more dependent on fewer varieties, there is greater vulnerability to catastrophic loss through disease. Similarly, as farm operations are more specialised and therefore more dependent, farmers respond conservatively to new 'alternative' technologies. The system discourages farmers from adopting agricultural practices which promote crop diversity, soil fertility through natural means and biological pest control. The practice of intercropping or companion planting is a case in point. It has been demonstrated that when two or more crops are grown in close proximity it is possible to increase the effectiveness of resource use and bring positive benefits in terms of soil fertility or pest management. Contract production does not foster companion planting.

While the operational flexibility of the farmer is being reduced by the contract system, that of the agribusiness corporation is increased. The corporation is able to traverse the globe for economic, political and ecological reasons; companies have a national and international horizon when it comes to searching for new markets and sources of supply. They will, where necessary, move to new regions and away from degraded land and water resources. As a consequence of the 'global reach' of transnational agribusiness, farmers growing pineapples in Queensland are in direct competition with growers in the Philippines and Thailand, while tomato growers in Victoria compete not just with growers in other states, but also with those in Thailand, Italy and Eastern Europe.

Standing at the apex of this system, orchestrating the production process, are the global agribusiness companies, whose ability to source from anywhere in the world means that the Australian pineapple farmer has to be able to match the lower production costs of the Thai farmer, or the subsidised production of a New Zealand or Canadian grower. As noted earlier, in attempting to compete under these terms, Australian producers have little choice but to further intensify production and, in the process, place greater environmental stresses on their productive resource.

Global sourcing means global competition for farmers and, as a consequence, heightened 'environmental competition'. While the expected degradation of productive land and water resources is a problem for individual farmers who are tied to one location, it is less problematic for companies as they can, and do, move across ecosystems. Virtually any location is expendable and firms will move from regions where degradation reduces crop yield. The short-term costs of soil and water conservation are usually more than the costs of moving with the promise of an early return on investment through increased output in the new location. Companies can move to another farm, region, country or continent. The dynamics of contract farmer, as structured by transnational agribusiness corporations, therefore has little room for traditional notions of stewardship where producers take responsibility for conserving the land and water resources upon which their crops and livelihoods are dependent.

The corporation is not site specific, and has no inherent attachment to any particular location (see Goodman 1991). Rather, the whole world is its production base and, as the following section shows, the corporation is fully prepared to exercise its freedom of movement when corporate interests dictate.

MOBILITY

The agribusiness corporation is characterised by its geographical mobility. George (1976) has documented how Castle and Cook (Dole) and Del Monte have moved entire facilities to countries where labour costs and input prices are lower. National companies similarly move across areas or regions to sustain or expand production; ecological and environmental factors are also important. The Australian potato industry provides an example. For many years, potato production was concentrated in Victoria, but with the growth of fast-food chains and the increased retail and domestic demand for frozen chips, Edgell–Birds Eye began to expand contract production in Tasmania. A principal reason was the preference expressed by McDonalds and other fast-food outlets for frozen chips made from the Russet Burbank variety, which grows well in Tasmania's soils and climate.

Edgell–Birds Eye have been the main suppliers of frozen chips to McDonalds. However, in recent years McCain, a Canadian-based multinational corporation, has been attempting to break into this lucrative market by encouraging the production of the Russet Burbank in the Ballarat area of Victoria. Although this area had previously grown potatoes under contract, these were the Kennebec variety used mainly for potato crisps, and it was found that the Russet Burbank did not perform well in the area. Yields were lower than for the Kennebec, and production was more costly and time consuming, yet McCain was not prepared to pay a price which would cover the increased costs of inputs. In the absence of a suitable variety that both satisfied the needs of the fast-food chains and gave Victorian growers an increased return, McCain began to undertake potato processing in Tasmania. In 1987, a plant was opened at Smithton, with a capacity for 50 000 tonnes (translating to an increase of 1000 hectares of land under Russet Burbank potatoes).

Clearly, this shift will have far-reaching impacts in the potato-growing areas of Victoria, where many farmers entered into production of the Russet Burbank in an attempt to meet changing patterns of demand on the part of the processing companies. Of equal significance is what the shift in location tells us about the environmental factors underpining the mobility of agribusiness corporations. Clearly, it was the unsuitability of the ecology of central Victoria to the cultivation of the Russet Burbank variety which persuaded McCain to undertake a major relocation and investment program in another more amenable environment.

From this perspective, agribusiness companies provide support for Schnaiberg's (1980: 15) view that 'humans have learned to operate across ecosystems, and to view their sociocultural production as nested in a different set of principles — economic, not ecological'. Agribusiness corporations have a national and international focus and need not be concerned about sustaining the resources on any given property or region. They are capable of moving their operations from one productive area to another, and choose farms and farmers which, according to their managerial judgement, are most capable of meeting the quality and quantity requirements demanded by mass markets. Farmers, on the other hand, are financially and emotionally tied to their land and are ultimately dependent upon conserving it.

COMPLEXITY

Food production and marketing has become more complex with vertical integration in agriculture and the expanding role of global agribusiness corporations. Products which at one time were marketed directly as final outputs (for example, fresh peas) are now the inputs to another link in the chain of food production and marketing.

Managing and controlling the contract system requires that agribusiness corporations must, as far as possible, eliminate uncertainty and minimise the scope for individual action and initiative. Solving problems of uncertainty of supply required companies, at least in their view, to exert formal control over farmers' decision making. However, as the flexibility and autonomy of farmers have declined due to corporate agriculture's widening span of control, problems of uncertainty have been replaced by problems of coordination which are associated with the structure designed to manage and control the behaviour and decision making of farmers. As with any large organisation, coordination is always a problem — and this is no less the case in contract farming. The use of agricultural chemicals is a case in point.

In conventional, market-based agriculture, the farmer normally makes decisions regarding the application of chemicals — whether to apply them at all, when, where, at what rate. In this situation, breaches of existing regulations governing the use of chemicals can only occur from ignorance or a disregard of regulations (both, of course, do occur). In the contract farming system, however, the farmer no longer makes all of the relevant decisions; many are made by the processing company's field officers or arise out of the processing plant's operating requirements. In this more complex situation, unique problems of coordination arise which can result in breaches of the regulations (Burch *et al.* 1992).

'Withholding' and harvesting dates provide a good example of this problem. For any given agricultural chemical, a withholding period is determined which specifies the period which must elapse between the application of the chemical and the subsequent harvesting and sale of produce which is safe for human consumption. Since market-based farmers retain control over these decisions, they can match their application and harvesting dates so as not to contravene the regulation. However, in the case of vegetable contracting, a farmer may be required to spray a crop on the direction of the processor's field officer, and in this case, the coordination of farm operations may break down. Harvesting is a processor responsibility and while approximate dates may be agreed at the beginning of a contract, these may be varied according to the weather, crop quality, plant requirements and so on.

In some cases, farmers may receive only 48 hours notice of harvesting, whereas withholding periods for commonly used pesticides are of the order of 3 to 14 days (Burch *et al.* 1992). Given this, it is clear that it is not always possible for a farmer to guarantee that the conditions of the withholding period have been met. Even more significantly, in interviews with farmers in the Lockyer Valley, instances were reported of farmers being told to spray crops by the processor's field officer, only to be told very soon after that the harvesting machinery would be arriving in the next day or so. Several farmers recounted instances where this had been their own experience, and others reported it as happening on neighbours' properties (Burch *et al.* 1992).

Obviously, there would be a major problem if this practice resulted in chemical residues remaining in the crops at harvesting. Industry sources suggested that such practices were unlikely in the case of beans, but could well occur with peas. This was not considered hazardous by the industry since the peas were protected by the pod and were sprayed with a non-systemic chemical (namely Endosulfin, with a 14-day withholding period). However, given the separation of ownership from control which is a feature of contract farming, no firm guarantees can ever be given that such practices are confined to one particular crop.

The practice of harvesting peas within the withholding period created other major problems for farmers. The existence of chemical residues affected the farmers' ability to harvest and market pea hay, which is usually sold as a fodder crop. Since the returns from this crop in 1990 were around $170–$210 per hectare (or about 40 percent of the returns from the pea crop), this would represent a major loss to farmers already experiencing falling incomes (Burch *et al.* 1992). This appears to have been a continuing problem. In 1991, it was reported that 120 tonnes of pea hay was burnt after it was found to contain chemical residues above maximum levels set by the National Health and Medical Research Council. The chemical had not apparently affected the processed peas, but would have built up in the fat layers of cattle if consumed as fodder (*The*

Courier-Mail 18 July 1991). The question inevitably arises, how much pea hay so treated has actually gone undetected and used as fodder?

This is but one example of the kinds of problems which emerge as contract farming spreads and adds to the complexities of a production and marketing system based on global organisation. While this system makes great claims to efficiency, these must be questioned. In particular, given the social and environmental impacts documented in this chapter, it is arguable that short-term 'efficiencies' can only be achieved by incurring long-term costs. In short, there are serious questions about whether the system is sustainable.

SUSTAINABILITY

Food production systems, when most effective and efficient, sustain the social and biophysical resources upon which rural community life and agricultural production are based. Although resource sustainability is a complex and controversial concept (Redclift 1987), defined simply it ensures 'the needs of the present without compromising the ability of future generations to meet their own needs' (World Commission on Environment and Development 1987: 8). In the context of agricultural production, 'sustainable agriculture does not deplete soil or people . . . [it] protects soil and water and promotes the health of people and rural communities . . . Sustainable agriculture needs to be ecologically sound, economically viable, socially just and humane' (*The Land Report* 1986: 8, cited in Milbrath 1989).

Given the nature of the environmental and social outcomes associated with contract farming, it cannot be seen as sustainable in the longer term. The agribusiness corporations which organise the production and marketing system are mainly concerned with short-term returns and, in the process, encourage operational practices which are drawing on the natural capital stock — the land, the biodiversity, the water resources and other elements of the ecosystem. The corporation's geographical mobility ensures its ability to traverse the globe in search of cheaper production sites and markets, with short-term 'efficiencies' achieved only at the cost of environmental degradation.

CONCLUSION

Industrial agriculture has changed the social organisation of farm production and thereby the status and role of the farmer. The farmer and the family farm are increasingly subordinate and dependent members of complex agribusiness corporations — organisations which have a national and international focus. Agribusiness, in serving national and international markets, has developed managerial systems which guarantee a steady supply of cheap, marketable produce. It has been in the interests of agribusiness companies to vertically integrate the farmer and farm family into the corporate structure and to coordinate the farmer's decision making about crop production and conservation with short-term corporate economic goals. A principal social impact has been the separation of land ownership from decision making.

Because of its promotion of capital-intensive and energy-intensive methods industrial agricultural has advanced a system of production that can continue only in the face of grave environmental costs of soil and water pollution and eventual declines in both the quantity and quality of farm produce. The advantages of a steady supply of relatively cheap agricultural produce to food processors and to consumer markets notwithstanding, the operational goals of maximising short-term profits, growth and efficiency by agribusiness corporations have had unfortunate consequences. The genetic diversity of agricultural ecosystems has declined. By promoting an intensive system of cultivation dependent upon energy-intensive technology, agribusiness corporations have indirectly promoted a system that has the capacity to degrade or destroy the land and the resources upon which it is based. Industrial agriculture, largely because of contract farming, has destabilised, and will continue to destabilise, modern agricultural ecosystems.

In this chapter, a model of the environmental and social impacts of the contract system has been developed emphasising seven basic concepts: responsibility, profitability, intensity, flexibility, mobility, complexity and sustainability. Its focus has been on the agribusiness corporation functioning as an organisation and, as all organisations must, attempting to manage and control external uncertainty so that supplies (farm produce, in this case) are guaranteed. From a corporate perspective coordination of farmers' decision making and corporate economic goals are essential to profit making. The way this is achieved (and a problem for the farmer) is for the company to cast the farmer into a subordinate, dependent position. This discussion has allowed us to address the questions of the power and dependence relationships between agribusiness corporations and farmers, and the consequences of these relationships for the stability of modern agricultural ecosystems. Environmental quality, social welfare and issues of equity, ethics and power cannot be separated when industrial agriculture is critically examined (Schnaiberg 1980). We are reminded of Bennett's (1976: 3) admonition that the study of the relationship between humans and nature requires the study of 'how and why humans use Nature, how they incorporate Nature into Society, and what they do to themselves, Nature and Society in the process'.

The last concept, sustainability, is perhaps the most important. Food production systems, when most effective and efficient, sustain the social and biophysical resources upon which both rural community life and agricultural production are based. Sustainable systems should degrade neither the soil and water used in production nor the people who are directly responsible for food production. They should conserve basic biophysical resources; all participants should receive an equitable economic return; it should be able to sustain the quantity and quality of production we need; they should respect and conserve critical human — as well as biophysical — resources.

REFERENCES

Bates, G. (1987) *Environmental Law in Australia*, Butterworths, Sydney.

Bennett, J. (1976) *The Ecological Transition*, Pergamon, New York.

Burch, D., Rickson, R. and Annels, R. (1992) 'Agribusiness in Australia', in K. Walker (ed.), *Australian Environmental Policy*, University of New South Wales Press, Sydney.

Burch, D., Rickson, R. and Thiel, I. (1990) 'Contract Farming and Rural Social Change', *Environmental Impact Assessment Review* 10 (1/2): 145–155.

Davis, J. (1980) 'Capitalist Agricultural Development and the Exploitation of the Propertied Labourer', in F. Buttel and H. Newby (eds), *The Rural Sociology of Advanced Societies: Critical Perspectives*, Croom Helm, London.

Gardner, E. (1985) 'Hydro-Salinity Problems in the Lockyer Valley — Real and Perceived, in *Landscape, Soil and Water Salinity — Lockyer/Moreton Regional Workshop*, Queensland Department of Primary Industries, Brisbane.

George, S. (1976) *How the Other Half Dies*, Penguin, Harmondsworth.

Goodman, D. (1991) 'Some Recent Tendencies in the Industrial Reorganisation of the Agri-food System', in W. Friedland, L. Busch, F. Buttel and A. Rudy (eds), *Towards a New Political Economy of Agriculture*, Westview, Boulder.

Lawrence, G. (1987) *Capitalism and the Countryside*, Pluto Press, Sydney.

Luckhurst, T. (1984) 'Integrated Contract Production: Some Pros and Cons', Paper presented to National Workshop: Marketing of Pigs and Pigmeat, Sydney.

Milbrath, L. (1989) *Envisioning a Sustainable Society: Learning Our Way Out*, State University of New York Press, Albany.

Minot, N. (1986) 'Contract Farming and Its Effect on Small Farmers in Less Developed Countries', Working Paper No. 31, Department of Agricultural Economics, Michigan State University.

Mooney, P. (1979) *Seeds of the Earth*, Inter Pares, Ottawa.

Nelson, A. and Murray, W. (1967) *Agricultural Finance* (5th edn), Iowa State University Press, Ames.

Redclift, M. (1987) *Sustainable Development: Exploring the Contradictions*, Methuen, London.

Reganold, J., Papendick, R. and Parr, J. (1990) 'Sustainable Agriculture' *Scientific American* 262 (6): 72–80.

Schnaiberg, A. (1980) *The Environment: From Surplus to Scarcity*, Oxford University Press, New York.

Talbot, R. *et al.* (1981) *Irrigation Quality of Lockyer Valley Alluvial Bores During the 1980 Drought*, Technical Publication No. 5, Department of Biology, Queensland Agricultural College, Gatton.

Thiel, I. (1985) 'As Like as Two Peas in a Can', unpublished M.Sc. dissertation, School of Science, Griffith University, Brisbane.

de Treville, D. (1987) *An Annotated and Comprehensive Bibliography of Contract Farming*, The Institute for Development Anthropology, Binghampton, New York.

Winson, A. (1990) 'Capitalist Coordination of Agriculture: Food Processing Firms and Farming in Central Canada', *Rural Sociology* 55 (3): 376–394.

World Commission on Environment and Development (1987), *Our Common Future*, Oxford University Press, New York.

16 AGRICULTURAL BIOTECHNOLOGIES: ECOSOCIAL CONCERNS FOR A SUSTAINABLE AGRICULTURE

RICHARD HINDMARSH

Genetic engineering is the most significant modern biotechnology. It emerged in the 1970s, at the same time as in-vitro fertilisation (IVF) techniques, and is a radical departure from 'classical' biotechnologies such as traditional plant breeding and fermentation science. A simple definition of genetic engineering (also known as recombinant or r-DNA technology) is the scientific manipulation of organisms at the cellular level in order to produce altered, or novel, organisms that carry out 'desired' or 'programmed' functions, invariably to facilitate industrial production processes. Technically, genetically engineered organisms (GEOs) are organisms whose genetic construct has been altered by insertion or deletion of small fragments of DNA. In the case of insertion, genetic material may be from a different strain of the same species or from a strain of a foreign species, or be synthetic (that is, designer genes engineered in the laboratory). In this way, not only can totally unrelated species share each other's genetic material, but totally novel organisms can be constructed.

The use of genetic engineering is now on the brink of widespread commercial application. While its potential benefits and hazards have been intensely debated for over two decades, they have neither been realised nor resolved. Prominent issues have included human ethics, the safety of containment facilities for genetic experimentation, the environmental release of GEOs, the identification and evaluation of any risks attached to such releases, the desirability and usefulness of emerging bioproducts, and the adequacy of monitoring and regulatory processes.

One issue now becoming more prominent is the potential of genetic engineering to contribute to sustainable agriculture. On one hand, proponent scientists and the chemical industry claim that genetic engineering will be a linchpin of sustainable agriculture, that it will ensure an environmentally safe cornucopian future. On the other hand, critics argue that not only will it exacerbate the problems of conventional agriculture, but it will also undermine ecological methods of farming (for background see Chapter 3).

To date, the debate and policy agenda of genetic engineering has been largely controlled by interested elements in the scientific community and the biotechnology industry itself. In that process, many social, political and environmental issues have been neglected. To correct this imbalance and to contribute to a public policy that is sufficiently well informed to formulate and generate policies in the Australian context, this chapter critically assesses the promises of genetic engineering for sustainability as well as delving 'behind' the policy agenda in order to reveal the more correct reasons for such promises — corporate hegemony over agricultural production and food supply.

THE TECHNOCENTRIC PROMISES OF GENETIC ENGINEERING

The genetic engineering industry is promoting a technocentric version of sustainable agriculture emphasising that increased productivity can be achieved through improved crop varieties, as well as decreased input costs and declining environmental problems. The proponents centre their claims on four major promises which are persistently advanced in order to gain universal support for a rapid deployment of the genetics industry:

1. the capability of herbicide-tolerant (or herbicide-resistant) crop research to replace hazardous herbicides with 'environmentally benign' herbicides;
2. the capability of pest-resistance research to reduce agrochemical usage, to counteract a growing resistance in insects to conventional pest control and to offer more precision than broad-spectrum insecticides (transgenic biopesticides are particularly relevant here);
3. the capability of nitrogen-fixation crop research to reduce chemical fertiliser usage; and
4. the low-risk projection of releasing GEOs into the open environment, where few or no environmentally adverse consequences eventuate.

HERBICIDE TOLERANCE

Highest on the agricultural genetic engineering research and development (R and D) agenda is the herbicide-tolerant plant variety. It appears that over 30 percent of proposals to field test genetically modified organisms comprise this particular product. Through herbicide-tolerance R and D, crops can be genetically adapted to so-called 'environmentally benign' herbicides. They can also be desensitised to withstand non-selective herbicides, more toxic herbicides and increased dosages of herbicides. Contrary to industry claims, herbicide usage is thus likely to increase. For example, cotton plants genetically desensitised to withstand the contact herbicide bromoxynil will allow much greater amounts of bromoxynil to be applied to cotton fields. Rissler and Mellon (1991: 8) estimate that 'if current uses of bromoxynil are maintained, the adoption of bromoxynil-tolerant cotton on only half the cotton acreage would more than double the use of bromoxynil' in the USA.

Proponents argue that bromoxynil is one of an 'environmentally cleaner' group of herbicides that are targeted for priority development. However, there are other R and D programs as well as recent data that contradict such claims. For instance, in Australia the CSIRO Division of Plant Industry has produced transgenic tobacco F1 progeny that survived spraying with dosages of the phenoxy herbicide 2,4-D that were four to eight times the recommended field application (Lyons *et al.* 1989). Although CSIRO scientists believe that 2,4-D is environmentally benign (D. Llewllyn pers. comm.) numerous concerns have been expressed over its safety, particularly with regard to carcinogenic and mutagenic effects. Moreover, Pimentel (1987) found that crops exposed to just the recommended dosages of 2,4-D became much more susceptible to insect infestation and disease, thus implying an increased need for higher dosages of complementary pesticides like insecticides and fungicides. Beneficial insects like bees have been found to suffer adversely from 2,4-D usage (Ritchie 1990) and detectable levels of 2,4-D residues have been found in the tissues of stressed and dying coral colonies (Glynn *et al.* 1984). A developing resistance in weeds to 2,4-D has also been confirmed (Georghiou 1989; Bureau of Rural Resources 1989).

CSIRO researchers have now spliced the gene expressing high 2,4-D tolerance into cotton (CSIRO 1990), which is extremely sensitive to this widely used herbicide. The CSIRO project, which is funded partly by the cotton industry, aims to desensitise cotton to the annual problem of spray drift from 2,4-D being applied to other crops (2,4-D easily evaporates and can be carried by the wind up to 30 km away). Desensitised cotton is advantageous not only for cotton farmers, but also for wheat farmers as restrictions can be eased on the use of the relatively cheap 2,4-D on wheat crops grown in close proximity to cotton crops.

Despite CSIRO's intentions to develop only desensitised cotton at this stage (D. Llewllyn pers. comm.) chemical manufacturers could hardly be expected to pass up the commercial opportunity to develop 2,4-D-tolerant seed for direct application of 2,4-D. Indeed, there are indications that by 1985 Dow Chemicals had developed 2,4-D-resistant tobacco, and Rhone-Poulenc 2,4-D-resistant carrots, and that both corporations were competing, along with Union Carbide, to develop other resistant crops including maize, rice and barley (Ruivenkamp 1989). As CSIRO scientists state, 'the introduction of the gene for 2,4-D mono-oxygenase into broad-leaved crop plants, such as cotton, should eventually allow 2,4-D to be used as an inexpensive post-emergence herbicide on economically important dicot crops' (Lyons *et al.* 1989: 533) including many fruits and vegetables. This indicates a wide market for both 2,4-D herbicide-tolerant crops (or seed) and 2,4-D. The Swiss pharmaceutical conglomerate Schering-Plough already holds a patent in the USA and Europe on the gene that expresses 2,4-D tolerance (D. Llewllyn pers. comm.).

Chemical-biotech-seed companies are moving more quickly to develop plants resistant to herbicides still under patent. Already, US seed multinational Pioneer Hi-Bred, which controls around 34 percent share of the US$1.33 billion seed corn market, is using a gene given to them by American Cyanamid for resistance to the imidizolinone family of herbicides. In return the company will sell the imidizolinone herbicides (New Zealand Department of Scientific and Industrial Research 1990). In another development, Calgene has patented the GlyphoTol herbicide-tolerance gene for cotton, and in 1987 acquired a cotton seed company to establish control over the breeding of cotton resistant to GlyphoTol (New Zealand Department of Scientific and Industrial Research 1990). While in Canada, atrazine-tolerant canola (a high quality rapeseed which is Canada's second most valuable crop and the third highest source of vegetable oil world-wide) has already been commercialised (Rissler 1990), even though over 55 species of 'weeds' are now resistant to the triazine group of herbicides to which atrazine belongs (LeBaron 1989). A herbicide like atrazine may also lead to an increased sensitivity in maize to Dwarf mosaic virus with, just like 2,4-D, the symptom worsening with an increase of dosage (Chaboussou 1986). Moreover, atrazine breaks down very slowly in the environment and is one of the two pesticides found most frequently in contaminated groundwater in the USA.

Even if herbicide-tolerance research were limited to newer, supposedly 'environmentally benign' chemicals, this would still pose environmental problems. For example, in 1989 the US Environmental Protection Agency cancelled and restricted various formulations containing bromoxynil on the grounds of potential birth defects in persons handling the products, as well as the induction of carcinogenic effects (Goldburg *et al.* 1990). Ecologically, it threatens most broadleaf plants as well as vegetation in wildlife habitats adjacent to crop plantations if misapplied, and is also highly toxic to some aquatic species (Goldburg *et al.* 1990).

Sulfonylurea, chlorosulfuron and imidazole are also among the newer, low-dose (or more concentrated) herbicides, but there are indications that their persistence in the environment harms subsequent crops. Sulfonylureas are also toxic to plants in minute quantities. While glyphosate, another more recent herbicide, degrades quickly in most soil types, it persists in runoff water and can be carried downstream in aquatic ecosystems. Some formulations of glyphosate contain allegedly 'inert' ingredients that are acutely toxic to some aquatic organisms (Goldburg *et al.* 1990). Significantly, the full range of environmental impacts of these herbicides is unknown owing to limited research and evaluation. Even so, plants that resist glyphosate, imidazole and the sulfonylureas have also been field tested (Goldburg 1989).

The rapid evolution of weeds resistant to some of the newer herbicides makes it unlikely that the older, more toxic herbicides could in fact be easily replaced. For instance, in Australia, ryegrass is already cross-resistant to most sulfonylureas, among other herbicides (LeBaron 1989). This problem is further compounded where at least 100 herbicide-resistant weed species have been identified, and weed populations tolerant to almost every known herbicide have been discovered (Comstock 1989). Therefore the emerging biotech-pesticide approach will most likely utilise a package or mix of newer, as well as older, herbicides. Such problems illustrate the flawed promise of biotechnology, or what Rissler (1990) has referred to as 'a promise betrayed'.

Further problems may arise with the possible transfer of herbicide tolerance from GEOs to weeds, for example, through hybridisation (see Teidje *et al.* 1989; Ellstrand and Hoffman 1990). Herbicides considered environmentally 'safer' would no longer be effective against weeds that had captured a gene for herbicide tolerance. Consequently, weed populations would increase causing the pesticide treadmill to accelerate.

World-wide, more than 79 corporate/state research programs are developing more than 23 herbicide-tolerant crop lines, including cotton, maize, corn, potato, rice, sorghum, soybean, wheat, tomato, alfalfa and sugar cane (Rural Advancement Fund International (RAFI) 1987; Sans 1988; Goldburg *et al.* 1990). These crops will further entrench the chemical approach, which, in turn, will further increase soil and water pollution, pest resistance and chemical residues in food. In the process, natural ecological processes will be further distorted and the erosion of biodiversity accelerated.

THE TRANSGENIC BIOPESTICIDE

More ecological risks are presented with the 'transgenic biopesticide'. This confers the plant with a built-in resistance to insects by transferring a foreign gene that expresses a naturally occurring toxin (a biotoxin) into its cells. This is achieved by genetically splicing the gene into bacteria that commonly colonise the plant. The toxin is then either expressed through the leaves and stems or the vascular system of the plant, and attacks the intestinal tracts of target insects. It is widely claimed to be harmless to non-target insects, birds and higher animals. The indications are that 20 to 30 percent of proposals for environmental release (or commercialising and testing) involve such products, which are optimistically forecasted to reach the marketplace during the next decade.

Because this technique does not rely on chemical insecticides it is claimed to be environmentally clean. It is also claimed to be a more precise approach than existing broad-spectrum insect control strategies. Again, however, there are hidden costs which undermine the environmental promise.

One problem is that just as chemical pesticides have induced the emergence of pesticide-resistant pests, so transgenic biopesticides can be expected to exert strong selection pressures for pests with a resistance to natural biotoxins. Over 500 species of insects have now developed resistance to one or more chemical insecticides, and many of these are major pests (Georghiou 1989). A secondary effect of that problem is demonstrated in the USA where, despite a tenfold increase in the use of insecticides from 1945 to 1988, annual crop losses to insects rose from 7 percent to 13 percent (Pimentel 1989). World-wide that is an overall loss of about 15 percent (Boulter 1989).

Bacillus thuringiensis (Bt) is the primary bacterium being genetically manipulated for the transgenic biopesticide, with an enormous projected market for its insect-resistant crops. Bt is a soil bacterium possessing a gene that produces natural protein insecticides. It has been used restrictively for over 30 years as a commercial biological control agent, and is especially important for many organic and other 'alternative' farmers. Yet when applied more intensively, as has been done in laboratory experiments, 10 insect species developed rapid resistance to Bt strains. Even more significantly, populations of Indianmeal moth and diamondback moth in the field have developed resistance to Bt (Georghiou 1989), even as the new Bt-adapted varieties are being field tested.

Some indication of how long it could take for widespread resistance to genetically engineered insecticidal crops to develop is given by monoculture crop geneticists. They estimate that 5 to 15 years after they introduce a new form of genetic resistance into a crop strain, that resistance collapses in the face of a newly evolved form of disease or pest (Myers 1979). As insects develop resistance to a strain of Bt, another strain will be utilised and then another one and so on — a biological treadmill will parallel the chemical one. Even proponents believe that there is some validity to such claims. The CSIRO is currently undertaking research to develop cotton varieties resistant to the *Heliothis* caterpillar, using the Monsanto Bt-toxin, and has stated: 'Given the chance, the *Heliothis* caterpillar will develop resistance to the Monsanto BT-toxin, just as it is now doing to the chemical pesticides being used to control it' (CSIRO 1990: 44).

To counteract this eventuality, and to ensure the long-term usefulness of the genetically engineered plants, the CSIRO is attempting to produce cotton plants containing multiple biotoxin genes ('stacked' genes), and contends that there will be only a small probability of insects gaining resistance to these genes simultaneously (CSIRO 1990). Yet, in the case of synthetic chemicals, multi-resistance has become an increasingly common phenomenon. Seventeen insect species can now resist five classes of chemicals simultaneously (Georghiou 1989).

A strategy proposed to extend the longevity and effectiveness of the biopesticide approach is to modify the Bt gene with a range of biotoxin genes from other species. Although numerous Bt strains exist, only a few have been found to be toxic enough to kill insects. Yet, this strategy again invites the question of how long will it be before insects gain a resistance to the wider pool of naturally occurring biopesticides?

Extremely serious questions are therefore posed concerning the alternatives that will be available when insects develop resistance to both synthetic and naturally occurring pest control strategies, and when other species and ecological processes are disrupted in the process. Widespread insect resistance to biotoxins would not only have adverse consequences for conventional agriculture, but also for 'alternative' agriculture, undermining its ecological methods of insect control through appropriate manual application of biopesticides, crop rotation, companion planting or intercropping, predator traps and so on. As the New Zealand Department of Scientific and Industrial Research (1990: 58) has recognised, resistance to Bt would 'reduce the efficacy of new resistant crops, and the efficacy of current uses of the toxin, and it could also change the role that the insecticidal protein plays in the natural ecosystem'.

The extensive use of Bt and other biotoxins could also cause a dramatic change in insect population dynamics which would disrupt pollinator and natural plant communities, both locally and regionally (see Burch *et al.* 1990). Furthermore, some strains of Bt have also been found to be detrimental to beneficial earthworms. Another potential hazard is the transfer of genes that express biotoxins from modified crops to weeds, making the latter less susceptible to their usual herbivores and thus allowing them greater reproductive success. Mutations could also occur, particularly if the beta-exotoxin is used. If Bt mutated it might switch from attacking caterpillars to attacking beneficial beetles which act as predators in controlling pests (Pimentel 1989).

Finally, there is another problem which impinges directly on human health. Naturally occurring toxins can be extremely dangerous so that the genetic engineering of plants for resistance to pests may produce metabolites in food that are even more harmful to humans than existing pesticides (Pimentel *et al.* 1984; Miller, cited in Hileman 1990). With the much faster pace of biotechnology development, new gene transfer techniques may allow a more rapid changing of toxin levels, the introduction of totally new ones, or the development of a secondary situation that invites the creation of a toxin (see Doyle 1988).

Because of these complexities in the chemical makeup of plants and the behaviour of agroecosystem dynamics, great care has to be taken in any development of the transgenic biopesticide as a pest control strategy. Contrary to industry claims, the concerns above indicate that the transgenic biopesticide may indeed fail on grounds of sustainability. Despite the above concerns, Ecogen, a US agrobiotechnology company, has been able to license rights to certain Bt-derived insecticidal genes to Pioneer Hi-Bred. The multinational plans to use the genes to develop new corn hybrids (Anon 1991).

ENVIRONMENTAL RELEASES

The preceding discussion of the role of biotechnology in agriculture presupposes the release of GEOs into the environment. When the issue of potential risks of this strategy is raised, proponents of genetic engineering claim that they are very low or non-existent (see von Weizsacker 1986; Davis 1987). Yet critics fear the possibility of pandemics caused by newly created pathogens, and the triggering of eventual significant ecological imbalances (Wheale and McNally 1988).

There are several reasons for this difference in risk perception. Most scientists responsible for low-risk statements are either molecular geneticists or microbiologists who specialise in biology at the molecular and cellular level of interaction. Often, they are directly involved with the genetic engineering industry by way of research contracts and other funding, or through their involvement in directly commercialising the technology. By way of contrast, many critics are ecologists who specialise in biology at the organism–ecosystem–biosphere levels of interaction, and who are generally independent of the industry. This means that ecologists are at least equally well placed to assess the potential consequences, while some would argue that they are better placed than the molecular geneticists or microbiologists.

The real world environment cannot be simulated in the laboratory and due to a limited understanding of many aspects of genetics, ecosystems and ecological processes, scientists cannot yet predict with any reasonable degree of certainty how altered organisms will 'behave' once released. As recently as 1988, the Biotechnology Consultative Group (1988: 14) stated that the scientific base of the bioindustry was very immature because there was an ignorance of '99% of the chemistry and biology of biological processes'.

At the same time, new evidence has surfaced that indicates the dangers of genetic engineering experimentation. For example, it was recently reported that much larger populations of viruses occurred in unpolluted water than had been previously established. Bergh *et al.* (1989) found densities of 250 million viruses per millilitre of water, and estimated that one-third of the bacteria in the water would suffer a bacteriophage (virus) attack each day. This has important implications for the potential to transfer genetic traits from waterborne GEOs into the indigenous bacterial population of aquatic ecosystems, and elsewhere.

Biotechnologists consistently advocate the precision of gene alteration using genetic engineering, claiming that the insertion or deletion of a single gene will result in a specific outcome. Yet a recent article in the influential journal *Science* reported that a single gene can control two totally unrelated processes (Levings 1990). In this instance, the gene for cytoplasmic male sterility was associated with southern corn leaf blight disease sensitivity in the 1970 US hybrid corn epidemic. While this eventuality may appear unusual to biotechnologists, it indicates the real need to proceed with extreme caution and to consider the possibilities of other outcomes. Indeed, Pollock (n.d.) suggests it 'seems to make it impossible to guarantee the safety of biotechnology'.

The effect of inserting a single gene into the genome (all the DNA of an organism in a single set of chromosomes) necessarily has an element of uncertainty associated with it because the biological characteristics of the organism are ordinarily determined by the complex interaction of groups of genes which have evolved together (see Royal Commission on Environmental Pollution (RCEP) 1989). The outcome of a single insertion depends both on the function of the inserted gene and on how it interacts with other genes in the genome (Hulsman 1991). If, for example, the gene inserted has a control or regulatory function, it may greatly alter the phenotypic expression (for example, leaf size) of other genes. Furthermore, small genetic changes can produce large effects especially if the altered gene affects embryonic development (Hulsman 1991). Consequently, it is important that the inserted gene adopts the correct pattern of expression during tissue and organ differentiation — something which is proving difficult to design (RCEP 1989). For instance, transcription errors in producing a protein from a gene may increase in frequency; small changes in the sequence of amino acids can greatly affect a protein's activity (RCEP 1989), and small changes in the genome may alter its physiological tolerance to ecological factors such as temperature or salinity, thereby influencing the organism's geographic range. Adverse ecological effects are the likely result (Simberloff 1990; Hulsman 1991).

Therefore, any distinction between the environmental release of a novel organism (for example, an indigenous organism where at least one gene has been modified or inserted) and an exotic organism (not indigenous in the environment into which it is released or introduced) is blurred. Because of the uncertainty of the outcome of a simple insertion of a gene and the way in which the novel organism subsequently responds to its new environment, and given the uncertainty of the way in which an exotic organism will respond to its new environment, one therefore does not pose less of a threat to the environment than the other (Hulsman 1991).

Given that vagueness in the distinction between novel and exotic organisms in practice, the environmental effects of exotic releases in the past do not auger well for future releases of GEOs. A study of 850 cases of introduced species in North America found that 104 (12 percent) caused the extinction of indigenous species (Simberloff 1981). Similarly, Williamson and Brown (1986) found that 10 percent of species investigated in a study of exotic organisms introduced into the UK had caused significant ecological effects. Therefore, even if the level of risk was in the lower region of around 1 percent for the release of GEOs — which some genetic engineers claim is quite acceptable — then that could amount to significant ecological damage given that we face the likely prospect of hundreds, even thousands, of 'batch' releases of GEOs over a long period of time.

At this stage, ecological damage from the introduction of novel organisms can only be 'guesstimated'. There exists great uncertainty over whether or not a transfer of spliced-in traits, such as resistance to pests, disease, salt or herbicides, will occur between modified organisms and non-target, naturally occurring organisms. Similarly, there is great uncertainty about how a gene-altered organism may adapt to conditions outside the laboratory; it may be quickly eliminated, cultivated safely, or encounter no natural controls to restrict its proliferation.

To overcome such uncertainties, biotechnologists advocate field tests. However, field testing is in itself an environmental release. Pimentel (1989: 73) concludes on the basis of previous experience with introduced pests that 'once genetically engineered organisms are released into the environment, the odds of ever controlling them is practically nil'.

Despite such risks, over 400 releases (small-scale field tests) have been conducted in OECD countries (United Nations Conference on Environment and Development 1991), with 10 occurring in Australia to date. So far, there appear to have been no 'escapes' or adverse consequences, yet serious questions are posed concerning the adequacy of post-release monitoring in field tests. For instance, in the case of tests with gene-altered microorganisms we need to know if the depth of the soil profile monitored is adequate and whether underlying groundwater is tested for contamination. The indications are that current assessment procedures are extremely inadequate.

Very soon, the odds of adverse consequences from GEOs will be shortened by the introduction of large-scale field tests, which are now being proposed. After three years of small-scale testing, Calgene USA has recently applied for permits to field test 2.3 million transgenic bromoxynil-tolerant cotton plants in 55 sites in 12 states (of the USA), of which 100 000 are also being tested for transgenic insect-resistance (Rissler and Mellon 1991).

The industry's promises that these uses of genetic engineering offer sustainability must be seriously questioned. This analysis shows such assertions to be flawed, naive and confused, or just 'doublespeak' aimed at deceiving the public in the interests of those who stand to profit from the new technologies. When evaluated ecologically they threaten to undermine the prospects of attaining a productive and long-term sustainable agriculture. Numerous large-scale releases of GEOs risk lessening natural species and genetic diversity, distorting natural ecological processes, and in the longer-term, significantly escalating agroecosystem disruption.

AGRIBUSINESS RESTRUCTURING: THE 'HIDDEN' AGENDA

Attracted by the commercial opportunities presented by agricultural biotechnology (variously predicted to be of the value of US$50 to $100 billion by the year 2000), as well as its potential to overcome environmental limits to industrial growth, transnational corporations (TNCs) began to seek control of biotechnology development, application and regulation from the mid-1970s. Since then there has been a growing concentration in the agricultural industry with new 'life-sciences' conglomerates interlocking corporate capital, seed (genetic supply) companies, small biotechnology firms, university or other research facilities, and chemical, pharmaceutical and petrochemical TNCs.

For instance, in 1989 a number of leading European chemical and food processing TNCs established a pressure group in the form of the Senior Advisory Group Biotechnology (SAGB) (Hodgson 1990; Anon 1990a). Its founding members include Hoechst, ICI, Monsanto, Rhone-Poulenc, Sandoz, Unilever and Ferruzzi. The group formed to influence and control the development and regulation of biotechnology in the new single European market which it hopes to promote as 'a supportive climate for biotechnology in Europe' (SAGB 1990). More recently, the broader genetic engineering industry formed the European Council for Bioscience to conduct a highly organised three-year campaign of education and information aimed at government in order to counter groups opposing genetic engineering (Lohr 1991).

The move to control the development of genetic engineering

is the most recent and significant stage in a restructuring of agribusiness that began in the 1960s, when petrochemical TNCs began diversifying from bulk chemicals into high-value speciality chemicals like pharmaceuticals and pesticides. The integration of the pharmaceutical and pesticide sectors was followed by the integration of the plant breeding sector, whereby the seed supply sector was transformed into a highly stratified and concentrated industry within just one decade (Hobbelink, Velle and Abraham 1990: 5). r-DNA technology offers the next step for further integration because of its capacity to forge interconnecting links between chemistry, pharmacology, energy, food and agriculture (Doyle 1985).

As DuPont's biotechnology director, John Hardinger, points out, 'the increasing application of molecular biology techniques is allowing the various segments of the world's largest industrial sector to form logical linkages that were never before practical' (cited in Klausner 1989). DuPont now collaborates with Holden's Foundation Seeds to develop improved hybrid corn varieties — the goal is to develop crop varieties that can resist disease, insects — and DuPont's herbicides (Anon 1989b). Food processing corporations are also using genetic engineering to improve backward linkages in the food chain from the supermarket to the seed. For instance, Nestlé has a joint venture with Calgene USA to develop a new soybean variety, and Campbell Soup has a contract with Calgene to develop high-solids tomatoes (Webber 1985).

OWNERSHIP OF THE SEED

The seed underpins the corporate agenda for genetic engineering — it is the 'vector' for biotechnological change. As the president of Agrigenetics (a US biotech-seed company purchased in 1985 by chemical giant Lubrizol) observed, 'The seedsman, after all, is simply selling DNA. He is annually providing farmers with small packages of genetic information' (Padwa 1983, cited in Kloppenburg 1988: 16). Through the seed, chemical conglomerates can thus genetically engineer the seed's DNA to facilitate the goals of their own research programs.

In this way, corporate seed ownership will intensify the dependency of farmers and society on chemical pest control regimes, create a new dependence of farmers on new pest control agents like the transgenic biopesticide, and increase the competitiveness of transnationals over independent seed companies. To consolidate such growth, chemical corporations spent more than US$10 billion buying up seed companies during the last decade (Pate 1989). Now, an estimated 10 TNCs control 50 percent of the pesticides market and the major part of the international seeds sector, thereby creating a new industrial sector — the genetics supply industry (Ruivenkamp 1987).

The ultimate danger of increased reliance on corporate r-DNA crop regimes is that eventually there will be few alternatives to genetically engineered monoculture seed. Farmers who want to use bromoxynil as a cotton herbicide will have to buy a 'package' of bromoxynil and bromoxynil-tolerant cotton seeds from Rhone-Poulenc — a major manufacturer of bromoxynil and a leading international seed manufacturer. Farmers who want to buy open-pollinated seed will find it increasingly hard to do so. Consequently, the current trend of farmers switching from conventional to ecological methods of farming, like permaculture, organic and biodynamic farming, could be seriously retarded.

The origins of the restructuring of the seeds sector are to be found in the introduction of the highly controversial Plant Variety Rights (PVR) legislation, which provides for new plant varieties developed by either selective breeding or by gene manipulation and transfer to be protected legally (Lawrence 1988). Such monopoly rights allow breeders to control the use and availability of a specific plant variety, levy and collect royalties, or extract increased surplus from the farm sector and fuel the growth of farm enterprises that are larger than family size (Lawrence 1988). The process of concentration in the genetics input industry overseas due to PVR is well documented (see O'Keefe 1981). For instance, following passage of the US Plant Variety Protection Act in 1970, corporate acquisitions were so extensive that 'the American Seed Trade Association [held a] . . . special symposium called "How to Sell your Seed Company" ' (Mooney 1979). By 1985 more than 1200 seed patents had been issued by the US Office of Plant Variety Protection, half of them to the subsidiaries of only 15 corporations (Doyle 1985).

One result of the widespread patenting of seeds has been the increasing marginalisation of public and farmer plant breeding programs. Plant breeding has become increasingly locked into commercial R and D priorities as the herbicide-tolerant plant indicates. Consequently, the development of diverse lines of plant varieties which offer more opportunity for sustainable agriculture are less likely to occur.

Another strategy for corporate control is to diffuse the biotechnological 'package' onto the market through contract farming. In the USA, 'roughly 32% of farm sales are concluded under some form of contract or are vertically integrated by business' (Doyle 1988: 69). This phenomenon is also spreading throughout other capitalist economies, like Australia (Burch *et al.* 1992). The future sustainability of agriculture will be directly affected by this practice, as Doyle (1988: 69) points out:

> In the future, biotechnology may give food processors and shippers a greater power of specificity in contracting with, or buying from farmers. And for those companies that supply farm inputs, gene-based products — whether in the form of seed, chemicals, or microorganisms — will certainly add a new dimension to their influence over agricultural productivity.

That potential power was recently signalled in Australia with a field test of a genetically engineered potato plant resistant to potato leaf roll virus by the CSIRO in conjunction with the Queensland Department of Primary Industries (CSIRO 1991). Significantly, Coca Cola Amatil, a major food processor and contractor for potatoes, partly funded the research. Undoubtedly, Coca Cola Amatil would specify that its contract growers purchase the company-preferred variety if it is successful.

The economic incentive to control genetic engineering is to be found in the global markets for commercial seed (about US$25 billion) and pesticides (US$20 billion) (Thayer 1991). By the year 2000 the seed market is estimated to be worth US$28 billion, which includes a US$12 billion opportunity for biotechnologically altered plant varieties (Fowler *et al.* 1988). Sales of herbicide-tolerant plants, which are based more on an extension of the market application of herbicides, could be near to US$6 billion (Goldburg 1989). Another US$6–8 billion market by the year 2000 is projected for the transgenic biopesticide market (Anon 1990b), and it is this fact, rather than a corporate concern for social and ecological benefits, which is directing research in this area.

ECOCENTRIST CONCERNS

Quite clearly, the corporate version of sustainable agriculture is to continue with conventional agriculture and to use biotechnology to overcome some of its central problems, such as declining productivity and international competitiveness, increasing pest resistance and genetic erosion, as well as widespread public opposition to chemicalisation. In other words, biotechnology is being used as a 'technological fix' to circumvent these problems, without questioning the flawed assumptions upon which that agriculture is based, and which gave rise to those problems in the first place.

One underlying assumption is the anthropocentric one — that all non-human lifeforms are in the service of humankind. Genetic engineering highlights this human-centred approach through its appropriation of non-human entities such as restriction enzymes, which act as an organism's natural defence system to invasion from foreign DNA. Infectious viruses are swarmed over by restriction enzymes which chemically disable or restrict them by 'slashing' the foreign DNA molecules to 'ribbons' (Suzuki and Knudtson 1989). Genetic engineers subvert this natural 'defence' process by taking restriction enzymes out of their natural context and into the laboratories to slice DNA to their purpose. Different DNA sequences can then be recombined to human design by using ligases (another type of enzyme) to splice together the ends of the separate DNA pieces.

This enables genetic engineers to circumvent biological, and evolutionary, limits to interspecies interbreeding in their drive to expand synthetically the environmental resource base for industrial production. However, this circumvention is occurring without any ecological understanding of why restriction enzymes act that way in the first place — is it for the long-term resilience and stability of the ecosystem? Who knows? Clearly the founder of Calgene has not considered this vital question when he stated that the plant genetic engineer's motto was 'any gene from any organism into plants' (cited in Kloppenburg 1988: 204).

From this perspective, biotechnology is not addressing the central issue in the development of a sustainable agriculture — the need for an ecologically sound *modus operandi*. In addition, it is clear that the environmental problems which the industry and its proponents claim will be resolved through genetic engineering are simply the outcomes of an earlier round of innovations which themselves were technological fixes attempting to overcome ecological limits. On this basis, the biotechnological approach will simply come to represent, not an ecologically acceptable alternative to conventional agriculture, but a 'new' form of conventional agriculture which will add to our environmental problems. In other words, in its capacity to expand synthetically the environmental resource base, r-DNA technology also has the capacity to diminish it ecologically.

The unacceptability of ecological and social risk presented by genetic engineering has been strongly expressed by environmental and consumer groups internationally. In what was West Germany, the Green Party called for a five-year moratorium on the commercial release of GEOs. Similarly in the UK, the UK Genetics Forum has called for a partial moratorium and a ban on environmentally irresponsible applications of biotechnology (Anon 1989a), while the USA has seen the formation of a number of groups that strongly oppose deliberate release, and which have been successful in legal actions to halt or delay releases of certain GEOs into the environment.

Similarly in Australia, the Australian Conservation Foundation (ACF) has called for a moratorium, pending the establishment of stringent laws to replace the existing system of voluntary self-regulation and until there is better understanding of Australia's complex and fragile environment (Toyne 1991). ACF's basic criteria for assessment of any release proposal is 'the maintenance of sustainability and biodiversity in both agricultural and natural environments' (ACF 1990: 35). The ACF has already singled out herbicide-tolerant plants for an outright ban.

THE AUSTRALIAN SITUATION

Of 65 biotechnology businesses in Australia (Playne and Arnold 1990) the Department of Industry, Technology and Commerce (DITAC) estimates there are now at least 50 Australian companies and research institutes actively developing and using genetically engineered organisms (DITAC 1990). Nineteen of these are involved in plant breeding. In the field of agrobiotechnology the emphasis has been on food processing technologies and pesticides (Biotechnology Consultative Group 1988). It is difficult to estimate how much has been spent by industry on agrobiotechnology, but in the broader area of biotechnology, expenditure grew from $5 million in 1982 (Australian Science and Technology Council 1982) to about $127 million in 1990, while public sector investment now amounts to some $100 million annually (Playne and Arnold 1990). One notable public sector player has been the CSIRO, which has emerged as a world leader in the field (CSIRO 1990). In 1988, the CSIRO had some 200 professionals involved in over 120 biotechnology projects of which some 84 (or 70 percent) were in agriculture (Biotechnology Consultative Group 1988). The CSIRO is now using genetic manipulation approaches in almost half its divisions (CSIRO 1990).

The creation of a favourable investment climate has proved attractive to the transnational agribusiness companies, who are responsible for the bulk of global investment in agrobiotechnology. Mooney (1989) estimates that US$12 billion was spent in the 1980s by TNCs in this area; most investment occurred through in-house research and joint ventures. A recent survey of 50 Australian biotechnology firms reported that 70 percent had formed some type of strategic alliance with an overseas partner, the largest number being with North American firms (Scott-Kemmis, Darling and Stark 1989). The reasons for such alliances included gaining access to larger markets and their established distribution channels, overcoming communication difficulties over long distances, and gaining resources for R and D programs. Overall, the general idea of Australian collaboration is to share biotechnologies, markets and profits. A more accurate view is that the Australian industry, and its research agenda, is becoming firmly entrenched within the global industry.

TNCs which are directly funding agricultural genetic engineering research in Australia include two subsidiaries of European and US companies affiliated to the SAGB — ICI (Aust) and Monsanto (Aust) — as well as Groupe Limagrain and Lubrizol. Other TNCs operating in Australia likely to apply the results of their parent company's genetic engineering programs include more subsidiaries of European companies affiliated to SAGB — Sandoz, Hoechst, Rhone-Poulenc and Unilever — as well as Bayer, Eli Lilly and Ciba-Geigy.

Foreign TNCs are thus actively developing bioproducts for the Australian market. The diffusion of these products will occur through their domination of the domestic seeds sector and the agrochemicals industrial and marketing sector. The extent of such domination can be seen from the following comment: 'seventeen of the world's top twenty [agribusiness] corporations and all but one of the top ten, have operations here. In all, nearly two-thirds of the fifty-nine companies producing or marketing agrochemicals in Australia are foreign transnationals' (Sargent 1985: 87–88).

Such has been the enthusiasm of the Australian government to link up with TNCs to develop the local industry that Lawrence (1989: 9) has charged that 'the state has introduced specific measures to expand and protect the development of a corporate-sector biotechnology industry in Australia'. Major incentives to the private sector have included: a voluntary self-regulatory regime; a narrow reliance on vested interests for policy advice; subsidisation for joint ventures and strategic alliances (particularly with foreign corporations), and several types of grant and assistance schemes (particularly for collaborative projects between public sector scientists and commercial groups); prioritisation of the field of molecular biology as a recipient of government research grants; the partial privatisation of plant material through the introduction of PVR legislation in 1987; the more complete privatisation of genetic material through the Patents Bill 1990; and finally, support for corporate penetration into public universities, as well as active participation in major corporate biotechnology assessment and development programs (see Lawrence 1989; and Persley 1989).

Perhaps the most outstanding example of the interlocking of the Australian public research base with the global corporate domination of genetic engineering is that of the joint venture between the CSIRO Division of Plant Industry and the French seed multinational, Groupe Limagrain, which has invested $22.5 million. The venture is to develop a genetic manipulation tool called 'gene shears' to control for viruses in plants and animals. Limagrain's goal is to produce hybrid seed resistant to diseases caused by viruses and viroids (O'Neill 1989).

Another avenue that interlocks Australian research with global companies is the privatisation of plant genetic resources. A similar restructuring to the one that occurred overseas during the post-PVR legislation period is now slowly developing in Australia. This is despite the earlier claims of PVR proponents that PVR would encourage the right of ownership and the development of small plant breeding firms by allowing them to compete on more equal terms with larger firms (O'Keefe 1981).

For example, after PVR legislation was introduced in 1987, Elders quickly established a seed network which stretched the length of the eastern seaboard with the acquisition of Fuller Seeds in Brisbane and West Seeds in Melbourne. These were added to its Hodder and Tolley seeds operations in Sydney, which Elders had purchased in 1985. The latter purchase had been made to take advantage of the impending introduction of PVR to develop a seed division in Australia (Anon 1988). Later, Elders also developed an interest in agrobiotechnology (Nugent 1988).

Foreign interests also became active in Australia at the same time. In 1987, Calgene Pacific P/L purchased three prominent Australian nursery companies. A major shareholder of Calgene Pacific P/L is Calgene USA. In Australia, Calgene Pacific is the only company genetically altering plants for export and is capable of producing over 30 million plants and commercial cuttings a year in its micropropagation laboratories and computer-controlled glasshouses (Osborne 1987). Plants of interest to the corporation include carnations, gerberas, roses and chrysanthemums. In 1990 Calgene Pacific formed a world-wide joint venture with Japanese TNC Suntory Ltd to develop and commercialise genetically engineered roses (Anon 1990c).

The year 1987 also saw Crompton-Hannaford P/L, an Australian seed and grain machinery manufacturer and a major supplier of seed treatment chemicals (as the Australian distributor for Bayer (Aust)), taken over by Uniroyal (Aust) P/L — a subsidiary of Uniroyal Chemical Company Inc. (USA). That acquisition expanded Uniroyal Chemical's business in seed treatment chemicals and the supply of new improved seed varieties (Anon 1987). In 1989, ICI Seeds (UK) purchased Pacific Seeds and ICI (Aust) purchased Incitec — a leading Australian fertiliser company and a leader in controlled release pesticide formulation technology. Along with Agrigenetics (Lubrizol), Incitec currently undertakes research into transgenic biopesticides.

It is most likely that the Patents Bill of 1990 will accelerate the emerging trends of privatisation of genetic material, and the integration of seed and agrochemical sectors within Australia.

THE OUTCOME

In the short term, the new life-sciences conglomerates will reap major rewards, just as their forerunners did through the introduction of industrialised agribusiness 'packages' throughout the world. Biotechnologies represent a new and very expensive agribusiness package. The integrated package will comprise brand agrochemicals together with herbicide-tolerant and multiple pest-resistant hybrid seed (as well as any other characteristics that the industry can build in). Through this biotechnological package, and with continued support from the state, transnational agribusiness will expand its hegemony in agricultural production and food supply, and thus sustain and expand its control politically, geographically, economically, socially and ecologically.

It is clear that if the bioindustry and bioscientific community continue to dominate local and global biotechnological policy processes, and to distort public sector research, the outcome will be an escalation of agricultural problems. The corporate genetic engineering plan to produce an environmentally cleaner, safer and more effective agriculture will fail, and at great cost to our society and fragile ecology.

Therefore, the risks of the various biotechnologies and their potential for developing sustainable industries must be carefully evaluated. All aspects of genetic engineering found to be ecologically unsustainable should be discarded. But how should assessment criteria be determined? One way would be to question the flawed anthropocentric assumptions of modern agriculture. Another way would be to question the impact of new (as well as existing) technologies on the stability of the ecosystem. What does that imply for biotechnological choice?

One major implication is that biotechnologies need to be re-evaluated in terms of whether or not they constitute ecotechnologies. Essentially, ecotechnological change reflects an holistic technological approach that helps society to form a deep ecological relationship with nature. In other words, the ingenuity of science should be harnessed to facilitate the development of future technologies that are based on ecological considerations, and that are designed to merge harmoniously human societal and environmental dynamics. With the adoption and rapid transition to this approach, long-term sustainability may be a realistic proposition.

For example, future energy technologies will be required to rely more on solar-based ecosystems, and future agriculture must be reliant upon ecologically sound regenerative practices and ecologically appropriate pest control strategies. Obviously, an integral part of this approach would be to direct much more public sector research into a better understanding of agroecological processes. It would also mean a thorough reappraisal of our values and lifestyle practices along with the development of a much deeper ecological consciousness than exists today. In other words, such a perspective implies a sophisticated and challenging approach for science given the complexity and diversity of ecosystems and ecological processes. In agriculture, it requires a comprehensive knowledge base and understanding of ecodynamics in order to find appropriate agroecotechnological structures and practices. Clearly, from the above discussion, genetic engineering has not yet proven to be one.

CONCLUSION

It is clear that certain emerging biotechnological R and D choices involving genetic engineering lack an ecologically sound perspective and therefore are unsustainable in agroecosystems. As such they should be discarded. To ensure long-term sustainability, the challenge for society is to ensure that only ecologically sound aspects of the biorevolution are researched and developed. Important and urgent challenges include countering the domination of biotechnological policy processes by corporations and their collaborative bioscientists, raising public awareness about the implications of biotechnological progress, supporting and developing ecotechnological change processes, developing a stronger network internationally to preserve and use open-pollinated plant varieties, and lastly, demanding a strict and effective regulatory regime that involves adequate and mandatory public oversight at all levels of biotechnological research and development.

NOTE

I would like to acknowledge the contribution made to this chapter by Dr David Burch (Griffith University) who provided useful comments on the initial draft and by Dr Kees Hulsman (Griffith University) who contributed to the section on environmental releases.

REFERENCES

Anon (1987) 'A Change for Hannaford', *Australian Seed Industry Magazine* 5 (4): 26–27.

Anon (1988) 'Hodder and Tolley open Sydney H.Q.', *Australian Seed Industry Magazine* 6 (1): 14.

Anon (1989a) 'Royal Commission calls for Tougher Controls', *Chemistry and Industry* 17 July 1989: 430.

Anon (1989b) *Genetic Technology News* January 1989.

Anon (1990a) 'Biotechnology Call', *New Scientist* 10 February: 9.

Anon (1990b) 'Alternative Pesticides Spawned by Biotechnology could grow to $8 Billion Market by 2000', *Biotechnology Bulletin* 8(5). Cited in United Nations Industrial Development Organisation (1990) *Genetic Engineering and Biotechnology Monitor* 27: 52–53.

Anon (1990c) *Genetic Technology News* June 1990: 9.

Anon (1991) 'Ecogen Licenses Genes to Pioneer Hi-Bred', *Chemical and Engineering News* 22 April 1991: 13.

Australian Conservation Foundation (1990) *Genetic Manipulation: Ecological, Social and Legal Issues*, Submission to the House of Representatives Standing Committee on Industry, Science and Technology Inquiry into Genetically Modified Organisms, September.

Australian Science and Technology Council (1982) *Biotechnology in Australia,* Australian Government Publishing Service, Canberra.

Bergh, O., Borsheim, K., Bratbak, G. and Heldal, M. (1989) 'High Abundance of Viruses found in Aquatic Environments', *Nature* 340: 467–468.

Biotechnology Consultative Group (1988) *Biotechnology in Australia*, Australian Government Publishing Service, Canberra.

Boulter, D. (1989) 'Genetic Engineering of Plants for Insect Resistance', *Outlook on Agriculture* 18 (1): 2–6.

Burch, D., Hulsman, K., Hindmarsh, R. and Brownlea, A. (1990) *Biotechnology Policy and Industry Regulation: Some Ecological, Social and Legal Considerations*, Submission to the House of Representatives Standing Committee on Industry, Science and Technology Inquiry into Genetically Modified Organisms, September.

Burch, D., Rickson, R. and Annels, R. (1992) 'Contract Farming, Social Change and Environmental Impacts', in K. Walker (ed.),

Environmental Policy in Australia, University of New South Wales Press, Sydney.

Bureau of Rural Resources (1989) *Submission to the Senate Select Committee on Agricultural and Veterinary Chemicals,* Department of Primary Industries and Energy, Canberra.

Chaboussou, F. (1986) 'How Pesticides increase Pests', *The Ecologist* 16 (1): 29–35.

Comstock, G. (1989) 'Is Genetically Engineered Herbicide-resistance (GEHR) Compatible with Low-Input Sustainable Agriculture (LISA)?', in J. MacDonald (ed.), *Biotechnology and Sustainable Agriculture: Policy Alternatives,* National Agricultural Biotechnology Council (NABC) Report 1, Ithaca.

CSIRO (1990) *Submission to the Inquiry into Genetically Modified Organisms by the House of Representatives Standing Committee on Industry, Science and Technology,* September 1990.

CSIRO (1991) *Genetically Engineered Plants — A Step Towards the Clever Country,* Media Release, CSIRO Division of Plant Industry, Canberra, 5 July 1991.

Davis, B. (1987) 'Bacterial Domestification: Underlying Assumptions, *Science* 253: 1329–1335.

Department of Industry, Technology and Commerce (1990) *Submission to the House of Representatives Standing Committee on Industry, Science and Technology Inquiry into Genetically Modified Organisms,* Canberra.

Doyle, J. (1985) *Altered Harvest,* Viking Penguin, New York.

Doyle, J. (1988) 'Potential Food Safety Problems Related to New Uses of Biotechnology', *Biotechnology and the Food Supply: Proceedings of a Symposium,* National Academy Press, Washington.

Ellstrand, N. and Hoffman, C. (1990) 'Hybridization as an Avenue of Escape for Engineered Genes', *Bioscience* 40 (6): 438–442.

Fowler, C., Lachkovics, E., Mooney, P. and Shand, H. (1988) 'The Laws of Life: Another Development and the New Biotechnologies', *Development Dialogue* 1/2.

Georghiou, G. (1989) 'Implications of Potential Resistance to Biopesticides', in D. Roberts and R. Granados (eds), *Proceedings of a Conference,* Boyce Thompson Institute for Plant Research, Cornell University, Ithaca.

Glynn, P., Howard, L., Corcoran, E. and Freay, A. (1984) 'The Occurrence and Toxicity of Herbicides in Reef Building Corals', *Marine Pollution Bulletin* 15: 370–374.

Goldburg, R. (1989) 'Should the Development of Herbicide-tolerant Plants be a Focus of Sustainable Agriculture Research?', in J. MacDonald (ed.), *Biotechnology and Sustainable Agriculture: Policy Alternatives,* National Agricultural Biotechnology Council (NABC) Report 1, Ithaca.

Goldburg, R., Rissler, J., Shand, H. and Hassebrooke, C. (1990) *Biotechnology's Bitter Harvest,* a report of the Biotechnology Working Group, USA.

Hansen, M. (1991) 'Biotechnology & Milk: Benefit or Threat?', *geneWATCH* 7 (1–2): 1–2.

Hileman, B. (1990) 'Alternative Agriculture', *Chemical and Engineering News* 5 March: 26–40.

Hobbelink, H., Velle, R. and Abraham, M. (1990) *Inside the Biorevolution,* International Organisation of Consumers' Unions and Genetic Resources Action International.

Hodgson, J. (1990) 'Growing Plants and Growing Companies', *Bio/Technology* 8: 624–628.

Hulsman, K. (1991) 'Some Issues Arising from the Roundtable Discussion 19 April 1991', House of Representatives Standing Committee on Industry, Science and Technology Inquiry into Genetically Modified Organisms, Canberra.

Klausner, A. (1989) 'Biotech Changing Agribusiness', *Bio/Technology* 7: 219.

Kloppenburg, J. (1988) *First the Seed,* Cambridge University Press, Cambridge.

Lawrence, G. (1988) 'Structural Change in Australian Agriculture: The Impact of Agri-Genetics', Paper presented at the Annual Conference of the Sociological Association of Australia and New Zealand, Australian National University, Canberra, 28 November–2 December.

Lawrence, G. (1989) 'Genetic Engineering and Australian Agriculture: Agenda for Corporate Control', *Journal of Australian Political Economy* 25: 1–16.

LeBaron, H. (1989) 'Herbicide Resistance in Plants', in J. MacDonald (ed.), *Biotechnology and Sustainable Agriculture: Policy Alternatives,* National Agricultural Biotechnology Council (NABC) Report 1, Ithaca.

Levings, C. (1990) 'The Texas Cytoplasm of Maize: Cytoplasmic Male Sterility and Disease Susceptibility', *Science* 16 November 1990: 942–947.

Lipton, M. and Longhurst, R. (1989) *New Seeds and Poor People,* Unwin Hyman, London.

Lohr, W. (1991) 'Biotech-Industrie plant europaweite Akzeptanz', *GID* 3/91: 3.

Lyons, B., Llewellyn, D., Huppatz, E., Dennis, E. and Peacock, W. (1989) 'Expression of a Bacterial Gene in Transgenic Tobacco Plants confers Resistance to the Herbicide 2,4-dichlorophenoxyacetic acid', *Plant Molecular Biology* 13: 533–540.

Mantegazzini, M. (1986) *The Environmental Risks from Biotechnology,* Frances Pinter, London.

Mooney, P. (1979) *Seeds of the Earth: A Private or Public Resource?* (revised edition), Inter Pares, Ottawa.

Mooney, P. (1989) 'An Informal Address by Pat Mooney', in *Beyond Biocides: People Linking for a Sustainable Future,* Third PAN International Meeting, Penang, Malaysia, 25–28 January.

Myers, N. (1979) *The Sinking Ark*, Pergamon, New York.

New Zealand Department of Scientific and Industrial Research (1990) *Genetic Engineering: A Perspective of Current Issues,* DSIR, Christchurch.

Nugent, M. (1988) 'Agribusiness Investment: Realising the

Opportunities', Paper presented to National Outlook Conference, Canberra.

O'Keefe, S. (1981) 'A Critical Review of Plant Patenting Legislation and the Australian Plant Variety Rights Bill', unpublished Honours dissertation, School of Australian Environmental Studies, Griffith University, Brisbane.

O'Neill, G. (1989) 'Genetic "shears" Cut Their Way to Market', *New Scientist* 29 July: 17.

Osborne, P. (1987) 'Biotech Nursery Blossoms', *Financial Review* 14 August, 1987.

Pate, M. J. (1989) 'Researchers Prepare Super Seeds of 1990s', *Agribusiness Worldwide* 4 (10): 6–12.

Persley, G. (ed.) (1989) *Biotechnology in Agriculture,* CAB International, Oxford.

Pimentel, D. (1987) 'Down on the Farm: Genetic Engineering Meets Ecology', *Technology Review* 90 (1): 24–31.

Pimentel, D. (1989) 'Biopesticides and the Environment', in J. MacDonald (ed.), *Biotechnology and Sustainable Agriculture: Policy Alternatives,* National Agricultural Biotechnology Council (NABC) Report 1, Ithaca.

Pimentel, D., Glenister, C., Fast, S. and Gallahan, D. (1984) 'Environmental Risks of Biological Pest Controls' *Oikos* 42: 283–290.

Playne, M. and Arnold, B. (eds) (1990) *Australian and New Zealand Biotechnology Directory 1990* (2nd edn), Australian Industrial Publishers, Adelaide.

Pollock, B. (n.d.) 'Cytoplasmic Male Sterility in Corn: A Problem for the US Science Community and Public', unpublished paper, Science Mediation Service, Boulder.

Rissler, J. (1990) 'Biotechnology Promise Betrayed', *Chemistry and Industry* 6 August: 500.

Rissler, J. and Mellon, M. (1991) *National Wildlife Federation Comments to the USDA APHIS on Two Applications from Calgene, Inc. to Field Test Cotton Plants Genetically Engineered to Tolerate the Herbicide Bromoxynil or Resist Insects and Tolerate Bromoxynil,* National Biotechnology Policy Center, National Wildlife Federation, Washington.

Ritchie, R. (ed.) (1990) *Australian Geography: Current Issues,* McGraw-Hill, Sydney.

Royal Commission on Environmental Pollution (1989) *The Release of Genetically Engineered Organisms to the Environment*, 13th Report, London.

Ruivenkamp, G. (1987) 'Social Impacts of Biotechnology on Agriculture and Food Processing', *Development* 4: 58–59.

Ruivenkamp, G. (1989) 'The Introduction of Biotechnology into the Agroindustrial Chain of Production', unpublished PhD thesis, University of Amsterdam, The Netherlands.

Rural Advancement Fund International (1987) *RAFI Communiqué Newsletters.*

SAGB (1990) *Community Policy for Biotechnology: Priorities and Actions,* Senior Advisory Group Biotechnology, Brussels.

Sans, M. (1988) 'Genetics Control Will Strengthen Big Firms', *Business Times* 23 February.

Sargent, S. (1985) *The Foodmakers,* Penguin, Ringwood.

Scott-Kemmis, D., Darling, T. and Stark, P. (1989) 'Strategic Alliances in the Australian Biotechnology Industry', *Australian Journal of Biotechnology* 2 (2): 122–126.

Simberloff, D. (1981) 'Community Effects of Introduced Species', in M. Nitecki (ed.), *Biotic Crisis in Ecological and Evolutionary Time,* Academic Press, New York.

Simberloff, D. (1990) 'Releasing Genetically Engineered Organisms: Introduced Species as a Model', a discussion held in the Zoology Department, University of Melbourne, 15 August.

Suzuki, D. and Knudtson, P. (1989) *Genethics: The Clash Between the New Genetics and Human Values,* Cambridge University Press, Cambridge.

Teidje, J., Colwell, R., Grossman, Y., Hodson, R., Lenski, R., Mack, R. and Regal, P. (1989) 'The Planned Introduction of Genetically Engineered Organisms: Ecological Consideration and Recommendation', *Ecology* 70: 298–315.

Thayer, A. (1991) 'Battling Back Financially, Ecogen Looks to Carve Niche in Pesticides', *Chemical and Engineering News* 11 March: 17–18.

Toyne, P. (1991) 'A Talk', in *Pan Pacific Conference, Veterinarians and the Environment,* Sydney, 13 May.

United Nations Conference on Environment and Development (1991) *Biotechnology Background, Part 1,* PC/67, Report of the Secretary General of the UNCED Conference, para. 21.

von Weizsacker, E. (1986) 'The Environment Dimensions of Biotechnology', in D. Danes (ed.), *Industrial Biotechnology in Europe*, Frances Pinter, London.

Webber, D. (1985) 'Calgene Strives to Lead in Plant Biotechnology', *Chemical and Engineering News* 29 April: 11–12.

Wheale, P. and McNally, R. (1988) *Genetic Engineering: Catastrophe or Utopia?* Harvester, England.

Williamson, M. and Brown, K. (1986) 'The Analysis and Modelling of British Invasions', *Philosophical Transactions of the Royal Society*, Series B, 314: 506–522.

17 COOPERATIVE LAND MANAGEMENT FOR ECOLOGICAL AND SOCIAL SUSTAINABILITY

PETER COCK

Does development for ecological sustainability require social change that is more radical than simply incorporating the biophysical environment into economic accounting? This minimalist approach has largely dominated the sustainable development debate (see Zarksy 1990) and amounts to no more than an attempt at revision, rather than transformation, of the social system. The most common approach to environmental thinking, management and action is directed towards a technical fix and/or support for the environmental victim. From the standpoint of sustaining long-term biodiversity it is not sufficient. The issue of the role that social reconstruction needs to play in redevelopment for ecological sustainability depends on the extent to which demands on the ecosystem need to be reduced by the materially overdeveloped world. Boyden (1989) suggests that as a start there is a need to reduce our energy/material demands to one-fifth of our present per capita usage (see also Boyden, Dovers and Sherlow 1990). Trainer (1991) argues that this reduction is insufficient and that our objective needs to be a reduction to one-tenth of present per capita usage. While the necessary percentage reduction in ecological demands is unclear, what *is* clear is that a radical reduction in the materially overdeveloped world is necessary (see Daly and Cobb 1989; Coombs 1990).

If a radical reduction is assumed to be necessary and if it is to be achieved, then we need to use a social structural perspective. How does the structuring of society shape environmental and social outcomes? Social structure is construed to mean the complex interrelationships of the component parts of society. If the structure of the social order is to be accountable to ecological and social — as well as economic — criteria, then the question of how it is to be restructured is critical. If we are to survive as a species, together with other species, then we have to postulate in a creative manner the kind of social redevelopment required for sustainability. What is the range of sustainable choices? Have we been so socialised in unsustainable living that we are incapable of adaptation within the necessary time-frame? The challenge is therefore one of developing visions and identifying constraints, while exploring mechanisms and processes for creative change.

In defining ecological sustainability we need to include the ongoing viability of human society as part of ecosystem diversity. Ecological sustainability requires that the organisation of human activities meets basic human needs and is compatible with biophysical constraints determined by other species' needs. When ecological sustainability is defined as 'the potential of human activities and encompassing ecosystems to remain viable indefinitely', it highlights the relationship between biophysical and social processes (Blake 1991: 1). The exclusive focus on the environmental outcomes of the social system has led to technical orientations that may suppress the symptoms. Yet in this way environmental problems are seen as being derived from the land itself rather than being linked to people's perceptions of the land — perceptions conditioned by social structures and processes. As Passmore (1974) said in the early stages of the environmental movement, ecological problems are social problems. However, a problem orientation to either social or environmental issues is inadequate to the task. What is required is a systematic approach that examines the environmental outcomes of how society is constructed. If it is accepted that at its base the environmental crisis is social and cultural, then the question shifts to an environmental sociology that analyses which modes of social construction contribute to the destruction and/or the reconstruction of the relations between nature and human society.

In the context of the need to reduce and change in a radical way the pattern of our demands on the biosphere, this chapter addresses the issue of the role of social structure in shaping land management's environmental outcomes. It examines the need for cooperative approaches to land use and care as an ingredient in the development of ecologically sustainable uses. The chapter critically reviews possibilities, barriers and realities involved in the development of cooperative approaches. Two Australian forms of cooperation, Landcare and land sharing, are examined. It is concluded that cooperative approaches are of themselves insufficient and that there needs to develop a partnership between cooperative institutions and government. Cooperative land management is taken to mean the joint organisation of human energy and material resources to contribute to the productivity and protection of the land. In particular, it refers to modes of social organisation that can reduce per capita non-human resource/energy use and can increase the participation of human resources in the productivity and care of the land. The cooperative approach to land management is an alternative to, or at least an addition to, the dominant mode of private, small individual lots on the urban fringe and the rise of corporate farming on broad hectares. In the context of a discussion of the need for an ecologically sustainable agriculture, the next section explores the benefits of a cooperative approach to land management.

COMMUNITY DEVELOPMENT THROUGH COOPERATIVE LAND MANAGEMENT

The discussion of the potential of cooperative land management needs to be seen in the context of the search for an ecologically sustainable agriculture. Ideas and practices for ecological farming have been around on the margins for a long time such as in the work of Rudolf Steiner and biodynamics (see also Faulkner, 1948). It is only with the increasing threat to productivity due to the dominant land uses that they have begun to be given mainstream attention. The development of an ecoperspective for agriculture involves transformation of land use practices into ones that are chemically free, have a better balance between capital and labour input, reduce external inputs (chemicals, fossil fuels) and increase the diversity of product types. This transformation is part of the creation of more productive niches as a structural base for the minimisation of vulnerability to weeds and pests, as well as the redefinition of what 'weeds' and 'pests' constitute.

The broad indicators of an ecologically sustainable agriculture are easier to see than to define and develop for each major land system or farm entity. Initial indicators for different situations are exemplified by the following: in cropping areas, retaining stubble and incorporating legumes in a rotation; and, in high rainfall grazing situations, securing effective pasture management and revegetation. In contrast, for irrigation areas using water efficiently would be more important (Campbell 1989; see also Owen 1987). More long-term indicators may well involve a shift from flood irrigation to drip feed or its replacement by another approach to farming. In semi-arid areas a change in the perception of appropriate land uses, from monocropping and sheep farming to harvesting kangaroos or the farming of emus would be beneficial (Grigg 1988). Kangaroo meat for human consumption could be legalised throughout Australia and promoted for its very low fat content. The development of an ecologically sustainable agriculture should mean that total system productivity over the long term is more likely to increase than to decline, contrary to what has been widely feared (see Wynen and Edwards 1988).

The search for modes of food production within ecological limits challenges the institutionally assumed answers to the questions of what types of food, how and in what areas food is produced. Ecologically sound land management involves a perspective that empowers a range of modes of management. This range would minimise dependence on any one person, species or function. The focus of attention needs to be as much on design for use complexity and cooperation between various dimensions as it is on improving the efficiency or productivity of any one part.

What Australians seek to do with the land — and the achievability of these aims — is as much determined by the organisation of social relations as it is by the nature of the technology and the inputs and outputs from the land. Increasing the diversity of modes of land tenure and management is an important contribution to increasing the flexibility to respond to changing environmental and market conditions. Cooperative ownership and formal cooperation between owners is a contribution to this diversity, as is increasing the mix of commercial farmers who work part time in the town and urban dwellers who work part time on the farm. For this new mix to develop and work, it needs a supportive social context.

Cooperatives can contribute to the integration of urban and rural cultures and to the matching of conservation with productivity. Cooperative land management can provide a new middle ground for a partnership between private and public land management. Cooperative social organisation for land management contributes to ecological sustainability by diversifying land uses in order to match land capability with differing human aspirations. Cooperative approaches are more likely to encourage a greater variety of animal husbandry modes, kinds of animals, and types of horticultural activities. There is a need not only for mixed land uses in terms of a range of produce but also for human settlement and for native flora and fauna habitat requirements.

From the standpoint of the individual's empowerment in the world, local cooperative development provides access to a human pool of so-called 'significant others' within one's immediate environment. Access to others willing to work cooperatively may reduce the need for machines to substitute for people and the need for explicit professional, paid relations. These structural circumstances can provide built-in, informal services and support.

Such cooperative development can provide a powerful bridge

between the individual household and the wider society. A clear sense of belonging and social identity is basic to human well-being. Social alienation, stress and their symptoms of suicide, drug use and acute anxiety tend to be minimised when each person is connected in intimate ways through a persistent network of other persons which extends beyond the nuclear family, but which does not grow far beyond the number of people a person can know. This is in contrast to the present high vulnerability of the individual to 'impersonal others' and machines, together with the instability and environmental impact derived from the high throughput of an individual's significant people and places — partners, friends and locality (see Cock 1980).

Community development involves the transformation of people from being disparate private producers and passive consumers to those having an active participation in community and public life. This approach stands in marked contrast to the traditional 'welfare' model of public provision of services to passive recipients, which serves to reinforce a relationship based upon control and dependency.

From an economic perspective, the economies of scale, together with the supportive context of cooperative land management, makes it potentially easier to live better with less. This is illustrated by the following:

- The cooperative pooling of capital through equipment sharing can lead to better utilisation, as well as increased capacity to purchase. This is especially so for large and little-used items. Furthermore, the increased availability of human resources makes the purchase and/or use of equipment and chemicals, particularly during periods of peak demand, less attractive. Cooperative approaches ease the burden of capital machine investment because there is a larger pool of investors.
- The organisation of cooperative labour can mean that each member is freed from the rural bind of always having to be on the farm, allowing members greater individual flexibility. There is a built-in social structural capacity to enable job rotation, for example in sharing the drudgery of necessary but often boring tasks, such as weed removal or fencing. There is an increased labour force for seasonal peaks.
- Cooperative land sharing may be the only accessible alternative to corporate ownership, hobby farm subdivisions and family inheritance. Its affordability is a key access point for the urban bred through shared and more equal ownership.

This is as important for the dignity of farm workers as it is for their commitment to the land.

From the perspective of farmer education and training, information access and use requires a social context that ensures interaction and interdependence, so that a multiplicity of factors are taken into consideration as part of the decision-making framework. Outside specialist expertise is limited in its knowledge of local conditions and in its capacity to be heard and applied. Useful, that is to say used, knowledge is dependent on a receptive social context. The lack of this receptivity has been at the bottom of the relative failure of agricultural extension support (see Brewin 1986). For knowledge to be developed and used involves a cooperative social reference group with direct input into the knowledge pool for decision making; it has been treated in the past as being merely a background reference point (see Chamala and Mortiss 1990). Cooperative land management groups provide easier access to external reference points and advice, as well as making regulation easier. Through peer group support, it also helps to set the social context for discussion about what is appropriate action or inaction.

Cooperative management provides a broader skill pool. This is especially needed given the range of skills required for ecologically sustainable land management, and the desire to modify the largely urban-based orientation of new rural landholders. Skill transfer is easier, and innovation more likely, in cooperative groups. As the degree of cooperation increases the costs of farmer education and skill transfer decrease because of an increase in the informal and formal channels of information flow (see Blake and Cock 1990).

ACCESS POINTS FOR COOPERATIVE LAND MANAGEMENT

Wherever there are lands at the margin, either at the urban–rural interface, at the public–private boundaries, or where productivity is threatened, holdings will often have become uneconomic. In these circumstances there are opportunities for restructuring for cooperative management. For example, land sharing cooperatives began in the 1970s with small, isolated commercially uneconomic holdings (see Cock 1981; Williams 1983). In contrast, Landcare groups began in the late 1980s, largely initiated by commercial landholders whose farm productivity had declined or who were under threat from such problems as rabbits, salinity or weeds.

The edges of cities is where there are the largest numbers and strongest growth of hobby or part-time small-scale farms.

It is here that there needs to be an assessment of the environmental balance sheet of this growth of hobby farms. On the credit side, there are more people with innovative ideas involved in rural land management, and the overall productivity of a catchment area may be increased including the regeneration of wildlife habitat (Beckwoldt 1983). On the debit side, the environmental impact of hobby farms on the urban–rural interface is clearly evident in the wasteful use of resources through overcapitalisation and underutilisation of equipment, the breaking up of the landscape, excessive servicing (for example, each property its own access road), introduced weeds, a dispersed settlement pattern, social isolation and high levels of absentee ownership. These impacts could be substantially reduced through cooperatives that share servicing in more clustered settlements.

At the interface between private and public lands there is the opportunity to develop a partnership between private land owners and public authorities to care for the land and maintain its productivity. Examples are having locals harvest surplus kangaroos from public lands, sharing weed removal, establishing habitat corridors between public and private land as well as across private land in return for rate relief for land taken out of production, and, finally, giving preference to local farmers as conservation volunteers and as paid rangers for the care of public lands. Where public interests need to override pre-existing uses, then legislation may be required to provide effective private compensation (see Neales 1990). These actions would help to engender a shared approach to bioregional management that confronts the ecological damage that has occurred and is occurring throughout the present system of human-made boundaries.

In the long term, for catchment planning and management to be effective, however, it has to encompass the political economy and the social structure of land tenure. In this way it will provide a middle ground between the present private–public dichotomy. One strategic area for action is the creation of cooperative environment zones around parks. These zones could provide the base for the building of a partnership between private and public land managers. Important public lands could be protected by a partnership between government workers and local volunteers if the adjacent areas to be privately developed were planned and populated by people committed to community development and environmental care. An example is Sherbrooke Forest, Melbourne, where there is a significant public–private interface. There is potential for the development of such a partnership, but there is at present insufficient mutually supportive environmental planning (Cock 1989; Aytan and Delacy 1989).

LAND CARING AND LAND SHARING AS MODELS FOR COOPERATIVE LAND MANAGEMENT

Before going on to discuss the nature and experience of cooperative land management it is important to consider the existing range of land management approaches. There is an increasing variety of different legal and social structures used for land management. To see these in context it is useful to put them along a continuum (see Figure 17.1). At one end there is the family farm. Within this type there is a subcontinuum of the nuclear family farm — itself varying from the full-time farm family with both partners and children running the farm, to the part-time farm family with one or more family members working part or full time off the farm — and the extended family farm, which varies from the migrant horticultural farmers to the Anglo-Saxon versions of the family trusts and companies. These farms are commonly involved in broadacre cropping and grazing.

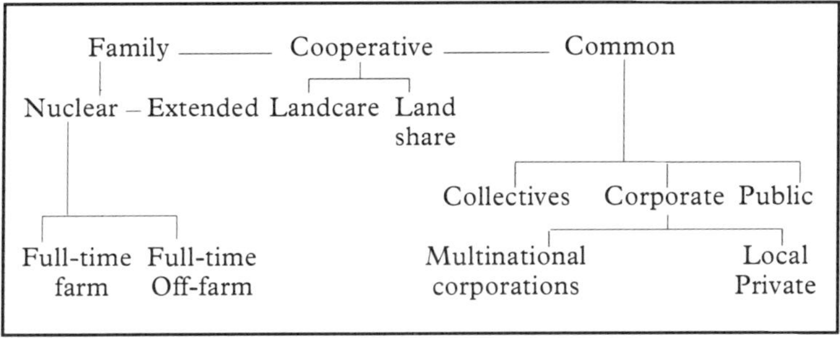

Figure 17.1 A private–public continuum of land management social structures

In the middle of the main continuum are cooperative farming enterprises. The subcontinuum here consists of land sharing and Landcare cooperatives. The land share cooperatives jointly own and cooperatively manage their land, whereas Landcare groups cooperate around a shared concern for their own and each others' land.

At the other end of the main continuum is common ownership of land, that is, ownership or membership by the many — those often not closely connected to the land or to each other. The social boundaries of ownership are large in comparison to the others in the continuum, while control is usually in the hands of the few. The subcontinuum consists, first, of collectives. State collective farms of the former USSR and China are examples. Secondly, there are corporate farms, whose owners vary from multinational companies involved in international multi-product activities (of which farming is one) to national public and private companies involved in producing one farm product in one location. Finally, there is public ownership of lands, varying from national parks to state parks and local reserves. The term common is used to highlight the impersonal detached ownership in contrast to the highly individualised management of the nuclear family farmer.

This schema fails to encompass fully the range of modes of social and legal structures used for land management, for example the kibbutzim of Israel and the tribal commons of indigenous peoples, which fit somewhere between cooperative and common land management. There are numerous variations throughout the world. For example, while public ownership of land is usually associated with the provision of public services and the protection of natural habitats, there is increasingly a mix of functions which can include agricultural activities such as the production of beef on Melbourne's Werribee sewerage farm.

Cooperative land management has a long history and is a well-established mode of social organisation for land management in many parts of the world. In Australia, cooperatives have been more concerned with the marketing of produce than with its production, or nowadays, with the repair of its consequences. This form has been marginalised by the twentieth century polarisation between public and private. This polarisation, represented by the two ends of the continuum, highlights the differences between private–public extremes both of which have significant social and environmental impacts. For example, the family farm is increasingly threatened economically by capital shortages and declining returns on investment. While farmers' commitment to their land is high, their ownership is declining, their control limited and they are largely locked into monocultural production regimes. While their corporate and public opposites command far more power in the marketplace, the scale and complexity of their operations inhibits responsiveness to ecological and human sensitivities, to the extent that the individual corporate worker or park ranger has little participation in land management decisions: the land is not mine, or ours, but theirs! In both public and corporate modes of social organisation there are very tenuous connections between the individual worker or park ranger, their controllers and the corporate shareholders or citizen owners.

In the light of this continuum, what types of land use are

best suited to ecologically and socially sustainable land management? How culturally constrained are we? The issue is to identify modes of land management that are not too distant from our own cultural heritage and yet move us towards caring for the land by sharing it. This should occur within a social context that is individually and community empowering. It is from this perspective that cooperative land management needs to be addressed, as a middle ground between the above extremes — one that can provide a social pathway to relearning how to live with, and from, our environment.

LAND SHARING COOPERATIVES

Within the land sharing model there is a variety of types. The most common of these is now called a multiple occupancy involving multiple households on the one jointly owned title. Usually the pattern of settlement is dispersed, with the productive land allocated to individuals or subgroups within the larger entity on a quasi-leasehold. These originated in the early 1970s, concentrated around northern New South Wales (Cock 1981). Hundreds were formed, most were illegal and they faced active local government opposition. Few of those begun in the 1970s lasted into the 1980s. These and newly formed cooperatives are now often legal in New South Wales and becoming so in other states (see Cock 1985; Woodward 1986; Metcalf 1986).

Most production on these cooperatives is for subsistence with only a small amount being sent to market. Common lands are usually those reserved for conservation or revegetation, for some equipment (tractor, truck) and for a building or shed. Initially purchased buildings are also owned in common (see Sommerlad, Dawson and Altman 1985).

The more organised and less dispersed type are environmental living cooperatives. These have the explicit purpose of seeking the settlement of bush and marginal lands in return for self-initiated strict environmental controls. For example, many prohibit dogs and cats and have an active commitment to weed eradication and flora and fauna regeneration. An illustration is the Around the Bend Cooperative on the outskirts of Melbourne, which is situated on 132 hectares and has planning permission for 32 houses. This registered community settlement society of 32 shareholders is located in a larger environmental living zone (see Watkins 1990).

Another example is Moora Moora Cooperative, Healesville, which owns 245 hectares, one-sixth of which is farming land. The rest is native forest in varying stages of regeneration.

Settlement is in six tight hamlets of five houses each. This cooperative's farming land is managed by a farm user's committee which is answerable to a Board of Directors. While every member has the right to farm the common at any given time only about one-sixth of the membership actually does so.

At Moora Moora, a complex system of rights and responsibilities has developed to accommodate the varying levels and types of interest in the productive land. For example, the land is managed as one entity, except for the subsistence gardens and orchards within each hamlet. There is a farm manager who is responsible for day to day management, answerable directly to the farm user's committee. There is a common barn with tractor and other equipment. Areas are set aside for horticulture, silviculture and grazing. Within the pasture, animals are rotated in groups. Stocking rates are set each year based on an assessment of the pasture carrying capacity. Individuals and small groups apply to run animals, which are privately owned. Members' responsibilities for farm work are based on their level of animal stocking. Each member has the choice of either a labour or a cash contribution.

There now exists a considerable number of land sharing cooperatives throughout Australia which have demonstrated the ability to survive over a decade and which have made a significant contribution both to bringing people back to the land and to the diversifying land management.

LANDCARE GROUPS

The Victorian Department of Conservation and Environment defines a 'LandCare' group (as it is termed in that state) as a 'voluntary organisation of local community members dedicated to combating land degradation, protecting wildlife and managing the land in a responsible way' (Department of Conservation and Environment 1990: 88). At its core are neighbouring family farmers who are cooperatively organised to tackle a shared land management issue that undermines their livelihood, such as salinity, pests or weeds. The Landcare movement began in earnest in the late 1980s, and has experienced rapid growth ever since. The initial thrust came from Victoria. In that state alone by the end of the 1980s there were nearly 70 registered LandCare groups, and at least as many developing groups. More than 3000 landholders working on 1.4 million hectares of land have been involved (McWaters 1990).

The notion of cooperative organisation for self-help has strong roots in the traditional culture of rural communities. Farmers have often assisted one another in responding to community needs in times of crisis. This culture, however, has been eroded gradually over a long period. Social changes reinforcing the seriousness of difficulties faced by rural communities are the declining levels of rural employment, the drift of young people to urban areas and the rise of poverty and stress (Lawrence 1987). These closely linked problems have tended to undermine people's confidence in the future of their communities.

Although a community development approach is well established in the theory and to a lesser extent in the practice of community services provision in Australia (see Grierson and Daddow 1985), it is much less widely accepted in the fields of agriculture and environmental management. Community action for environmental conservation is, however, becoming increasingly common.

Landcare groups are significant in that they draw upon social values and networks of support that constituted a more prominent aspect of rural life in the past. They provide an opportunity for community redevelopment focused on a group approach to care of the land, while at the same time empowering the individual farmer's capacity to act. They offer a structure for more efficient use of departmental, private and community resources, and a social context for moral support and the development of a new ethic of land management. Farmers who participate in cooperative groups exert a collective social pressure upon non-members who are either ignorant of land degradation issues or reluctant to become involved in land repair, as well as exerting a continuing supportive pressure upon fellow members.

Furthermore, Landcare groups encourage exploration of the central issues involved in land management. Due to the complexity of these issues, a variety of inputs from different perspectives is critical to making wise and — in the long term — effective decisions. Landcare groups also provide a structure within which accountability for the expenditure of public monies is increasingly possible. The potential of a cooperative approach has been acknowledged at the federal level through the government's acceptance of the Victorian LandCare Program as the model for a national program (see Blake and Cock 1990).

AN ASSESSMENT OF COOPERATIVE LAND MANAGEMENT

Land degradation issues are not of the highest priority for farmers. This prevailing attitude appears to reflect an interaction between perceptions of land degradation, basic value orientations and resource constraints. Three key factors can be identified.

1. A widespread fatalism in attitudes towards salinity and other land degradation problems (Williams 1979; Barr and Cary 1985).
2. 'Uncertainty about the applicability and cost-effectiveness of conservation measures to conditions on their farms' (Rickson *et al.* 1987: 193).
3. A preoccupation with short to medium term financial aspects of farming as opposed to longer-term environmental dimensions. Efficiency is defined by both farmers and consumers in terms of the narrow economics of inhouse, short-term cost/benefits of specialisation in one activity rather than in terms of an ecosystem view of multi-niche creation for societal effectiveness over the long term (see Daly and Cobb 1989).

Prices do not adequately reflect the environmental costs of food production. Household expenditures on food have declined historically relative to other expenditures. While farmers directly confront the environmental issues and the cost-price squeeze of rural production, urban populations demand low-cost produce and remain largely ignorant of — and are unaffected by — rural environmental pressures. Public expectations of low food prices constrain the farmers' ability to incorporate extensive conservation measures into their land management practices. Yet it is the urban public that demands levels of land and animal care that farmers often cannot afford or which make food production economically unviable.

While urban and international pressures drive the exploitation of the land (see Chapter 3) there is still the barrier of the Australian culture of privatism and consumerism, a culture that views freedom and empowerment only in terms of individual ownership and control. For example, the ingrained belief that farmers on their own decide what they will produce and how they will produce it overlooks the control held by banks, marketing boards and government regulation and the dependence of rural land managers on urban and international market forces.

Within the farming community, there is still widespread resistance to confronting the reality of land degradation, particularly on individual farmers' own properties (Cary, Beel and Hawkins 1984; and see Chapter 6). While psychological and economic factors are involved here, this resistance in part reflects the dominant Australian culture of self-reliant individualism and privacy (see Horne 1986). These cultural traits also erect a barrier to developing cooperative relationships between private and public management, as well as between neighbours. There is still a common tendency to focus on symptoms rather than causes and to blame someone else or, wherever possible, to export the problem to some other place. More concretely, there is also resistance to revegetation measures that encroach on actual or potentially productive land, unless it can provide an income (for example, lucerne or, in the longer term, pines). This resistance is heightened when the percentage of land in need of revegetation is a significant component of the total farm, as is the case in some dryland recharge areas. More generally, farmers' desires for public financial support are not always matched by a willingness to accept either significant accountability in relation to public concerns or the voluntary role in Landcare of the non-farming sector of the community.

While these factors represent major barriers to change, preparedness to implement conservation management measures is linked with farmers' levels of education and access to finance, and, more particularly, with their willingness to invest resources for longer-term returns, as well as the extent of their participation in a community's social life (Chamala, Keith and Quinn 1983).

It is clear that farmers' environmental awareness and their priorities need to change. This, however, is unlikely unless they are better resourced, both socially and materially. The crucial connection between the economic and social contexts of land protection is the need to create a new balance away from the overwhelming dominance of a cultural orientation towards private self-interest and individual self-reliance. The cultural alternative is an increased emphasis on shared community interests and mutual reliance.

SUSTAINING LANDCARE GROUPS

The tremendous upsurge in the development of Landcare groups has been a source of hope and enthusiasm in the face of despair. This upsurge is derived from the explosion of awareness of the severity of land degradation in all its varying and multiplying forms. The concept of Landcare groups is sound and the progress in their numbers is impressive.

Interviews with members from a number of LandCare groups in the Goulburn/Broken region of Victoria, however, revealed significant difficulties in group participation and management, thereby pointing to the need for advisory support which is skilled in community development and cooperative skills. Farmer education programs have failed to address, in any systematic or effective way, the issues of long-term group management. Nor have the appropriate structures and roles of government extension services to these groups been properly examined (see Blake and Cock 1990). Professionals skilled in group processes can be of great assistance in helping group leaders to develop explicit management structures and processes for decision making, action and accountability. Such structures and processes can clarify the rights and responsibilities of all members, and so relieve pressures, especially those experienced by more dedicated individuals. The development and implementation of basic structures and processes for group management need to occur during the early 'honeymoon' phase of the group, because later, when they are more needed, they will be difficult to develop. The recent work of Chamala and Mortiss (1990) provides a way of tackling these deficiencies.

Clearly, while Landcare groups need to emerge from within the farming community, the active encouragement of extension officers, and the allocation of sufficient and appropriate financial support from government is equally important. Government support for cooperative land protection action needs to facilitate the ongoing development of effective group organisation, especially to ensure equity of effort towards achievement of shared purposes. Government funding of groups to employ part-time coordinators is an important aspect of support for the core leadership of a group. Each group's coordinator and chairperson are key catalysts for group action as well as liaison with extension personnel. When there is a recession, pressures on funding are a significant challenge to the strength of Landcare groups, especially if their prime catalyst is access to government funding.

Cooperative Landcare groups are most likely to be sustainable over the long term if they are locally initiated and well organised, with a broad base of participation. In the absence of an adequate organisational structure, there is an inevitable over-reliance on the time, energy and, often, emotional resources of the few. Consequently, the overall performance of a group is likely to fluctuate with the capacities and commitment of these individuals. Non-participation by some local farmers and token participation by others are concerns which need to be continually addressed by any cooperative. There is a real risk that a group may collapse once the initial collective enthusiasm has worn off while the salinity, rabbits and other obvious signs of failure still remain despite the effort expended.

Those cooperative Landcare groups which were among the first to form, or which are now relatively large, often have special characteristics shaping their success which are not readily transferable to other groups. Groups which have acquired a reputation as highly innovative or effective, built through the enthusiasm and imagination of core members and promoted through media attention, can tend to be overstated as models. Established groups which have displayed a high degree of enthusiasm, organisation and networking skills have also had the ability to attract relatively high levels of support from the public sector — and, in some cases, also the non-government sector — as well as being able to influence the development of state or regional policies. A good understanding of the means by which external resources and assistance can be attracted is, in fact, an important factor shaping group performance. While a highly innovative and committed group may require relatively little assistance, other groups will typically require much more.

There is yet to be a full-scale evaluation of the environmental and social performance of Landcare groups. This is partly because it is still too early to conduct such a comprehensive evaluation. The member attitude survey by O'Brien and Pennicuik (1989) reveals little of their performance apart from a high level of shared commitment to constructive action for land repair. They conclude that the backbone of the scheme is the Department of Conservation, Forests and Lands (O'Brien and Pennicuik 1989). Government extension services and private consultants are valuable sources of assistance to be called upon. Apart from the impact on farmers of land degradation itself, however, it is the influence of respected individuals within local networks that is the most powerful motivating factor.

Campbell (1989) points to the issues involved in the assessment of Landcare groups' performance. This assessment needs to take into account the initial ecological and sociological context in which the groups formed, and the time scale for their social development as groups and to evaluate their actual contribution to land repair.

A draft report of the Ecologically Sustainable Development Working Groups (1991) recommended that governments continue to support Landcare groups as they represented a principal means of encouraging information transfer among farmers. It was considered that whole farm planning would be a logical extension of attempts by Landcare groups to address local environmental problems. The role of women in Landcare was singled out as a major attribute of the Landcare 'model'. (The most recent evaluation of Landcare can be found in Chapter 11.)

There is considerable potential for Landcare groups to extend the scope of their activities to other areas of cooperative activity, including the sharing of expensive agricultural machinery and the bulk purchasing of materials. Such cooperation has an obvious economic rationale in cutting production costs. In some circumstances economic cooperation may represent a viable response to the pressure to 'get big or get out'. Government programs can facilitate cooperation by providing incentives for joint purchases. By reinforcing the links between economic and environmental concerns, Landcare groups are likely to achieve a more secure organisational base. The additional complexities involved in defining rights and obligations within such arrangements, however, require careful attention.

Landcare groups have the potential to draw in the participation of local people not directly engaged in rural production, including urban residents with an interest in the conservation of the environment and the well-being of rural communities. People who might ordinarily be excluded from access to extension activities, agricultural associations or similar rural organisations may become involved. Encouragement of tree planting in urban settings may have a spin-off effect in support for rural action, and vice versa. Service clubs, ethnic associations, social welfare groups, women's groups, environmental organisations and churches have a potential role in contributing to community mobilisation for land protection. While some organisations may have an interest in activities directly related to land protection, others may be more concerned with the human consequences of economic dislocation and psychological stress associated with land degradation.

The broader the social base and scope of concern, the more likely the response is to gain momentum and to achieve an enduring significance. Unless the scope of land protection groups can be expanded to encompass explicitly the needs and mutual support of people within affected communities, the groups may eventually wither or collapse when the enthusiasm of their leadership wanes. There needs to be a greater emphasis on the human needs of rural communities suffering from land degradation and concomitant social pressures on the agenda of land restoration.

COOPERATIVE LAND SHARING

Cooperative land management requires more time to be given to decision making, conflict resolution, land use monitoring and the organisation of people. It is easiest to own and manage land in common for conservation and recreation. For example, with a bush block, it is largely a matter of 'lock it up and leave it'. In the case of cooperative farming much more management is required. In farmer cooperatives it is easier to undertake permaculture than animal production. The hardest is that which requires the greatest level of management of people and machines, such as with annual crop production. The latter is, in cooperative terms, the least successful except perhaps for the cooperative provision of infrastructure such as fencing, water supply and large equipment items.

The clarity of organisational form is a requisite foundation for the evolution of sustainable cooperatives and for the development of constructive informal community dynamics. The naive vision of the simple life and self-sufficiency has tended to suppress the capacity to develop and affirm the complex interdependencies that land sharing involves. The rejection of corporate bureaucracy requires an alternative form of organisation — one with explicit form and function — to develop as a first step towards strengthening the customs, rituals and symbols behind the growth of a sharing community. It takes a long time to rediscover and to evolve appropriate shared cultural realms which can nourish the community and sustain it during crisis (see Kanter 1972).

There is much to learn from the anarchists' tradition of participatory models of consensus building and decision-making processes. There is equally much to learn from corporate modes of socioeconomic organisation. That is, once a decision has been made through participatory modes of decision making, then there needs to be in place effective organisation that has the authority and capacity to carry out the agreed action or inaction. The environment movement — and the rural resettlement movement (Williams 1983) — have been strong on participatory decision making but weak on organisational clarity, efficiency and accountability. Afraid of any hint of hierarchy, they have often stripped themselves of the collective authoritative capacity to act — something particularly needed in crises or simply when small, but contentious, difficulties arise.

A lack of effective social organisation leads to the following processes being set in train. In order to function at all there is a privatisation of decisions by the few, often in an informal way, that is difficult to access or challenge. Alternatively, or additionally, this disorganisation results in the formal monopoly of a particular group of the most able, who become exhausted and burnt out in a few years. (In one sense the most committed tend to be consumed by the least committed.) The lack of explicit group power generates confusion, withdrawal and powerlessness, and adds to the burnout of those previously committed. Decision-making meetings are no longer valued because they are so exhausting, further weakening the capacity for participatory democracy.

The lack of clarity of structure — arising from fear of an emergent bureaucracy — results in confused lines of communication and coordination between components. For example, the documentation of agreed actions is often inadequate. As a result, no one knows what was decided three months ago, because the minutes were not valued or written up or were lost. This generates structurally induced disagreement about the history of agreement.

The ongoing struggle to develop an effective mix and balance between individual–household and cooperative–state rights and responsibilities is a vexing issues that continues to trouble even successful cooperatives. Cooperatives that fail to grasp the nettle of being clear about their land management practices and how this is reflected in the organisation of rights, responsibilities and accountability have a limited capacity and questionable viability in the long term (see Kanter 1972). The experience of the cooperative rural resettlement movement in the 1970s is illustrative of these social processes and of the socioecological consequences of such disorganisation (see Cock 1985).

Where there is pressure for redevelopment such as for hobby farming, the Moora Moora approach described above is preferable as a settlement pattern. The multi-occupancy approach to land sharing may be best used in a situation of intensive horticulture. Access to collective wisdom has to be balanced with the need for leadership and authority to initiate and give focus, particularly in the case where one's livelihood is dependent on the land's sustained output. The record of land sharers as agricultural producers is patchy. Innovative in concept, in Australia it rarely generates commercially marketable produce as an entity, as distinct from that produced by a few individuals within a multi-occupancy (see Sommerlad *et al.* 1985). This is in contrast to the highly productive kibbutz. Where the production from the land is a central consideration, then placing private activity within the context of cooperative organisation is probably the most successful approach in the present Australian context. It is here that the Landcare model is most appropriate in pre-existing commercial farming circumstances.

In order to increase the scope and sustainable levels of cooperation the cultural context needs to be supportive and the social organisation systematic. Yet it must have the flexibility and responsiveness to be able to take account of, and utilise the differences in, temperament and skills of the members. Clearly, as the levels of cooperation increase, so too does the role that local formal and informal social structures and processes can play in shaping environmental outcomes.

The above assessment applies to both types of cooperatives discussed above — particularly in relation to their common desire to evolve a cooperative approach to land management. The issues of particular concern differ between each of the two types discussed. What they share is an attachment to the land and its well-being and the tenuous use of cooperative modes of social organisation to improve its management. Their variations of origin, of social composition and of reference points makes little difference to the social issues that any cooperative group faces.

THE NEED FOR A PARTNERSHIP BETWEEN COOPERATIVE GROUPS AND GOVERNMENTS

Active commitment to cooperative development can only arise where there exists a high level of community awareness that land management issues relate directly to the organisation of social interests. This involves community education in order to cultivate appropriate levels of understanding and concern within different segments of the population. Government is in a position both to assist community action and to promote its own policy objectives through strategic allocation of public resources, channelled through community organisations. Regulatory measures still have a valuable function where there exists sufficient community support to achieve effective implementation. Strategic planning is important to guide resource allocation to propose and support regulatory measures which will help to realise policy objectives appropriate to regional or local circumstances.

From the 1960s to the 1990s there has been something of a conversion in the public perception of cooperative action from deviance to trendiness. Before too long cooperation may well be the dominant mode. The Landcare movement has been a vital catalyst in bringing cooperative group action into focus as a legitimate and vital part of the struggle for land restoration. Government support, as we have seen, is necessary to achieve sustainable cooperative land management. But government should not impose arbitrary limits on the scope of action. Just as difficult, but equally important, is the involvement of groups which traditionally have little influence or decision-making power in the community — for example women and migrants from non-English-speaking countries.

A major role of government is to promote awareness of important issues and to assist local action. Nevertheless, it should be up to local people to assume predominant responsibility for the future of their own area. This is not to say that individuals should be able to do with their land as they wish, or that government sanctions do not have an important role in some circumstances. Government is responsible for establishing policy frameworks, resolving conflicts and providing coordination between different sectors, although each of these activities should involve community participation.

There are numerous isolated individual cases of private conservation ventures. Their habitat size, and their social and political support base are too small to be sustainable. It is vital that they are clustered together, covering a large enough area to provide for significant human settlement, habitat protection and agricultural production. Developing an agreement between the local authority and a local cooperative that is explicit and enforceable is an important role for local government.

The clamour for local empowerment to engender social mobilisation needs to be developed in partnership with an acceptance of external accountability. The difficulties of land sharing and the need for external accountability to ensure the tough decisions are made (decisions that affect the short-term interests of some or all of the local community, but ensure sustainability) are great. What is needed are externally produced guidelines of accountability which can be used as a reference, thereby overcoming local attachments and hostilities that might otherwise cloud perceptions. Purely local decision-making responsibilities and accountabilities can cripple the capacity to make ecologically sustainable decisions or to challenge damaging land practices, let alone ensure the necessary land healing.

Local authorities strongly endorse the need for planning approval. Up until now, however, they have generally opposed cooperative settlement. When such settlements are approved, their authorities have been weak on regulatory follow-through. Regulation without sanctions against rule-breaking quickly becomes meaningless — a semblance of authority that often masks an interest in generating income via planning and building licence fees.

To complement a cooperative's own policies, there needs to be support from government as a community educator and as a regulator of environmental use. In general, planning regulation without local support has failed because it is basically not enforceable. In the face of environmental rule-breaking government is often forced, as in the case of illegal tree removal, to turn a blind eye. However, merely to leave responsibility to the initiative of groups of local individuals is insufficient. In short, a local community often has the responsiveness but not the authority; the state has the authority but not the necessary sensitivity to local conditions. The conflict between those who believe in purely local community power and responsibility and those who look to government as protector and enforcer is not to be resolved by these extremes, but by carefully designed interactions.

CONCLUSION

The two case types examined focus on cooperative approaches to the management of particular properties or groups of properties. This is distinct from the more general question of the broader mobilisation and development of rural communities. For cooperative approaches to land management to be sustainable, they not only need government support but need to be part of a broadly based transformation of the cultural and political framework.

A community-based strategy for rural action and change must aim to develop the collective capacities of communities to assume more responsibility for their own future well-being by focusing on the following: awareness raising; motivation of appropriate forms of action; the organisation of effective social structures; and the legitimation of the roles of all parties.

The development of a collaborative relationship between landholders, community organisations and government agencies is a basic requirement, complementing a cooperative approach within the respective organisations. Cooperatively organised, voluntary action should form a core of the sustainable land management response. If it is to be effective over the long term this will need to be both supported by appropriate financial incentives and advice and reinforced by suitable regulation.

Only as community understanding and cooperative action are developed will the seemingly intractable become tractable. The rapid growth in the establishment of new Landcare groups lends weight to the view that cooperative landcare is a creative, soundly based innovation which indicates the potential for a broad shift toward cooperative farming systems which are more sustainable. Additionally, a tremendous potential within the urban margins exists for the mobilisation of cooperative action for the provision of basic necessities. These include neighbourhood gardens, street permacultures, cooperative transport and housing.

The gradual reduction of levels of cooperation and mutual accountability within land sharing cooperatives, however, supports the view that the initial enthusiasm for cooperative land repair needs to be used to develop the organisational power necessary for sustainable cooperative action, together with ongoing active support of the wider community — local, public and official. Cooperatives work best when they exist within a culturally supportive environment, are clearly organised, have a skilled membership and operate within known natural limits.

This action cannot be separated from the necessity of an approach to social redevelopment that encompasses cultural transformation, grassroots social organisation, political support and global awareness and, not least, the cultural transformation in our ways of looking at and valuing the world. The mobilisation of each person's sustained responsibility for planetary well-being is dependent on the development of a supportive cultural context, local community development and political will. Each of the above elements is necessary to ensure the sustainability of the others and the development of our capacity to live better with less.

REFERENCES

Aytan, J. and Delacy, Y. (1989) *Living in Sherbrooke* (2nd edn), Victorian Association for Environmental Education, Melbourne.

Barr, N. and Cary, J. (1985) *Farmers' Perceptions of Soil Salting: Appraisal of an Insidious Hazard*, School of Agriculture and Forestry, University of Melbourne, Melbourne.

Beckwoldt, R. (1983) *Wildlife in the Home Paddock: Nature Conservation for Australian Farmers*, Angus and Robertson, Sydney.

Blake, T. (1991) 'Social Structures for Ecological Sustainability: A Critical Social Theory Perspective', in P. Cock (ed.), *Social Structures for Sustainability*, Centre for Resource and Environmental Studies, Australian National University, Canberra.

Blake, T. and Cock, P. (1990) *Salinity and Community: Awareness and Action for Salinity Control in the Goulburn/Broken Region: Towards a Community Development Approach*, Graduate School of Environmental Science, Monash University, Melbourne.

Boyden, S. (1989) 'An Energy Scenario for an Ecologically Sustainable Australia', unpublished paper, Centre for Resource and Environmental Studies, Australian National University, Canberra.

Boyden, S., Dovers, S. and Sherlow, M. (1990) *Our Biosphere Under Threat: Ecological Realities and Australia's Opportunities*, Oxford University Press, Melbourne.

Brewin, D. (1986) *Group Conservation Programs and Soil Conservation Practices*, School of Agriculture and Forestry, University of Melbourne, Melbourne.

Campbell, A. (1989) 'Evaluation of Landcare Groups: Performance Indicators for Different Stages of Group Development', unpublished paper, School of Agriculture and Forestry, University of Melbourne, Melbourne.

Cary, J., Beel, A. and Hawkins, H. (1984) *Farmers' Attitudes Towards Land Management for Conservation,* School of Agriculture and Forestry, University of Melbourne, Melbourne.

Chamala, S., Keith, K. and Quinn, P. (1983) 'Australian Farmers' Attitudes Towards, Information Exposure to, and Use of, Chemical and Soil Conservation Practices', *Tillage Systems and Social Science* 3 (1).

Chamala, S. and Mortiss, P. (1990) *Working Together for Land Care: Group Management Skills and Strategies,* Australian Academic Press, Brisbane.

Cock, M. (1989) 'Sustainable Development: Planning for the Interface of Urban Settlement and Environmentally Sensitive Areas', unpublished thesis, Graduate School of Environmental Science, Monash University, Melbourne.

Cock, P. (1980) 'The Alienating Consequences of the Nuclear Family's Size and Mobility', in D. Davis *et al.* (eds), *Living Together: Family Patterns and Lifestyles*, Centre for Continuing Education, Australian National University, Canberra.

Cock, P. (1981) 'The Centre of Alternative Australia: The Rainbow Region', *Current Affairs Bulletin* November: 4–14.

Cock, P. (1985) 'Sustaining the Alternative Culture: The Drift Towards Rural Suburbia', *Social Alternatives* 4 (4): 12–16.

Coombs, H. (1990) *The Return of Scarcity,* Cambridge University Press, Cambridge.

Daly, H. and Cobb, J. (1989) *For the Common Good,* Beacon, Boston.

Department of Conservation and Environment (1990) *Tree Victoria Action Plan,* Land Protection Division, Department of Conservation and Environment, Melbourne.

Ecologically Sustainable Development Working Groups (1991) *Draft Report — Agriculture*, Australian Government Publishing Service, Canberra.

Faulkner, E. (1948) *Ploughing in Prejudices,* Michael Joseph, London.

Grierson, D. and Daddow, N. (1985) *You Can Do It: Overcoming Youth Unemployment,* The Joint Board of Christian Education of Australia and New Zealand, Melbourne.

Grigg, G. (1988) 'Kangaroo Harvesting and the Conservation of the Sheep Rangelands', *Australian Zoology* 24: 124–128.

Horne, D. (1986) *The Public Culture*, Pluto Press, London.

Kanter, R. (1972) *Commitment and Community: Communes and Utopias in Sociological Perspective,* Harvard University Press, Cambridge, Massachusetts.

Lawrence, G. (1987) *Capitalism and the Countryside*, Pluto Press, Sydney.

Metcalf, W. (1986) 'Dropping out and Staying in: Recruitment, Socialisation and Commitment Engenderment Within Contemporary Alternative Lifestyles', unpublished PhD thesis, Griffith University, Brisbane.

Neales, S. (1990) 'Our Farmland Fight', *Australian Farmer* August/September.

McWaters, V. (1990) 'The Year of Land Care: An Opportunity to

Put Land Protection on the Map', *Trees and Natural Resources* 32 (3).

O'Brien, B. and Pennicuik, M. (1989) *An Evaluation of Rural Community Participation in the Landcare Program,* The Research Network, Melbourne.

Owen, C. (1987) *A Resource Guide for Sustainable Agriculture,* Wagga Wagga Cottage, Wagga Wagga.

Passmore, J. (1974) *Man's Responsibility for Nature: Ecological Problems and Western Traditions,* Duckworth, London.

Rickson, R., Saffigna, P., Vanclay, F. and McTainsh, G. (1987) 'Social Bases of Farmers' Responses to Land Degradation', in A. Chisholm and R. Dumsday (eds), *Land Degradation: Problems and Policies,* Cambridge University Press, Cambridge.

Sommerlad, E., Dawson, P. and Altman, J. (1985) *Rural Land Sharing Communities: An Alternative Economic Model,* Australian Government Publishing Service, Canberra.

Trainer, T. (1991) 'Thinking about the Nature of the Required Conserver Society', in P. Cock (ed.), *Social Structures for Sustainability,* Centre for Resource and Environmental Studies, Australian National University, Canberra.

Watkins, S. (1990) 'Living with the Environment: A Melbourne Experiment', *The Age* 20 November: 19.

Williams, M. (1979) 'The Perception of the Hazard of Soil Degradation in South Australia: A Review', in R. Heathcote and B. Thom (eds), *National Hazards Symposium,* Academy of Science, Canberra.

Williams, S. (1983) *Low Cost Rural Resettlement,* Australian Rural Adjustment Unit, University of New England, Armidale, New South Wales.

Woodward, J. (1986) *Multiple Occupancy Development in the Shire of Tweed,* Ministry for Planning and Environment, Sydney.

Wynen, E. and Edwards, G. (1988) *Towards a Comparison of Conventional and Chemical Free Farming in Australia,* Economic Discussion Papers, School of Economics, La Trobe University, Melbourne.

Zarksy, L. (1990) *Sustainable Development: Challenges for Australia,* Commission for the Future, Melbourne.

INDEX